The Subnuclear Series • Volume 52

Proceedings of the International School of Subnuclear Physics

STATUS OF THEORETICAL UNDERSTANDING AND OF EXPERIMENTAL POWER FOR LHC PHYSICS AND BEYOND

50th Anniversary Celebration of the Quark

THE SUBNUCLEAR SERIES*

Series Editor: ANTONINO ZICHICHI, *European Physical Society, Geneva, Switzerland*

20. 1982 GAUGE INTERACTIONS: Theory and Experiment
21. 1983 HOW FAR ARE WE FROM THE GAUGE FORCES?
22. 1984 QUARKS, LEPTONS, AND THEIR CONSTITUENTS
23. 1985 OLD AND NEW FORCES OF NATURE
24. 1986 THE SUPERWORLD I
25. 1987 THE SUPERWORLD II
26. 1988 THE SUPERWORLD III
27. 1989 THE CHALLENGING QUESTIONS
28. 1990 PHYSICS UP TO 200 TeV
29. 1991 PHYSICS AT THE HIGHEST ENERGY AND LUMINOSITY: To Understand the Origin of Mass
30. 1992 FROM SUPERSTRINGS TO THE REAL SUPERWORLD
31. 1993 FROM SUPERSYMMETRY TO THE ORIGIN OF SPACE–TIME
32. 1994 FROM SUPERSTRING TO PRESENT-DAY PHYSICS
33. 1995 VACUUM AND VACUA: The Physics of Nothing
34. 1996 EFFECTIVE THEORIES AND FUNDAMENTAL INTERACTIONS
35. 1997 HIGHLIGHTS OF SUBNUCLEAR PHYSICS: 50 Years Later
36. 1998 FROM THE PLANCK LENGTH TO THE HUBBLE RADIUS
37. 1999 BASICS AND HIGHLIGHTS IN FUNDAMENTAL PHYSICS
38. 2000 THEORY AND EXPERIMENT HEADING FOR NEW PHYSICS
39. 2001 NEW FIELDS AND STRINGS IN SUBNUCLEAR PHYSICS
40. 2002 FROM QUARKS AND GLUONS TO QUANTUM GRAVITY
41. 2003 FROM QUARKS TO BLACK HOLES: PROGRESS IN UNDERSTANDING THE LOGIC OF NATURE
42. 2004 HOW AND WHERE TO GO BEYOND THE STANDARD MODEL
43. 2005 TOWARDS NEW MILESTONES IN OUR QUEST TO GO BEYOND THE STANDARD MODEL
44. 2006 THE LOGIC OF NATURE, COMPLEXITY AND NEW PHYSICS: FROM QUARK-GLUON PLASMA TO SUPERSTRINGS, QUANTUM GRAVITY AND BEYOND
45. 2007 SEARCH FOR THE "TOTALLY UNEXPECTED" IN THE LHC ERA
46. 2008 PREDICTED AND TOTALLY UNEXPECTED IN THE ENERGY FRONTIER OPENED BY LHC
47. 2009 THE MOST UNEXPECTED AT LHC AND THE STATUS OF HIGH ENERGY FRONTIER
48. 2010 WHAT IS KNOWN AND UNEXPECTED AT LHC
49. 2011 SEARCHING FOR THE UNEXPECTED AT LHC AND THE STATUS OF OUR KNOWLEDGE
50. 2012 WHAT WE WOULD LIKE LHC TO GIVE US
51. 2013 REFLECTIONS ON THE NEXT STEP FOR LHC
52. 2014 STATUS OF THEORETICAL UNDERSTANDING AND OF EXPERIMENTAL POWER FOR LHC PHYSICS AND BEYOND

Volume 1 was published by W. A. Benjamin, Inc., New York; 2–8 and 11–12 by Academic Press, New York and London; 9–10 by Editrice Compositori, Bologna; 13–29 by Plenum Press, New York and London; 30–52 by World Scientific, Singapore.

*For the complete list of titles in this series, please go to
http://www.worldscientific.com/series/ss

The Subnuclear Series • Volume 52

Proceedings of the International School of Subnuclear Physics

STATUS OF THEORETICAL UNDERSTANDING AND OF EXPERIMENTAL POWER FOR LHC PHYSICS AND BEYOND

50th Anniversary Celebration of the Quark

Edited by

Antonino Zichichi

European Physical Society
Geneva, Switzerland

World Scientific

NEW JERSEY • LONDON • SINGAPORE • BEIJING • SHANGHAI • HONG KONG • TAIPEI • CHENNAI • TOKYO

Published by

World Scientific Publishing Co. Pte. Ltd.
5 Toh Tuck Link, Singapore 596224
USA office: 27 Warren Street, Suite 401-402, Hackensack, NJ 07601
UK office: 57 Shelton Street, Covent Garden, London WC2H 9HE

British Library Cataloguing-in-Publication Data
A catalogue record for this book is available from the British Library.

The Subnuclear Series — Vol. 52
STATUS OF THEORETICAL UNDERSTANDING AND OF EXPERIMENTAL POWER FOR LHC PHYSICS AND BEYOND
50th Anniversary Celebration of the Quark
Proceedings of the International School of Subnuclear Physics 2014

ISBN 978-981-3148-63-5

Desk Editor: Rhaimie Wahap

Printed in Singapore

Professor Bruno Zumino lecturing at the
International School of Subnuclear Physics in 1969.

PREFACE

In June/July 2014, a group of 132 physicists from 24 countries met in Erice to participate in the 52nd Course of the International School of Subnuclear Physics. The countries represented by the participants were: Armenia, Azerbaijan, Brazil, Canada, Cuba, Denmark, Egypt, France, Georgia, Germany, Greece, Hungary, India, Italy, Japan, the Netherlands, Pakistan, Poland, Russia, Slovak, Switzerland, Ukraine, United Kingdom and the United States of America.

The School was sponsored by the Academies of Sciences of Estonia, Georgia, Lithuania, Russia and Ukraine; the Chinese Academy of Sciences; the Commission of the European Communities; the European Physical Society; the Italian National Institute for Nuclear Physics; the Italian Ministry of Education, University and Scientific Research; the Sicilian Regional Government; the Weizmann Institute of Science; the World Federation of Scientists and the World Laboratory.

Following the three years dedicated to the celebrations of the 50th anniversary of the School, this year's Course was dedicated to the 50th Anniversary Celebration of the Quark with Murray Gell-Mann and George Zweig. In fact, during the last half century the School has been involved in all crucial steps of our Physics. Few examples: SU(3) flavour and SU(6) (with SU(2)-spin coupled with SU(3)-flavour) dismantled by the "No-Go-Theorem", the battle between S-Matrix and Field Theory, the Universality of the weak forces (started with the ε-parameter and the non existence of the "flavour changing-neutral currents" solved by the existence of "charm"), the experimental search for the 3rd lepton in the early sixties before the discovery of the CP breaking, the birth of the Electroweak Unification and SSB (Spontaneous Symmetry Breaking) mechanism, the discovery of the negative sign of the β-function and of asymptotic freedom, the triumph of non Abelian field theories (QCD and QFD) with all consequences (including Instantons), the discovery of Supersymmetry (many years — and not few days — after the "No-Go-Theorem"). Now we would like LHC to give us the first sign for the existence of the Superworld. Finally we should not forget that, since complexity exists at the fundamental level, a totally unexpected discovery should be given to us by LHC or by some other Lab engaged at the frontier of our knowledge.

As usual, the lectures dealt with the latest and most significant achievements in both theoretical and experimental Physics, with several of them devoted to reflections about the next steps for LHC. Moreover, three groups of problems were open for competition: the first group was "Mixings", the second was "Elementary and composite states" and the third one was "Symmetry breakings".

An original feature of the School, introduced in 1996, is a series of special sessions devoted to "New Talents", with the aim of encouraging and promoting young physicists to achieve recognition at an international level. This is a serious problem in Experimental Physics, where collaborations count several hundreds of participants and it is almost impossible for young fellows to be known. Even if with much less emphasis, the problem exists also in Theoretical Physics. So it was decided to offer 36 young fellows the possibility to give an open presentation of the results of their studies, followed by a discussion. Among them 26 students were awarded a diploma and 4 were given the New Talent Awards. It was also decided to award a special scholarship for the best student in honour of Prof. Bruno Zumino, to whom this year's School is dedicated.

During the organization and the running of the Course, I enjoyed the collaboration of my colleague and friend, Gerardus 't Hooft, who shared with me the Directorship of the School. I would like to thank him, together with the group of invited scientists and all the people who have contributed to the success of the Course.

I hope the reader will enjoy the book as much as the students attending the lectures and the discussion sessions, which, as every year, have been the focal point of this School activity. Thanks to the work of the Scientific Secretaries, the discussions have been reproduced as faithfully as possible. At various stages of my work I have enjoyed the collaboration of many friends whose contributions have been extremely important for the School and are highly appreciated. I thank them most warmly. A final acknowledgement to all those in Erice, Bologna and Geneva who have helped me in so many occasions and to whom I feel very indebted.

Antonino Zichichi

Geneva, October 2014

CONTENTS

STATUS OF EXPERIMENTAL POWER

DISCUSSIONS

CLOSING CEREMONY

«ETTORE MAJORANA» FOUNDATION AND CENTRE FOR SCIENTIFIC CULTURE
TO PAY A PERMANENT TRIBUTE TO GALILEO GALILEI, FOUNDER OF MODERN SCIENCE AND TO ENRICO FERMI, THE "ITALIAN NAVIGATOR", FATHER OF THE WEAK FORCES

INTERNATIONAL SCHOOL OF SUBNUCLEAR PHYSICS
52nd Course: *STATUS OF THEORETICAL UNDERSTANDING AND OF EXPERIMENTAL POWER FOR LHC PHYSICS AND BEYOND*
Erice, 24 June – 3 July 2014

OPENING LECTURE

A LESSON FOR THE FUTURE OF OUR SCIENCE

MY TESTIMONY ON LORD PATRICK M. S. BLACKETT

ANTONINO ZICHICHI
Pontifical Academy of Sciences, Vatican City
INFN and University of Bologna, Italy
CERN, Geneva, Switzerland
Enrico Fermi Centre, Rome, Italy
World Federation of Scientists, Beijing, Geneva, Moscow, New York

Paul A.M. Dirac Lecture Hall – Patrick M.S. Blackett Institute
Erice, 25 June 2014

OPENING LECTURE

A LESSON FOR THE FUTURE OF OUR SCIENCE

MY TESTIMONY ON LORD PATRICK M. S. BLACKETT

ANTONINO ZICHICHI
Pontifical Academy of Sciences, Vatican City
INFN and University of Bologna, Italy
CERN, Geneva, Switzerland
Enrico Fermi Centre, Rome, Italy
World Federation of Scientists, Beijing, Geneva, Moscow, New York

CONTENTS

A Lecture on "My Testimony on Lord Patrick M.S. Blackett" has been delivered at the Clore Lecture Theatre – Imperial College, London, 30th April 2014

(Picture courtesy of the Imperial College London)

The Blackett Institute in Erice.

«ETTORE MAJORANA» FOUNDATION AND CENTRE FOR SCIENTIFIC CULTURE
INTERNATIONAL SCHOOL OF SUBNUCLEAR PHYSICS
52nd Course: *STATUS OF THEORETICAL UNDERSTANDING AND OF EXPERIMENTAL POWER FOR LHC PHYSICS AND BEYOND*
Erice, 24 June – 3 July 2014

Opening Lecture

A LESSON FOR THE FUTURE OF OUR SCIENCE
MY TESTIMONY ON LORD PATRICK M. S. BLACKETT

I – INTRODUCTION

First of all let me welcome all of you to this 52nd Course of the Subnuclear Physics School. This is the first time that I am obliged not to be with you. During these many past decades more than hundred thousands scientists, from all over the world have participated to the Erice Schools, the majority being engaged in activities very far away from Physics.

Let me show the exact data.

DATA ON ACTIVITIES SINCE 1963

124 SCHOOLS,
1.704 COURSES,
118.302 PARTICIPANTS(*)
COMING FROM **932** UNIVERSITIES
AND LABORATORIES OF **140** NATIONS.

My Opening Lecture is the answer to the question posed by some of the many fellows mentioned in the Table above: Why the Institute where we are now is dedicated to Lord Patrick M.S. Blackett?

(*) **129** Nobel Laureates (80 of them were awarded the Nobel Prize after their participation to the Ettore Majorana Schools and 49 were already Nobel when they started to take part in the Centre activities).

Let me thank, Professor Mike Duff, who organized at Imperial College – few weeks ago – a formidable set of events and invited me to recall what did happen when I was 67 years younger than now.

In fact, the first time I learned about the existence of the Blackett group was in 1947 when the "strange particles" called Vs were discovered by a group of physicists whose leader was the same fellow who gave to Sicily the privilege of having two years of peace (1943-1945) instead of the terrible two last years of war.

1947 was the last year of my pre-University School in Trapani (Ximenes Liceum) and when I read about this totally unexpected discovery I could not imagine that seven years later I would have been the youngest member of Professor Blackett physics group, whose detectors (Wilson cloud chamber and associated equipment) were installed in the highest Lab of Europe (3580 meters a.s.l.) the Sphinx Observatory at the Jungfraujoch Lab. Professor Blackett was the leader of the most powerful group of experimental physics in the world.

An incredible sequence of unexpected events allowed me to become a pupil of Professor Blackett. This has been of vital importance not only for my career in Physics but also for my activities including the establishment of the **Ettore Majorana Centre for Scientific Culture in Erice**, the **WFS**, the **World Lab** with its projects implemented in many developing countries and the **Scientific Culture** endeavours to let people know the role of scientific discoveries in everyday life.

Professor Blackett was convinced that it is us, the physicists, who must be engaged directly to let people, outside our labs, know what is the role of Science for the progress of our Civilization.

The scientific discovery at the 1st level of Galilean Science is the real motor for the progress in technological inventions that allow the quality of life to be at the level it is today.

It is thanks to Lord Blackett that I had the privilege to spend an evening with **Bertrand Russell** and know his views on us physicists engaged at the frontiers of human knowledge in order to understand the Logic of Nature.

It is always thanks to Professor Blackett that I learned the existence of the

"**Blackett effect**" which allowed, during the 2nd World War, the British Navy to win the Mediterranean battle thus giving to Sicily the privilege of having two years of Peace (1943-1945) instead of the terrible two last years of war, as said before.

In this lecture I want **to pay tribute** to a physicist who has played a vital role in the discovery of the Subnuclear Universe and in the promotion of Scientific Culture.

Today at the centre of everybody's attention are the problems concerning the **role of Science** in order to be the source of high quality life instead being the source of high precision and high power weapons.

In the years after the 2nd World War these were problems far from being of interest to the great public.

When I had the privilege of being in the physics group, whose leader was Professor Blackett, I learned a lot about these problems.

Professor Blackett was engaged not only in Physics frontiers but also in problems concerning the role of Science in the Culture of our Time with his friend Bertrand Russell.

The year 1947 was a great year. As said before, this is the year when I first knew the existence of the Blackett group where the "strange particles" had been discovered. In 1947 I had finished the pre-University studies and when the discovery of the V-particles was announced, I could not imagine, being in Sicily at that time, that in 1955 I would have succeeded to be the youngest fellow of the Blackett group. **In 1947** no-physicist would have been able to predict the existence of the Subnuclear world.

When the V-particles were discovered by the Blackett group at the Jungfraujoch Lab (3580 meters a.s.l.), Enrico Fermi said: «*These strange particles are the price we have to pay for having discovered* ***probably*** *everything needed to understand the Logic of Nature*».

In fact with all particles discovered (p, n, e, π, γ) it was possible to understand the molecular, atomic and nuclear structure of all types of matter (mountains, oceans, the Earth, the Sun, the Moon, Galaxies). And this was not all. It was also possible to predict the existence of "Virtual Phenomena".

Blackett with his young fellow Occhialini discovered (1932) the associated production of (e^+e^-) pair, thus giving experimental support to the existence of the "vacuum polarization" process theoretically envisaged by Dirac thanks to the discovery of his equation (1929). The "vacuum polarization" is the starting point of "Virtual Physics" which is now the frontier of High Energy Physics. Before 1947 no one was able to predict the discovery of the Lamb-shift, a "Virtual Physics" effect much more simple than the "vacuum polarization". We all work thanks to the existence of "Virtual Physics". Let me try to explain what it is.

Imagine the most powerful technology. No matter how powerful and precise it is, this technology will be unable to "directly" observe (measure, determine) "Virtual Phenomena". You could say: these phenomena do not exist. Here is the great novelty of those times. A virtual phenomenon produces rigorously predictable and experimentally observable "effects". Example: the so-called "anomalous" magnetic moment of the "heavy lepton" called "muon".

Today we can study the Unification of all Fundamental Forces of Nature, thanks to the fact that we can mathematically compute all possible "Virtual Phenomena" produced by the three Fundamental Forces of Nature: the Electroweak, Strong and Gravitational. All these new frontiers of Physics started in 1932 with Blackett and Occhialini discovery of the simultaneous production of (e^+e^-) confirmed by the "Lamb-shift" in 1947.

In the same year another wanted particle the "nuclear glue", the π–meson, was discovered. The π–meson allows the existence of many protons and neutrons in a nucleus, since it acts as the "nuclear glue".

And now comes the incredible sequence of unexpected events which allowed me to become the pupil of **Professor Blackett**.

II – AN INCREDIBLE SEQUENCE OF UNEXPECTED EVENTS

The year was 1955 and I was with a fellowship of the INFN in the "Guglielmo Marconi" Physics Institute of the Rome University La Sapienza, going up and down to and from the Cervinia Lab, since cosmic rays were the only source of High Energy Particles in Europe.

It was 12.00 p.m. – Sunday – and I was working with cosmic rays Wilson cloud chamber pictures taken by our group in the Cervinia Lab (3480 meters a.s.l.). A phone call by Bruno Brunelli, the Deputy Group Leader, was asking me if I could leave next day to participate at the International Conference since he was sick (40° fever). These were times when each Country had a quota for participants. Minus one implied that, at the next Conference, the Country had his quota reduced. This was the reason for "last minute call".

I did go to the Conference where a great event was announced.

A new discovery by the same group who in 1947 discovered the V–particles. This was the first name originally given by the group leader, Professor Blackett, to this unexpected discovery.

Why V-particles? Because they appeared as inverted V. Late one two V_S^0 were identified to be a baryonic event ($\Lambda^0 \to p\pi^-$) and a heavy mesonic event ($\theta^0 \to \pi^+\pi^-$). The discovery of the Vs was the starting point of the "strange particle" physics which no-one in 1947 would have imagined to be the opening of a new horizon towards the Subnuclear Universe.

The new discovery presented by a member of the Blackett group (G.D. James) needed a reaction which was "forbidden". Let me show a copy of the Proceedings where this is reported [G.D. James, *Some Notes on the Production of V-particles*, Suppl. Nuovo Cimento Vol. IV, X n. 2, page 328 (1956)].

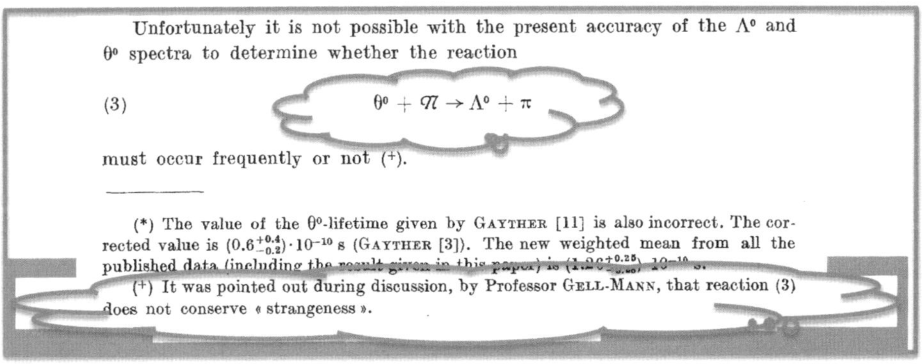

Unfortunately it is not possible with the present accuracy of the Λ^0 and θ^0 spectra to determine whether the reaction

(3) $$\theta^0 + n \to \Lambda^0 + \pi$$

must occur frequently or not (+).

(*) The value of the θ^0-lifetime given by GAYTHER [11] is also incorrect. The corrected value is $(0.6^{+0.4}_{-0.2})\cdot 10^{-10}$ s (GAYTHER [3]). The new weighted mean from all the published data (including the result given in this paper) is $(1.26^{+0.25}_{-0.20})\cdot 10^{-10}$ s.

(+) It was pointed out during discussion, by Professor GELL-MANN, that reaction (3) does not conserve « strangeness ».

As reported by G.D. James, it was pointed out by Professor Gell-Mann that reaction (3) did not conserve the strangeness quantum number. This was a great event for my career. Professor Butler, the co-discoverer of the first strange particles was the Chair of the Session.

(Picture courtesy of the Imperial College London)

Back row left to right:

Prof. P.T. Matthews (Theoretical Physics), Prof. M. Blackman (EXSS), Prof. J. Mc Gee (Instrumental Technology), Prof. Sir J. Mason (Cloud Physics)

Front row left to right:

Prof. A. Salam (Theoretical Physics), Prof. C.C. Butler (HENP), Lord P.M.S. Blackett (Head of Physics), Prof. W.D. Wright (Optics), Prof. H. Eliott (Cosmic Ray and Space Physics)

G.D. James did not answer to Gell-Mann objection. In my broken English I said that at production (10^{-23} sec) the θ^0 has $S = +1$ but, according to the most recent developments of our understanding the physics of strange particles, the θ^0–meson, later becomes a mixture of $(\theta^0 \pm \bar{\theta}^0)$. The correct reaction is therefore

$$\bar{\theta}^0 + N \rightarrow \Lambda^0 + \pi .$$

The only difficulty is that there was no experimental evidence for the production of heavy mesons with positive and negative strangeness.

A.Z. and Murray Gell-Mann talking about heavy mesons with positive and negative strangeness.

Europe had only **Cosmic Rays.**

No SC – now 60th anniversary celebration at CERN

PS + SPS

LEP

LHC.

After the 1955 Conference Professor Blackett invited me to join his group (1955), and this is how I became the youngest member of his group. Two years later (1957) the associated production of heavy mesons (whose name from θ became K) with positive and negative strangeness was discovered, as illustrated below.

After 2 years: 1955 + 2 = 1957

Associated production of

$$\begin{pmatrix} S = +1 \\ S = -1 \end{pmatrix} \Rightarrow \text{K mesons}$$

$$K^+ + \overline{K}^0$$

$$K^0 + \overline{K}^0$$

was discovered by the Blackett group.

The photos, taken at the Jungfraujoch Lab (3580 meters a.s.l), **were at least 3 years old**. How it happens that in the photos taken at Jungfraujoch there was so much Physics and in the photos taken at Cervinia there was nearly never an interaction producing many particles?

Professor Blackett was the author of the Electronic Trigger System with threshold Energy

$$E_{Rays}^{Cosmic}\ Interaction\ \geq 10\ GeV$$

The Cervinia Trigger was at an energy level ten time lower E $\gtrsim$ 1 GeV.

The Jungfraujoch Lab where the Blackett group had its powerful detectors placed in the Sphinx Observatory.

Another view of the Jungfraujoch Lab.

Before going to the next Chapter let me give a synthesis of my sequence of events in the Table 1 below.

1947 – The end of my studies at the pre-University level.
1947 ① Lamb–shift *wanted* ② π–meson *wanted* ③ Strange particles – *Totally Unexpected*
1948 + 5 ⇒ 1953: end of University studies and becoming [1955] the youngest member of the Blackett group.

Table 1

Thanks to Blackett the same research group has been involved successfully with the three (1947) great achievements: Virtual Physics (Lamb-shift), Nuclear Glue (π–meson) and Strange Particles.

The most spectacular one being the **Strange Particles** illustrated in **Figure 1**.

The final states of the first two Vs discovered by the Blackett group.

$$\Lambda^0 \to p + \pi^-$$

$$\theta^0 \to \pi^+ \pi^-$$

implied the existence of the nuclear glue (π–meson), with all consequences illustrated in **Figure 2**.

The discovery of the Lamb-shift was the simplest example of Virtual Physics (vacuum polarization effects) with all consequences illustrated in **Figure 3**.

From now on the Figures will report in a synthetic way the developments up to the present day of the new horizons opened by the discoveries originated in the Blackett Lab.

IIIA – BLACKETT AND THE ORIGIN OF THE SUBNUCLEAR UNIVERSE

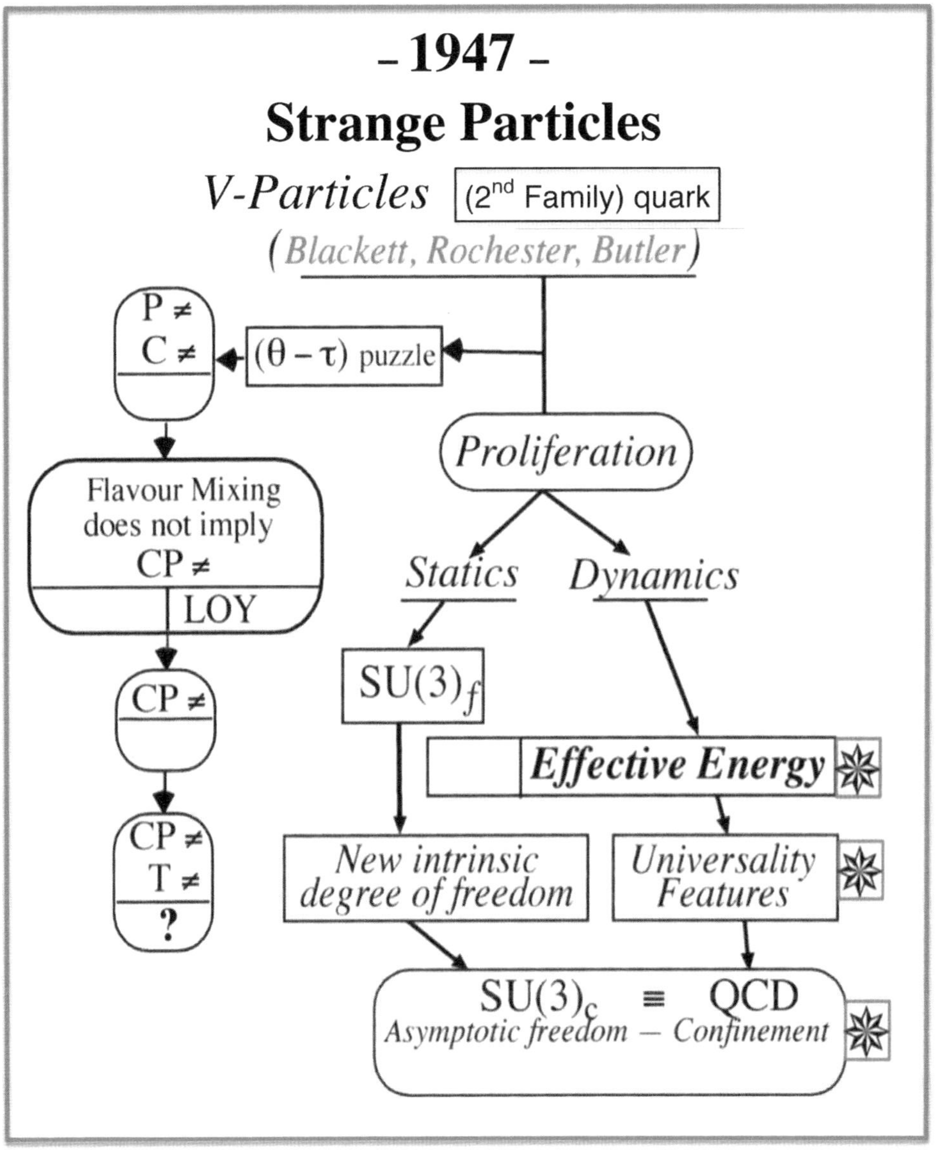

Figure 1

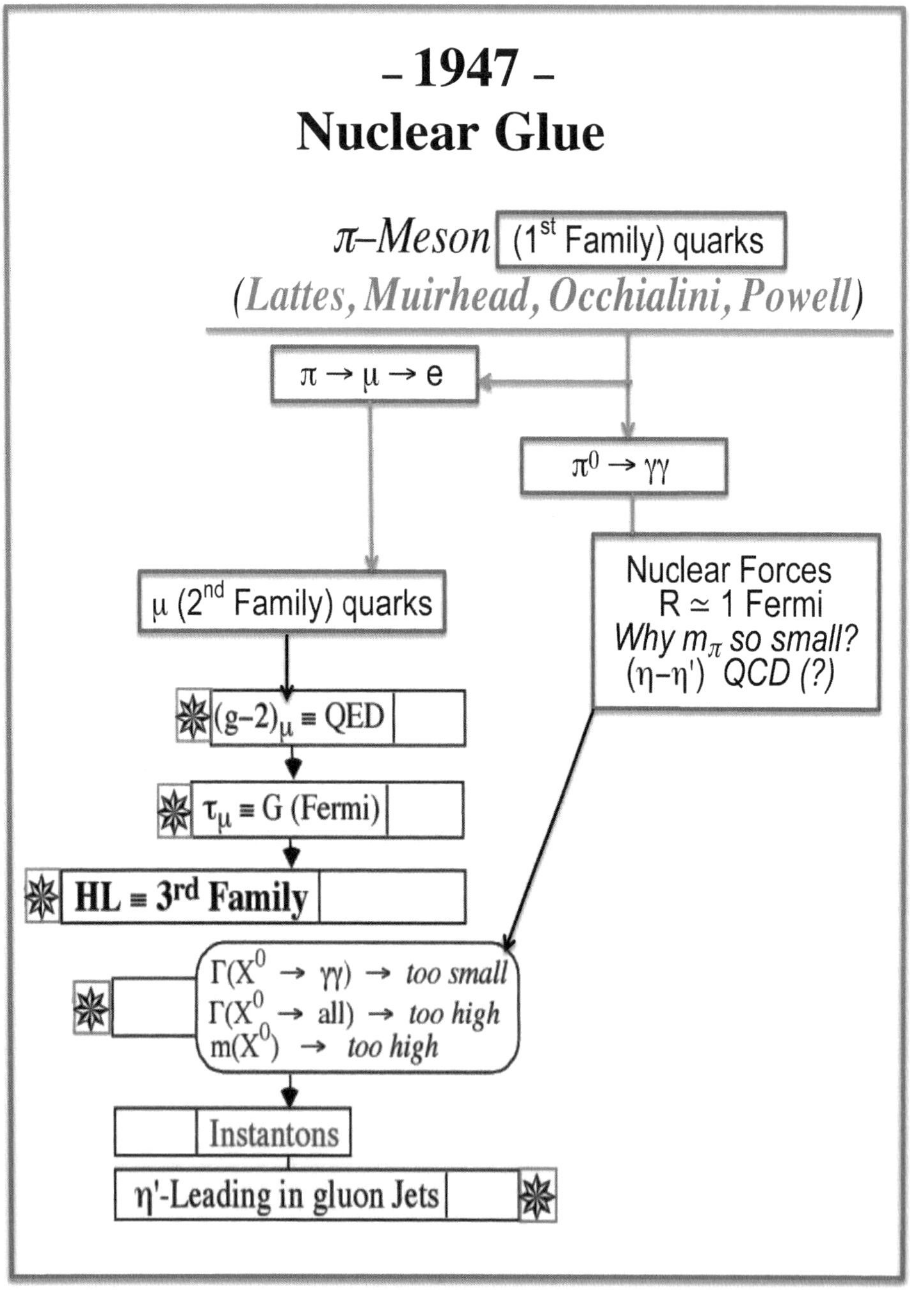

- 1947 -
Nuclear Glue
π–Meson (1st Family) quarks
(Lattes, Muirhead, Occhialini, Powell)
π → μ → e
π0 → γγ
μ (2nd Family) quarks
Nuclear Forces
R ≃ 1 Fermi
Why mπ so small?
(η−η') QCD (?)
(g−2)μ ≡ QED
τμ ≡ G (Fermi)
HL ≡ 3rd Family
Γ(X0 → γγ) → too small
Γ(X0 → all) → too high
m(X0) → too high
Instantons
η'-Leading in gluon Jets

Figure 2

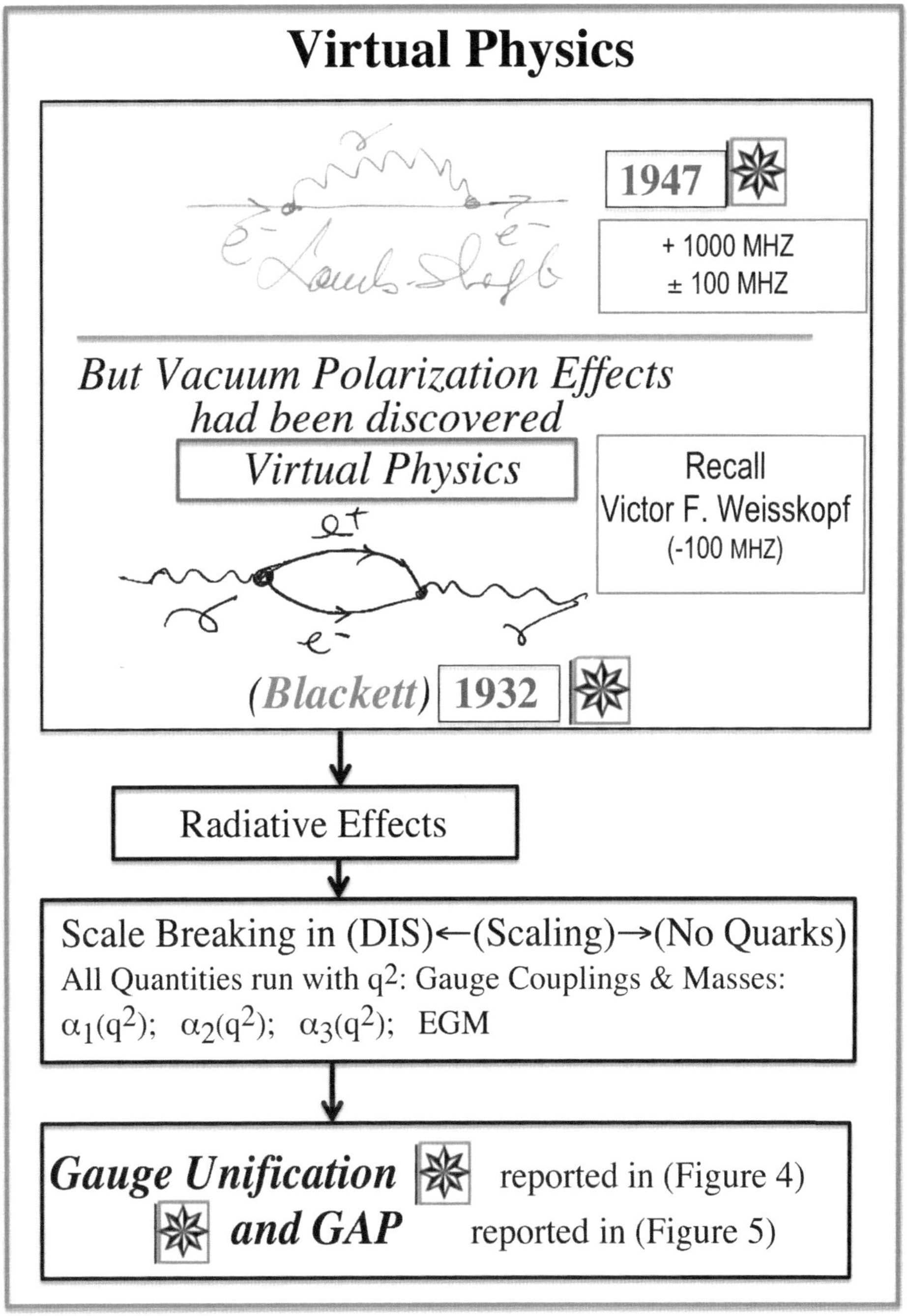
Virtual Physics
1947
+ 1000 MHZ
± 100 MHZ
But Vacuum Polarization Effects
had been discovered
Virtual Physics
Recall
Victor F. Weisskopf
(-100 MHZ)
(Blackett) 1932
Radiative Effects
Scale Breaking in (DIS)←(Scaling)→(No Quarks)
All Quantities run with q^2: Gauge Couplings & Masses:
$\alpha_1(q^2)$; $\alpha_2(q^2)$; $\alpha_3(q^2)$; EGM
Gauge Unification reported in (Figure 4)
and GAP reported in (Figure 5)

Figure 3

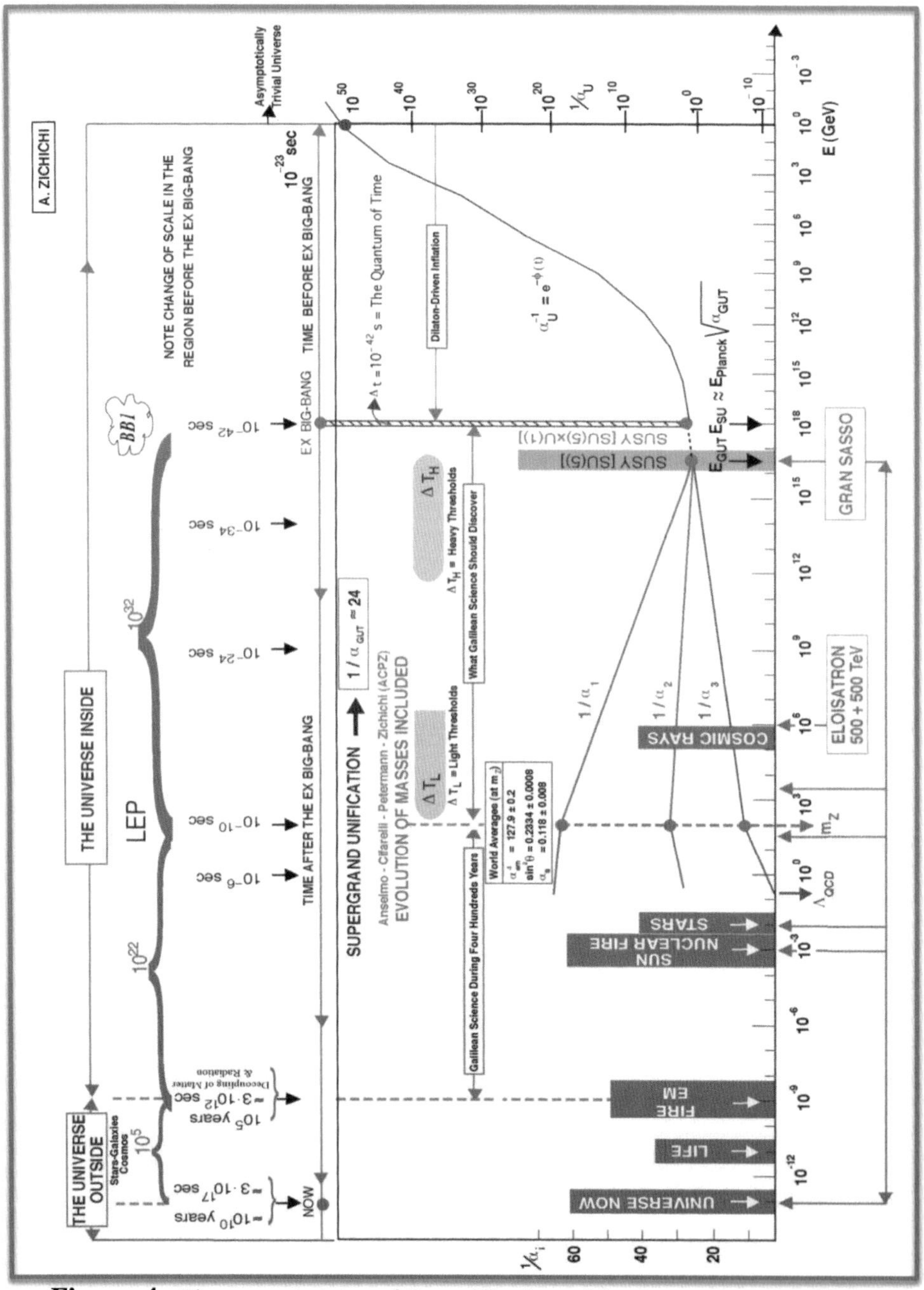

Figure 4: The present status of the unification of the three gauge couplings α_1 α_2 α_3 is reported. This is the most accurate description of the unification whose mathematics is in Figure 6.

THE GAP BETWEEN E_{GUT} AND E_{PLANCK}

In Figure 5 is reported the GAP which exists between the Planck energy, E_{Planck} , and the two energy levels, E_{GUT}, where the three gauge couplings converge and E_{SU}, the energy where the RQST (Relativistic Quantum String Theory) puts the origin of the Gravitational Forces. The existence of this GAP is, at present, being studied.

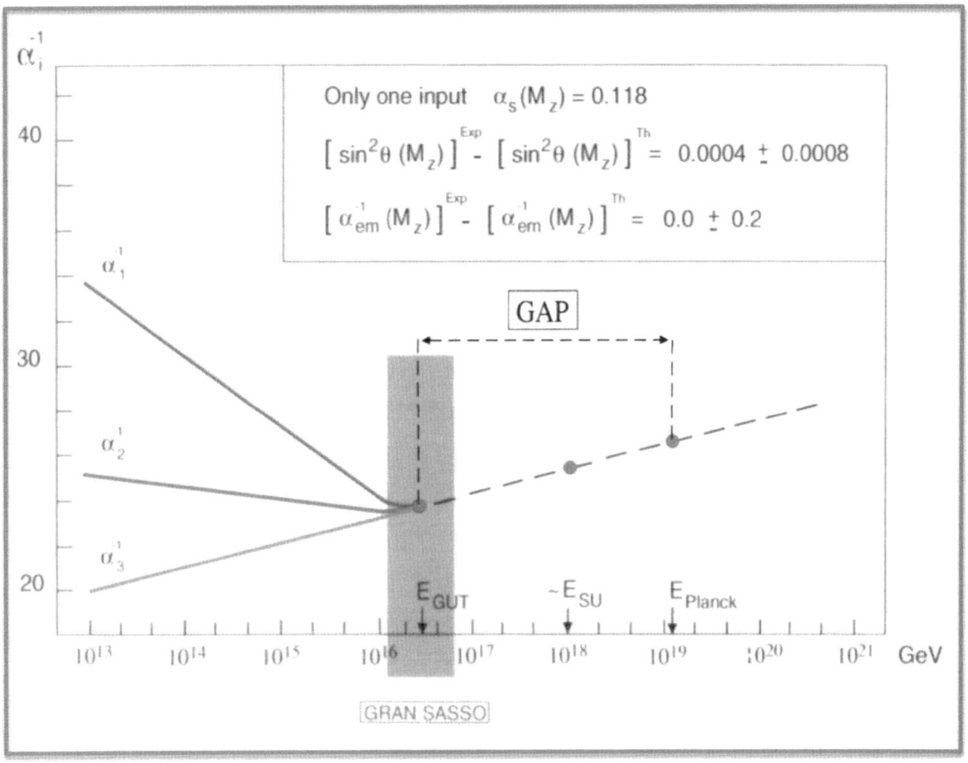

Figure 5

GUT (Grand Unified Theory): the Mathematics (Figure 6).

THE UNIFICATION OF ALL FUNDAMENTAL FORCES

The three lines (α_1^{-1}, α_2^{-1}, α_3^{-1}) in Figure 4 result from calculations executed with a supercomputer using the following system of equations:

$$\mu \frac{d\alpha_i}{d\mu} = \frac{b_i}{2\pi}\,\alpha_i^2 + \sum_j \frac{b_{ij}}{8\pi^2}\,\alpha_i\alpha_j$$

α_i , α_j (with $i = 1, 2, 3$; and $J = 1, 2, 3$ but $i \neq j$),

This is a system of coupled non-linear differential equations where the existence of the Superworld is included. This system describes how the gauge couplings (α_1, α_2, α_3) vary with "μ", the basic parameter which depends on the energy of the elementary process, from the maximum level of Energy (Planck Scale) to the energy level of our world. **During more than ten years (from 1979 to 1991), no one had realized that the energy threshold for the existence of the Superworld was strongly dependent on the "running" of the masses.**

This is now called: **the EGM effect** (from the initials of Evolution of Gaugino Masses). This effect produces a factor 700 in the threshold for the lightest supersymmetry particle. No one knows the value of the threshold Energy for the production of the lightest supersymmetry particle. The EGM effect lowers this value by a factor 700. Suppose that somebody would say that the threshold is at 700 TeV, thanks to the EGM effect this value is going to be 1 TeV.

Figure 6

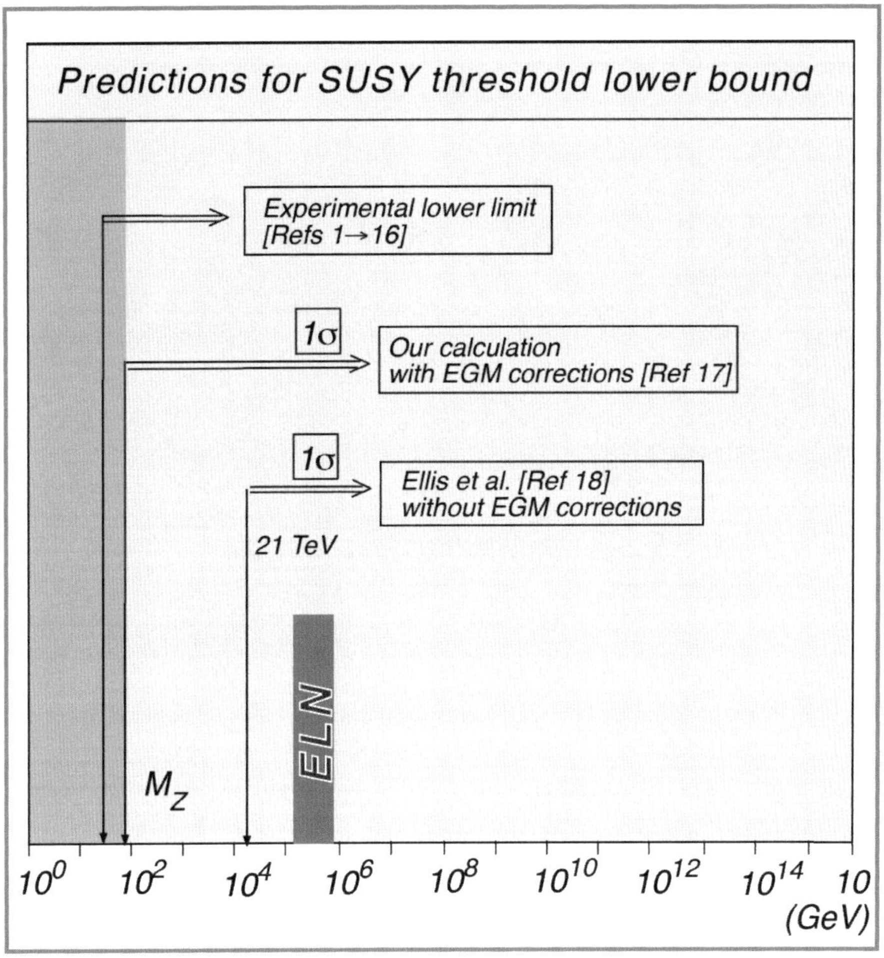

Figure 7

This Figure illustrates the EGM effect for the case when the prediction was at 21 TeV.

IIIB – BLACKETT AND THE WHOLE OF OUR KNOWLEDGE

– 1929 –

(From left) Lord Patrick Maynard Stuart Blackett, Pyotr L. Kapitza, Paul Langevin, Lord Ernest Rutherford, Charles Thomson Rees Wilson outside Cavendish Laboratory (1929). **This photo – dated 1929 – is a gift in 1955 of Professor Blackett to the youngest member of his group** (A.Z.). On this occasion Professor Blackett said that on the same year (1929) the best synthesis on the future of Physics had been given in Florence by **Orso Mario Corbino**, the founder of what would have been the famous "Panisperna Group" where Enrico Fermi invented the "slow neutron technology" which allowed all elements of the Mendeleev Table to become "radioactive". This is how Fermi discovered the "Weak Forces". Here are the words of Professor Corbino(*): "*The only possibility of great discoveries in Physics dwell in the eventuality of succeeding in modifying the inner nucleus of the atom. This will be the task truly worthy of the future in Physics*".

(*) «*La sola possibilità di grandi scoperte in Fisica risiede nella eventualità che si riesca a modificare il nucleo interno dell'atomo. E questo sarà il compito veramente degno della Fisica futura*».

Let me say few words on each physicist in the photo.

Blackett [discovery of (e^+e^-) production in cosmic rays and of the first examples of strange particles. A picture was with the "mesonic state" (θ^0) and the "baryonic state" (Λ^0). This is the reason for giving opposite values of the new quantum number called "strangeness": S = +1 to the meson and S = −1 to the baryon].

Kapitza [discovery of superfluidity (no secret Labs, Erice Statement) and hero in Science for his courage to refuse the directorship of the nuclear-fusion-bomb (called now H–bomb) thousands times more powerful than the nuclear-fission-bomb **called atomic** (Hiroshima and Nagasaki). Why called Atomic? Eugene Wigner (Time Reversal Invariance) told me the reason. Before 1940 the word **"Nuclear" did not exist in our language**. People would have been confused. The most advanced scientific word was Atomic. This is why the first nuclear bomb was called "Atomic"](*).

Langevin [discovery of paramagnetism (Radioactivity) and inventor of the "twin paradox". He gave to Becquerel the correct suggestion which allowed the discovery of radioactivity].

Rutherford [discovery of the atomic nucleus, i.e. of the fact that atoms are composite objects and that more than 99% of their mass is concentrated in a volume which is a million of a billion times smaller that the atomic volume].

Wilson [invention of the "**cloud-chamber**": the instrument which allowed during more then five decades the enormous number of discoveries using the "cosmic rays" as primary high energy particles. **Nature was very kind with us: all new particles have a life-time of $\simeq 10^{-10}$ sec** (perfect for centimeters flights). The production time for the same particles is $\simeq 10^{-23}$ sec].

(*) Nuclear Fission → [1945 – Uranium Bomb (Hiroshima);
1945 – Plutonium Bomb (Nagasaki);

Nuclear Fusion → [1952 – H-Bomb (USA);
1953 – H-Bomb (URSS).

VIRTUAL PHYSICS AND VIRTUAL HISTORY

The historians have invented "Virtual History" following the Physics discovery of "Virtual Physics".

As said before Professor Blackett is the physicist who discovered in 1932 the simultaneous production of (e^+e^-) pairs in cosmic rays thus giving the needed experimental evidence for the existence of the first example of "Virtual Physics", the "vacuum polarization effects", theoretically predicted by Dirac in 1929.

"Virtual **Physics**" and "Virtual **History**" operate in the two asymptotic limits of Complexity (minimum and maximum levels) but share a common property: **Evolution**.

In fact History is "Evolution of the World in its Real Life" ≡ (**EWRL**), and Science is "Evolution of our Basic Understanding of the laws governing the world in its Structure" ≡ (**EBUS**).

The two evolutions (EWRL) and (EBUS) are dominated by Unexpected Events with Enormous Consequences (**UEEC** events) as illustrated in Table 2.

The world would not be as it is if any of the UEEC events reported in the History columns of Table 2 did not take place. The Virtual world, which would come from "Virtual History", would not be like our world, but "Virtual Physics" is exactly our Physics.

Virtual History could not have been invented if "Virtual Physics" had not been discovered thanks to Professor Blackett. In Table 2, Science and History are compared. The Logic of Nature allows the existence of the lowest limit of Complexity, **Science**, and of the highest limit of Complexity, **History**. How to go from the lowest to the highest limit is illustrated in Table 3.

UEEC EVENTS IN HISTORY AND IN SCIENCE

	In History ≡ EWRL		In Science ≡ EBUS
I	What if Julius Caesar had been assassinated many years before?	*I*	What if Galileo Galilei had not discovered that $F = mg$?
II	What if Charles VII had not been able to win the 100 years war?	*II*	What if Newton had not discovered that $F = G\frac{m_1 \cdot m_2}{R_{12}^2}$?
III	What if America had been discovered a few centuries later?	*III*	What if Maxwell had not discovered the unification of electricity, magnetism and optical phenomena, which allowed him to conclude that light is a vibration of the EM field?
IV	What if Napoleon had not been born?	*IV*	What if Becquerel had not discovered radioactivity?
V	What if Louis XVI had been able to win against the 'Storming of the Bastille'?	*V*	What if Planck had not discovered that $h \neq 0$?
VI	What if the 1908 Tunguska Comet had fallen somewhere in Europe instead of Tunguska in Siberia?	*VI*	What if Lorentz had not discovered that space and time cannot both be real?
VII	What if the killer of the Austrian Archduke Franz Ferdinand had been arrested the day before the Sarajevo event?	*VII*	What if Einstein had not discovered the existence of time-like and space-like real worlds? Only in the time-like world, simultaneity does not change, with changing observer.
VIII	What if Lenin had been killed during his travelling through Germany?	*VIII*	What if Rutherford had not discovered the nucleus?
IX	What if Hitler had not been appointed Chancellor by the President of the Republic of Weimar Paul von Hindenburg?	*IX*	What if Hess had not discovered cosmic rays?
X	What if Pyotr Kapitza accepted to be the leader of the USSR H-bomb Project as wanted by Stalin?	*X*	What if Dirac had not discovered his equation, which opens new horizons, including the existence of the antiworld?
XI	What if Nazi Germany had defeated the Soviet Union?	*XI*	What if Fermi had not discovered weak forces?
XII	What if Karol Wojtyla had not been elected Pope, thus becoming John Paul II?	*XII*	What if Fermi and Dirac had not discovered the Fermi–Dirac statistics?
XIII	What if Gorbachev had not been defeated by Yeltsin?	*XIII*	What if Yukawa had not proposed the existence of a "meson" in order to have the nuclear glue?
XIV	What if the USSR had not collapsed?	*XIV*	What if the 'strange particles' had not been discovered in the Blackett Lab?

Table 2

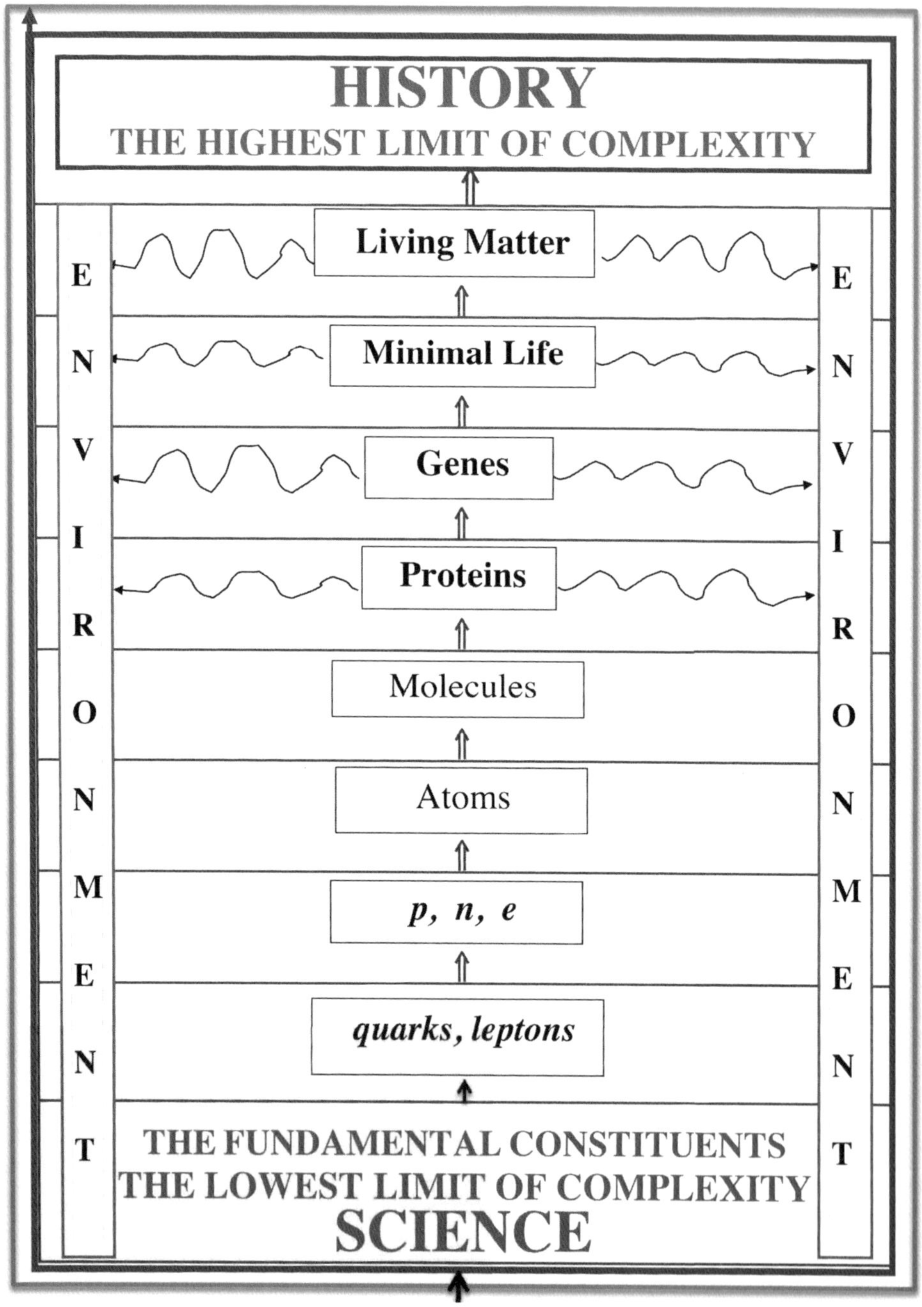
HISTORY
THE HIGHEST LIMIT OF COMPLEXITY
Living Matter
Minimal Life
Genes
Proteins
Molecules
Atoms
p, n, e
quarks, leptons
THE FUNDAMENTAL CONSTITUENTS
THE LOWEST LIMIT OF COMPLEXITY
SCIENCE
E N V I R O N M E N T
E N V I R O N M E N T

Table 3

Science and History bring us to consider the "whole of our knowledge", reported in Figure 8. Let me say few words on Figure 8.

The **First Big Bang** (BB1) is needed in order to describe how we go from the Vacuum to the Universe, made with inert matter.

The **Second Big Bang** (BB2) is needed in order to describe how we go from inert matter to the matter endowed with life. This is being studied in many laboratories. Hundreds of scientists are fully engaged in studying what is called the "problem of minimal life": i.e. how many pieces of inert matter are needed in order to produce the most elementary piece of "living matter".

The **Third Big Bang** (BB3) is needed in order to describe how we go from the matter endowed with life (and no Reason) to the most elementary form of living matter endowed with the privilege of having also Reason.

Attention please: with the term Reason it is not meant the most simple form of Reason needed to the living matter in order to guarantee life.

There are hundreds of thousands forms of living matter (vegetal and animal, small and big) but only one form of living matter is able to invent "**Language**" (which gives rise to Permanent Collective Memory, better known as Written Language), "**Rigorous Theoretical Logic**" (known as Mathematics) and "**Rigorous Experimental Logic**" (known as first level Galilean Science).

These three Big Bangs are indicated in Figure 8, where the knowledge of the Universe plays a central role.

However, if the Third Big Bang would have not occurred, we could not be here discussing our problems and I could not have written this lecture.

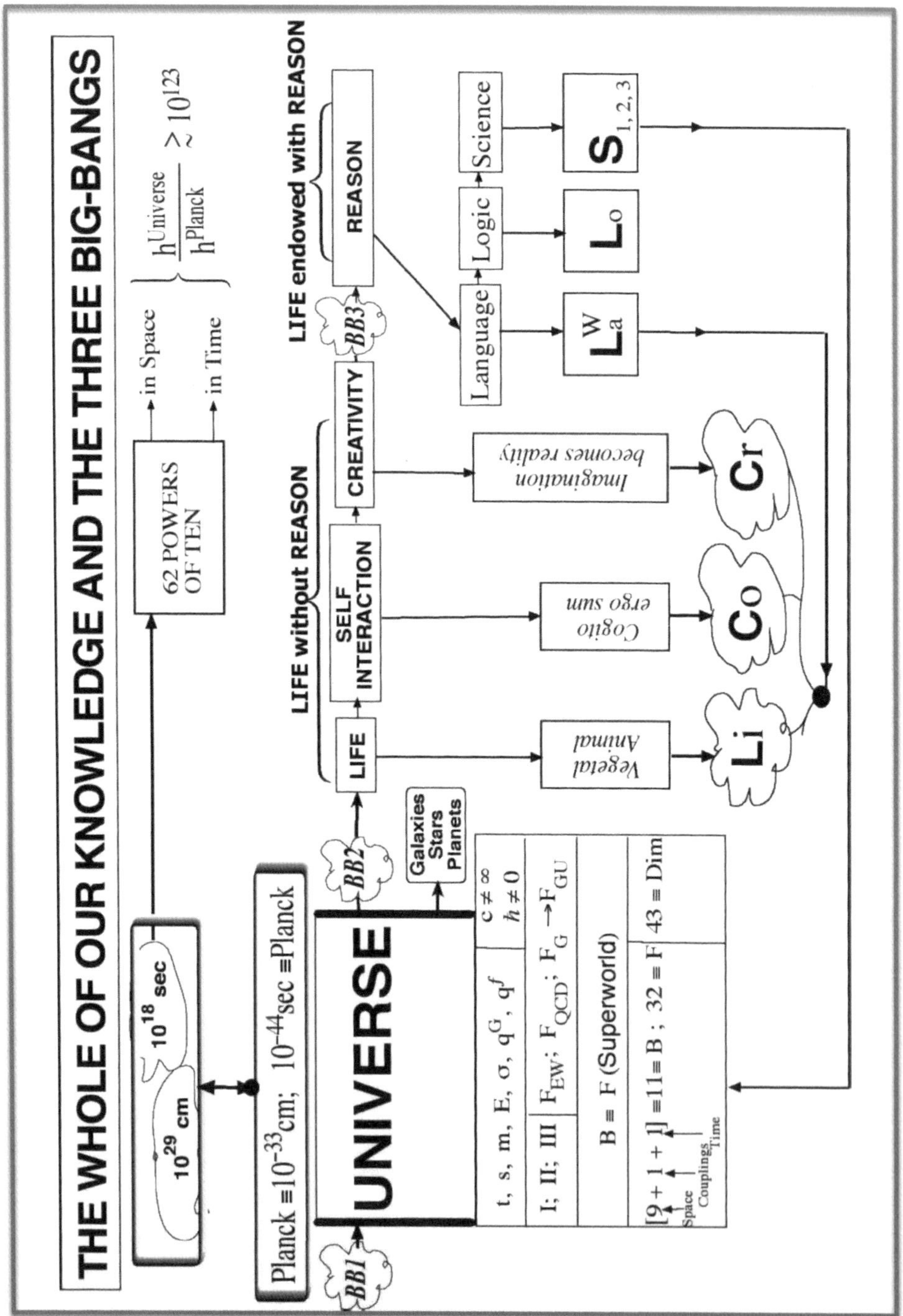
THE WHOLE OF OUR KNOWLEDGE AND THE THREE BIG-BANGS
10^{29} cm
10^{18} sec
62 POWERS OF TEN
in Space
in Time
$\frac{h^{Universe}}{h^{Planck}} \gtrsim 10^{123}$
Planck ≡10^{-33}cm; 10^{-44}sec ≡Planck
BB1
UNIVERSE
BB2
Galaxies Stars Planets
LIFE without REASON
LIFE
SELF INTERACTION
CREATIVITY
BB3
LIFE endowed with REASON
REASON
Vegetal Animal
Cogito ergo sum
Imagination becomes reality
Language
Logic
Science
Li
Co
Cr
L_a^W
L_o
$S_{1,2,3}$
t, s, m, E, σ, q^G, q^f
c ≠ ∞ ħ ≠ 0
I; II; III
F_{EW}; F_{QCD}; $F_G \to F_{GU}$
B ≡ F (Superworld)
[9 + 1 + 1] ≡11≡ B ; 32 ≡ F
43 ≡ Dim
Space Couplings Time

Figure 8

IV – BLACKETT AND RUSSELL (GALILEI, EINSTEIN, GÖDEL)

It is thanks to **Lord Blackett** that I had the privilege to spend an evening with **Bertrand Russell** and know his views on us physicists engaged at the frontiers of human knowledge in order to understand the Logic of Nature.

Blackett invited me at his place along with his friend, Bertrand Russell, who said: «*You, physicists, would sell your souls to the devil just to be allowed to implement your experiments*». And turning to me: «*What about you, my young fellow, what do you think of it?*».

Actually, that meeting at Blackett's house had the purpose of discussing **Relativity** and the role played by **Galilei**. Blackett was not happy of what Russell had written in his book the **ABC of Relativity**; Galilei was only cited once despite the fact that Galilei's Principle of Relativity was so well formulated that included Electromagnetic Phenomena (unknown to Galilei). I was there because, from childhood, I had read all of Galilei's writings.

From the Galilei-Einstein theme, the conversation turned to a subject that was **gnawing at Russell's** mind: *Physicists, that vile cursed race of beings*. He did not say it out loud, but he certainly harboured the thought. At one point he even exclaimed: "**The day there will be a slowdown in the race for armaments you physicists will be in trouble**. We do not know for how many decades this will continue. Therefore, your future is secure. However, if I could think of a world without any more conflicts between great powers, if I were you I would change job. I don't know when the "**cold war**" between the two super powers will become "**hot**". We should do all we can to avoid this and you should know that, in a planet without the two opposing super powers, the race for armaments will subside and so will the financing of frontiers research. The **political powers are not financing your projects to better understand the Logic of Nature. This belongs to us and the truth will be in our hands. You physicists are getting all that attention from governments because of the potential applications of your research to war technologies**. The day that interest will wane, there will be no more financing for your projects".

Later on, coming back to Russell's rather shocking point of views, Blackett told me: «*You know, Russell is mad at us because, after all,* ***Kurt Gödel*** *is one of us*».

This Era of Peace, for which Blackett and Russell strongly disagreed as to the way and means to achieve it, has now become reality; there is no more need for a race for armaments. We are now at the crossroads: will the political powers of free and democratic countries follow the road predicted by Russell's pessimism and feared by all the free worlds' scientists? Or will they go the way **Blackett had shown**, during the difficult times following the end of the war, by putting in place the **first European Scientific Institution, CERN**, which he wanted to be beyond all ideological, political and racial barriers, without secrets or frontiers, and entirely dedicated to fundamental research without any kind of warlike technological applications?

Russell

The further away the scientific discovery is from what the culture believes having understood **the more unpredictable the technological applications**, for good or evil, will be. You are talking now about Fundamental Forces and elementary particles without having understood what is at play. After the Nuclear Universe you are working to discover another type, more sophisticated and microscopic, of Universe. The governments, **as long as the secret laboratories will continue to exist**, will keep financing you out of the reciprocal fear of the new technologies which could turn out to be applicable to arms that would be ever more precise and ever more potent.

V – THE "BLACKETT EFFECT" IN THE 2nd WORLD WAR

It is always thanks to Professor Blackett that I learned the existence of the "**Blackett effect**" which allowed the British Navy to win the Mediterranean battle during the 2nd World War. The Blackett effect refers to the problem of going from A to B: the best solution is to send all navies distributed in concentric circles. The maximum loss cannot exceed 70%. The proof was given by the fact that the British Navy succeeded to go on liberate Malta with few per cent losses. Professor Blackett was Chief Advisor on Operational Research in the Admiralty during the 2nd World War.

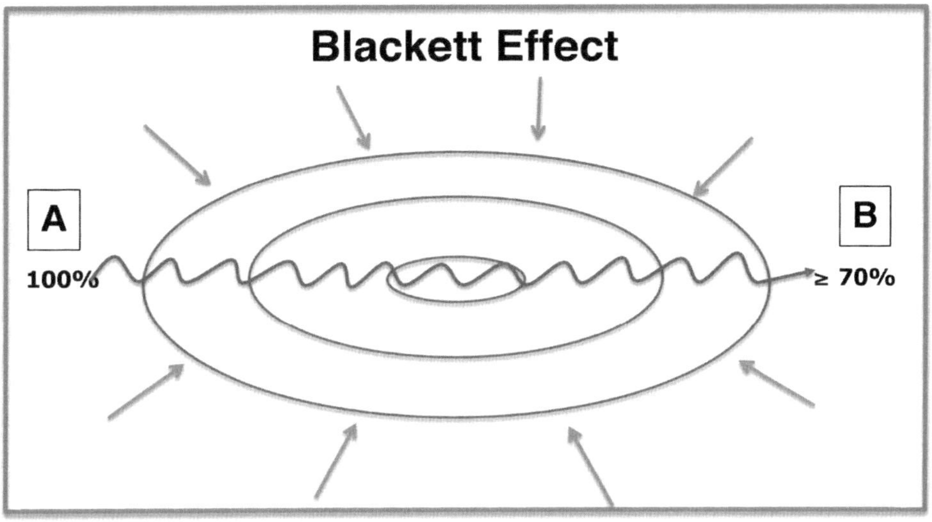

Note: The surrender of the Island (Malta) was considered inevitable.

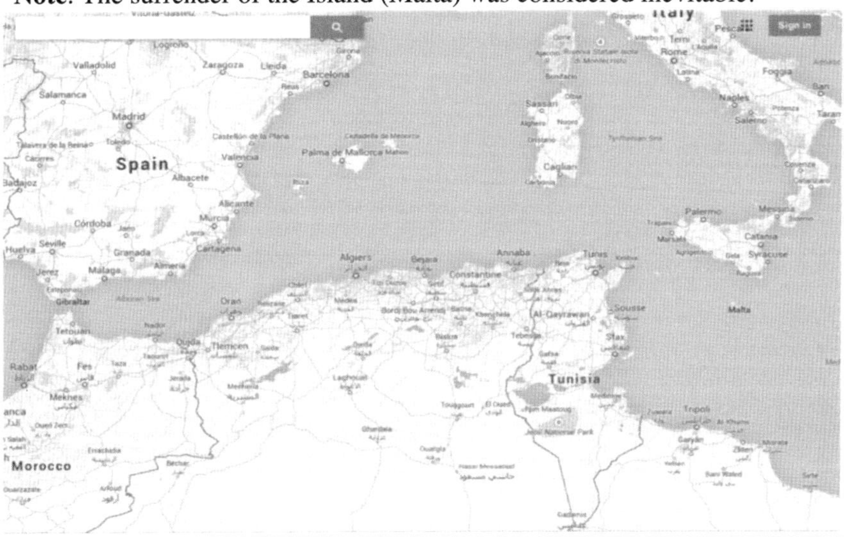

Thanks to the Blackett Effect the war in Sicily ended in 1943. The 2 years of terrible tragedies (1943-1945) **did not occur in the place where I was living.**

VI – NEW INSTITUTIONS FOUNDED

Being a pupil of Blackett has been of vital importance not only for my career in Physics but also for my activities including the establishment of the **Ettore Majorana Centre for Scientific Culture in Erice**.

Professor Blackett was convinced that it is us, the physicists, who must be engaged directly to let people, outside our labs, know what is the role of Science for the progress of our Civilization.

The scientific discovery at the 1st level of Galilean Science is the real motor for the progress in technological inventions that allow the quality of life to the be at the level it is today.

From left: Luigi Dadda, Pierre A. Piroué, Enrico Bignami, Yuval Ne'eman, Richard L. Garwin, John C. Eccles, Eugene P. Wigner, the author, Edward Teller, Paul Adrien Maurice Dirac, George Charpak (1981).

The constitutive act of
the Ettore Majorana Foundation and Centre for Scientific Culture (EMFCSC)

The constitutive part and the name of the new institution.

We, here at CERN, now and for the years to come, we have decided to establish a new institution named after the great Sicilian physicist Ettore Majorana.

This institution has to be called "foundation and centre for scientific culture".

The foundation is the central body to provide through our joint action and collaboration the strongest possible support to the centre for scientific culture. We will do our best to convince universities, laboratories and other institutions, the world over, to contribute to the activities of the

"Ettore Majorana Foundation and Centre for Scientific Culture"

in the most efficient way. We do want, as it is presently at CERN, that this new institution operates with a minimum of bureaucracy and we will support with our promotion, the world over, its activities.

Purpose of the EMFCSC.

The purpose of the EMFCSC is to implement the spirit of this great european laboratory, CERN, where all the activities are without secrecy and where every physicist, no matter his nationality, his religion, his political ideas, can freely take part in the challenges of modern galilean science, as the founders of CERN wished.

The activities of the EMFCSC must be open to all the persons who want to learn, exactly like in the early times, when the first university was established in Bologna.

The EMFCSC must not issue certificate or diploma but only and exclusively learning. Any formal burden should be avoided.

The EMFCSC has to implement, not only in physics but in all fields of scientific and cultural knowledge, the great ideals of the founders of CERN, some of whom are also among the founders of this new institution (P.M.S. Blackett, I.I. Rabi). Our purpose is to contribute, through the EMFCSC, to establish, the world over a new trend in the international collaboration.

1/2

Power of action and responsibility.

We want that this new institution — EMFCSC — is to be governed by only one of us who has to be the president of the foundation and the director of the centre.

We give all powers of action and all responsibility to him which we decide to be Antonino Zichichi.

We do emphasise that his action has to be for the EMFCSC to become, in the years come, an example of how we firmly believe that collaboration between scientists and thinkers has to be implemented the world over.

Location of the head quarters of the EMFCSC.

The EMFCSC must have its head quarters in the city of the ancestors of Ettore Majorana and of one of us (Antonino Zichichi). Other sections can be established in other places, if needed.

The Future of the EMFCSC.

The new institution EMFCSC must keep the present document as the basic act of establishing its legal existence and the only basic law for all its activities in all fields of advanced human knowledge for the years to come.

Note of legal value.

V.F. Weisskopf

In my position as CERN DG I have decided to allow this document to be signed here in this extra-over national geographical area and to have legal value without any further bureaucratical burden.

Signed J.S. Bell

P.M.S. Blackett

I.I. Rabi

V.F. Weisskopf

A. Zichichi

Geneva, May 8th 1962

2/2

The World Federation of Scientists

Organisation & Status

The World Federation of Scientists (WFS) was founded in Erice, Sicily, in 1973, by a group of eminent scientists led by Isidor Isaac Rabi and Antonino Zichichi. Since then, many other scientists have affiliated themselves with the Federation, among them T. D. Lee, Laura Fermi, Eugene Wigner, Paul Dirac and Piotr Kapitza.

The WFS is a free association, which quickly grew to include more than 10,000 scientists drawn from 110 countries. All members share the same aims and ideals and contribute voluntarily to uphold the Federation's Principles. The Federation promotes international collaboration in Science and technology between scientists and researchers from all parts of the world - North, South, East and West. The Federation and its members strive towards an ideal of free exchange of information, where scientific discoveries and advances are no longer restricted to a select few. The aim is to share this knowledge among the people of all nations, so that everyone may experience the benefits of the progress of Science.

The creation of the World Federation of Scientists was made possible by the existence, in Erice, of a centre for scientific culture named after the physicist Ettore Majorana, the *Ettore Majorana Centre for Scientific Culture*. This Centre, which has been dubbed "The University of the Third Millennium", has attracted over 100,000 scientists from all over the world since its founding in 1963. The Ettore Majorana Centre was a precursor of the World Federation of Scientists and its action to mitigate planetary emergencies.

The World Federation of Scientists rapidly identified 15 classes of Planetary Emergencies and began to organise the fight against these threats. One of its main achievements was the drawing up of the Erice Statement, in 1982, by Paul Dirac, Piotr Kapitza and Antonino Zichichi, clearly setting out the ideals of the Federation and putting forward a set of proposals for putting these ideals into practice. Another milestone was the holding of a series of International Seminars on Nuclear War which have had a tremendous impact on reducing the danger of a planet-wide nuclear disaster and have ultimately contributed to the end of the Cold War.

In 1986, through the action of a group of eminent scientists (most of whom were members of the WFS) the International Centre for Scientific Culture ICSC-World Laboratory was founded in Geneva to help achieve the goals outlined in the Erice Statement. To achieve these, specific pilot projects have been implemented to overcome the Planetary Emergencies. The ICSC-World Laboratory works on the principle that one of the better ways of helping Developing Countries is to support the participation of their scientific elite in projects aimed at the solution of their particular problems, working in collaboration with their peers in Developed Countries and contributing to the advancement of Science and human knowledge as a whole.

Other achievements have been the establishment of the Erice Prize, the Gian Carlo Wick Gold Medal Prize, the formulation of the Farnesina Statement and the Lausanne Declaration and the National Scholarship Programmes.

The ICSC – World Laboratory

Geneva – Moscow – Beijing – New York

What is the World Laboratory

The International Centre for Scientific Culture, ICSC - World Laboratory, is an international non-governmental organisation recognised by the United Nations and granted, by the federal government of Switzerland, the status of "Public Utility Organisation". It has established 5 World Laboratory branches, 42 research centres and signed formal agreements with 20 countries and 65 universities, research institutes and laboratories. It has implemented successfully over 100 pilot projects, in all scientific fields, in developing countries and granted access to higher institutes of learning and research laboratories to more than 5'000 young scientists from developing countries. The World Laboratory is unique in that it is able to obtain the scientific cooperation of its members and Project Directors on a totally voluntary basis.

The International Centre for Scientific Culture - World Laboratory was founded in CERN, Geneva, in 1986 by a group of eminent scientists, to help achieve the goals outlined in the Erice Statement. written by Professors Paul A.M. Dirac, Piotr Kapitza and Antonino Zichichi at the Ettore Majorana Centre for Scientific Culture in Erice, Italy in 1982, and subsequently endorsed by 10,000 scientists the world over. The World Laboratory has demonstrated that one of the better ways to aid developing countries is to encourage collaboration between their scientific elite and those of developed countries, in projects aimed at solving their particular problems, whilst at the same time advancing scientific and human knowledge as a whole. - thus contributing actively to the fight against the Planetary Emergencies defined by the World Federation of Scientists.

The World Laboratory Structure

Since the basic principles of the World Laboratory are high-reactivity guaranteed by the highest expertise available, with a minimum of bureaucracy, it is organised in a way to avoid the burden of an overly centralised administrative structure.

The ICSC – World Laboratory is managed by the President, who reports to the General Assembly, and who relies on the Executive Committee to set the overall policies of the organisation. Various scientists are appointed by the President to become members of the Scientific Committee and the President's Office, to serve in an advisory capacity and advise him on the selection of projects and project directors. The Project Directors Committee handles the co-ordination between the various projects. The General Secretariat, in Headquarters, handles the administrative, financial and control services.

Founding Institutes and Scientists

Academia Sinica
African Academy of Sciences
"Ettore Majorana" International Foundation and Centre for Scientific Culture
"Galileo Galilei" Foundation.
Institutes of Fundamental Studies – Sri Lanka
International Centre for Theoretical Physics
International Centre of Insect Physiology and Ecology
Permanent Inter-state Committee for Drought Control in the Sahel (CILSS)
Third World Academy of Sciences

Professor Sir John Eccles	*Nobel Laureate*
Professor M. Gourdin	*University of Paris*
Professor R. Hofstädter	*Nobel Laureate*
Professor H. Isliker	*University of Lausanne*
Professor G. Motta	*University of Naples*
Professor D. Palumbo	*University of Palermo*
Professor H. Rieben	*Jean Monnet Foundation for Europe*
Professor A. Salam	*Nobel Laureate*
Professor K. Siegbahn	*Nobel Laureate*
Professor S.C.C. Ting	*Nobel Laureate*
Professor B. Vittoz	*President Swiss Institute of Technology*
Professor A. Zichichi	*CERN, Geneva*

President

Professor A. Zichichi	*CERN, Geneva*

Vice President

Professor T.D. Lee	*University of Columbia, New York*

Executive Committee

Professor Pietro Adonnino	*University of Rome*
Professor Giuseppe Guarino	*University of Rome*
Professor Blaise Knapp	*University of Geneva*
Professor Guido Piragino	*University of Torino*
Professor Henri Rieben	*University of Lausanne*
Professor Kai M.B. Siegbahn	*University of Uppsala*

A Summary of Results and Achievements

(in collaboration with the World Federation of Scientists)

One hundred pilot projects and dozens of programmes were dedicated to the fight against planetary emergencies. Some results follow:

Basic Science and Technology –

Hundreds of bright young scientists from developing countries were given training in CERN, to help in the development of highly sophisticated detectors for the LAA, the ALICE and the R&D for Advanced Technologies experiments.

National Scholarship Programmes established in 38 developing or newly emerging countries with over 2'000 scholarships granted every year.

An International Scholarship Programme, in association to Pilot Projects, granted over one thousand international scholarships to bright young scientists from Developing Countries, for training in the best institutes and laboratories in industrialised countries.

Young scientists from developing countries were involved in the development of frontiers research on spectroscopy equipment in our research centre in Uppsala.

A large number of young scientists from developing countries were involved in the cosmic rays research programmes in the Gran Sasso underground laboratory.

A centre for advanced Science and technology was established in Beijing, in collaboration with the Chinese Academy of Sciences. It was the first established National Scholarship Programme and, over the last 18 years, it consistently introduced whole frontiers of Science areas in China, with 484 one-month-long domestic symposiums and 18 international symposiums training more than 30'000 scientists, 168 volume publications and nearly 4150 papers published in international journals.

A research centre for high-energy physics and cosmology was established in Islamabad.

A computer-assisted design centre was established in the University of Buenos Aires and training provided to young scientists.

A technological park centre was established in Buenos Aires to improve the interaction between universities, industry and research centres in Argentina. Training was provided to young scientists.

A total of 20 agreements were signed with governments from Developing Countries.

A total of 42 World Laboratory Research Centres were established in developing countries with the collaboration of national institutions.

A total of 65 scientific collaboration agreements were signed with governmental institutions and research institutes.

Climate and Environment –

Study and elaboration of mathematical models for drought and flooding on a regional and global basis.

The damming effect of geopotential formations on monsoons and the ensuing droughts was demonstrated.

Protection of coastal marine environment in the southern Mediterranean Sea.

The Monitoring of Planetary Emergencies –

Since 1998, a Permanent Monitoring Panel (PMP) was established for each group of Planetary Emergency, with the scope of:

- Monitoring the scientific results of research conducted in the relevant fields of Science;
- Organising scientific workshops to reconcile conflicting views on scientific issues and channel research efforts;
- Helping disseminate relevant scientific data and information;
- Organising yearly Group Meetings and reporting on group activities to the WFS Annual General Meeting in Erice; writing recommendations for use by governments and international agencies.

The 16 PMPs are:

- Biotechnology
- Brain and Behaviour
- Climatology
- Defence Against Cosmic Objects
- Desertification
- Energy
- Floods and Extreme Weather Events
- Information Security
- Limits of Development
- Missile Proliferation
- Mother & Child
- Pollution
- Motivations for Terrorism
- Mitigation of Terrorist Acts
- Transmissible Spongiform Encephalopathy
- Water

Forty-four Sessions of the International Seminars on Nuclear War and Planetary Emergencies were held, since 1981, in Erice.

A national competition was established in Italian high schools, to increase the awareness of the coming generation of the Planetary Emergencies.

A large project was launched for high schools in Italy to assemble and operate muon telescopes, using simple material developed as part of our ALICE experiment in CERN. The high school students and their teachers will thus be exposed to an ongoing physics experiment, to which they can effectively contribute.

Desertification –

Two research centres were established in China. Long-term studies and field applications showed that desertification could be forecasted; it can be stopped and even made to regress.

Energy and Pollution –

A research centre for Fusion was implemented in China and training provided to young scientists of South-East Asia.

A long-term programme for the development of nanocrystalline photovoltaic energy was implemented in China, with the collaboration of the Swiss Institute of Technology.

Another long-term programme was implemented in China for the development of an energy-efficient and pollution-free carbon slurry network and research centre.

A solar research and training centre was established in Senegal for the photovoltaic applications. Training was provided to instructors and graduate students.

Extreme Weather Events –

The method for the forecasting of extreme local weather events, using simple means, was developed and made available to many developing countries. Dedicated research and operational centres were established throughout the Mediterranean Basin and training provided to the local staffs.

Floods –

A vast system for the forecasting and the management of floods was implemented in the Yellow River Basin. The population living between the dikes, numbering in millions, is now given warning and evacuated in case of forecasted catastrophic events.

Food –

A new source of protein for developing countries was derived from the selective cultivation of a common species of bean.

A wide variety of fruit trees was shipped to China. A research centre with laboratory and nursery was provided along with an experimental canning factory. Training was provided to the local scientists and staff both in China and in European laboratories and institutes.

Global Monitoring of the Planet –

Numerous Seminars were dedicated to the study of the search for solutions to the problems of Global Warming and the Ozone Hole.

A sophisticated seismological system was put in place in the Mediterranean Basin, with stations on all sides of the Mediterranean and linked to worldwide networks. In addition, regional seismological networks were implemented in countries prone to earthquakes.

Infectious Diseases –

A network of 15 research institutes focused their studies and field research on AIDS in 14 African countries, and two dedicated research centres were established in Uganda and Nigeria.

Medicine –

A vast programme for advanced biotechnology was undertaken in China. Training was provided in European institutes and three centres were equipped and established in Beijing and Shanghai. Two of these centres have since become the leading centres in China.

An epidemiological study and survey for the prevention and the treatment of heart diseases was implemented in Kenya. A specialised research and clinical centre was established and training was provided to medical personnel of various hospitals and university centres.

A vast survey of the elderly population was undertaken in China, the first of its kind. A specialised research and clinical centre was established and training provided to local medical personnel.

A long-term programme, called Kangaroo Mothers, was started in 1989 to conduct studies and seek scientific proof of the validity of a new method to cater for premature babies, in developing countries lacking the necessary incubators infrastructure. Once the studies proved the method to be beneficial, a vast programme of training was implemented worldwide. The method has now even been adopted in various developed countries as it proved to be superior to the classical incubator method.

Another pilot programme for teaching deaf-mute children was disseminated in developing countries. Training was provided to local medical personnel and specialised training centres were established in three locations. That very successful programme will help mitigate the psychological, social and economic consequences for the disabled children.

Information Security –

The dedicated PMP investigated the emerging threat to the functioning of information and communication technology (ICT) systems and made appropriate recommendations. The Report and Recommendations of the PMP on Information Security, *Toward a Universal Order of Cyberspace: Managing Threats from Cybercrime to Cyberwar* (Report), is part of an ongoing effort that has been undertaken by the WFS to address threats in this arena. The vital importance for emerging and developing countries to ensure the security of Information was stressed, thus avoiding manipulations and economic disaster. Co-operation was developed with ICT and WSIS and a joint publication issued with ITU.

Motivations for Terrorism –

The Permanent Monitoring Panel on the Motivations for Terrorism was established to contain and possibly eliminate the growing scourge of terrorism and its impact, monitoring research that is being conducted, organising scientific workshops to channel research efforts and reconcile conflicting views on scientific and related ethical issues, helping disseminate relevant scientific data and information, developing recommendations for use by governments and international agencies, elaborating project proposals to conduct research in collaboration with scientists and others from relevant countries, and seeking to operationalize the results of scientific research in the cause of deeper understanding and for the purposes of cause of peace rather than war.

Mitigation of Terrorist Acts –

The Permanent Monitoring Panel on the Mitigation of Terrorist Acts has considered nuclear and biological megaterrorism, first to determine and to explain the magnitude of the hazard and then, in the case of biological megaterrorism, to identify the great benefit potentially available in the case of a terrorist-initiated pandemic of contagious disease such as smallpox by non-pharmaceutical interventions such as hand sanitation, improvised masks, air purification, use of household bleach - all to reduce the reproduction factor R of the disease from one generation of illness to the next "serial interval." Evidently, rapid development, production and use of an appropriate vaccine for the disease will minimize the serious economic and social burden of draconian public health measures.

The PMP Terrorism-Mitigation continues to maintain an overview of the capability in principle and in practice of detecting the flow of fissile material that might be used by terrorists to build nuclear weapons; of the monitoring capability for the early detection and identification of disease that might be terrorist-induced or for that matter a natural outbreak, taking into account that terrorist-induced disease might have many foci. Mitigation involves rapid assessment of the hazard - e.g., the extent and nature of radiological contamination in case of the dispersal of radioactive materials, to minimize further contamination and exposure of the population, as well as the tools and procedure to decontaminate where possible and warranted. The response involves not only systems analysis informed by physics and chemistry, but notably the hard work of defining and sharing "best practices" and of sketching materials that would most effectively limit the damage from a pathogen.

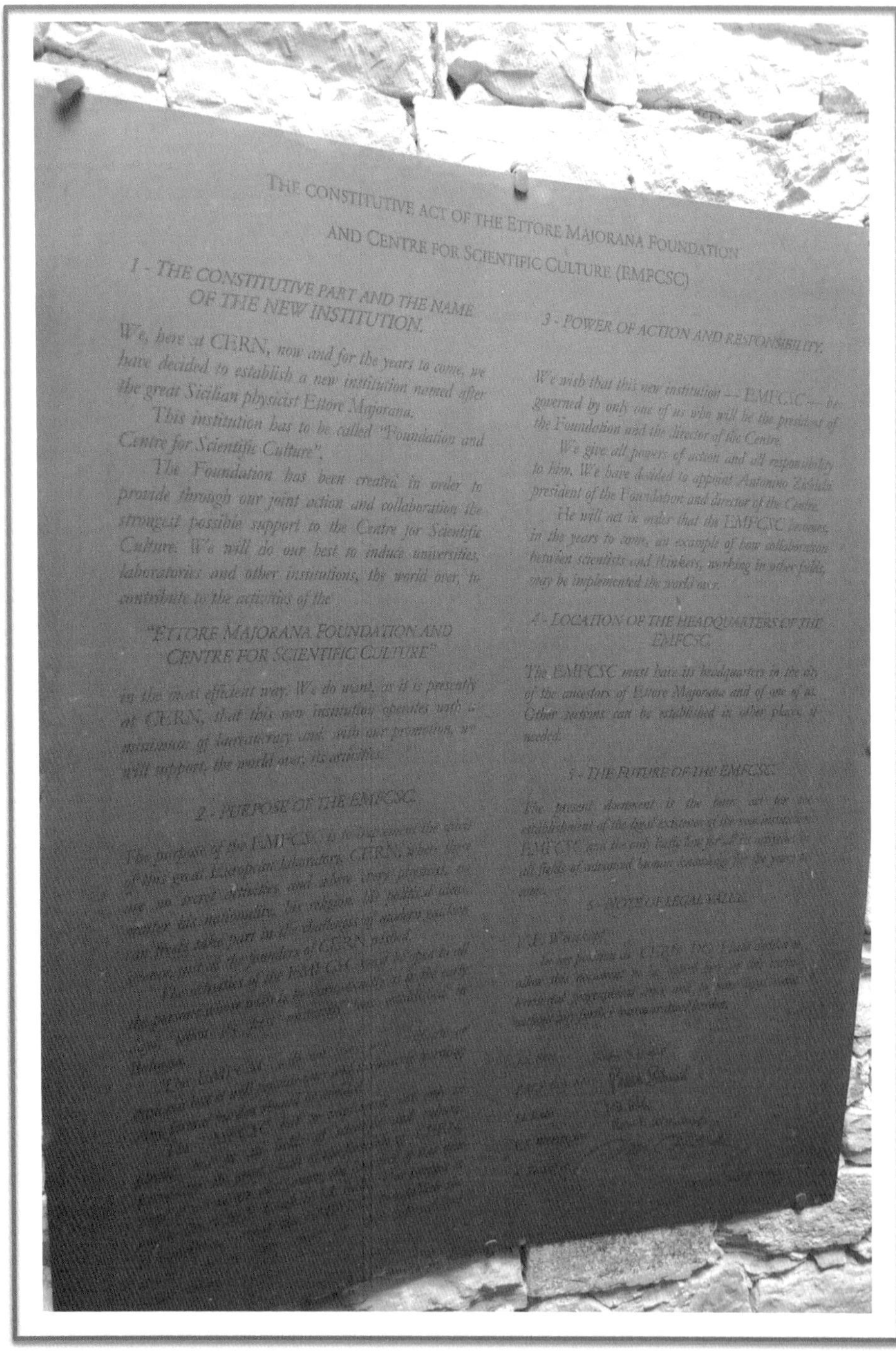
THE CONSTITUTIVE ACT OF THE ETTORE MAJORANA FOUNDATION
AND CENTRE FOR SCIENTIFIC CULTURE (EMFCSC)
1 - THE CONSTITUTIVE PART AND THE NAME OF THE NEW INSTITUTION.
We, here at CERN, now and for the years to come, we have decided to establish a new institution named after the great Sicilian physicist Ettore Majorana.
This institution has to be called "Foundation and Centre for Scientific Culture".
The Foundation has been created in order to provide through our joint action and collaboration the strongest possible support to the Centre for Scientific Culture. We will do our best to induce universities, laboratories and other institutions, the world over, to contribute to the activities of the
"ETTORE MAJORANA FOUNDATION AND CENTRE FOR SCIENTIFIC CULTURE"
in the most efficient way. We do want, as it is presently at CERN, that this new institution operates with a minimum of bureaucracy and, with our promotion, we will support, the world over, its activities.
2 - PURPOSE OF THE EMFCSC.
3 - POWER OF ACTION AND RESPONSIBILITY.
We wish that this new institution — EMFCSC —
governed by only one of us who will be the president of the Foundation and the director of the Centre.
We give all powers of action and all responsibility to him. We have decided to appoint Antonino Zichichi president of the Foundation and director of the Centre.
He will act in order that the EMFCSC becomes, in the years to come, an example of how collaboration between scientists and thinkers, working in other fields, may be implemented the world over.
4 - LOCATION OF THE HEADQUARTERS OF THE EMFCSC.
The EMFCSC must have its headquarters in the city of the ancestors of Ettore Majorana and of one of us. Other sections can be established in other places, if needed.
5 - THE FUTURE OF THE EMFCSC.

The inauguration in 1991
of the Blackett Institute by Professor Kibble

P.M.S. Blackett's Statement and the Director of the Blackett Laboratory at Imperial College in London, Thomas W.B. Kibble.

«We experimentalists are not like theorists: the originality of an idea is not for being printed in a paper, but for being shown in the implementation of an original experiment.»

Lord Patrick Maynard Stuart Blackett, 1962.

This unveiling Ceremony recalls me the other one where Frascati **Lab AdA-ADONE** were given the **EPS Historic site**.

On that occasion I recalled the famous **statement by Enrico Fermi**:

«Without Memory neither Science
nor Civilisation could exist».

INFN - EPS
Ceremony

European Physical Society – EPS Historic Site
The AdA Storage Ring at the INFN Frascati National Laboratories

AdA, ADONE, (J/ψ)
AND THE
3rd LEPTON

Antonino Zichichi
INFN and University of Bologna, Italy
CERN, Geneva, Switzerland
World Federation of Scientists, Beijing, Geneva, Moscow, New York

INFN Frascati National Laboratories
Thursday, December 5th, 2013 – 11 a.m.

Let me show page 3 of my Opening Lecture at the AdA-INFN-EPS unveiling Ceremony.

In the next graph it is shown the position of the great event (J/ψ) and the ADONE–SPEAR energy range.

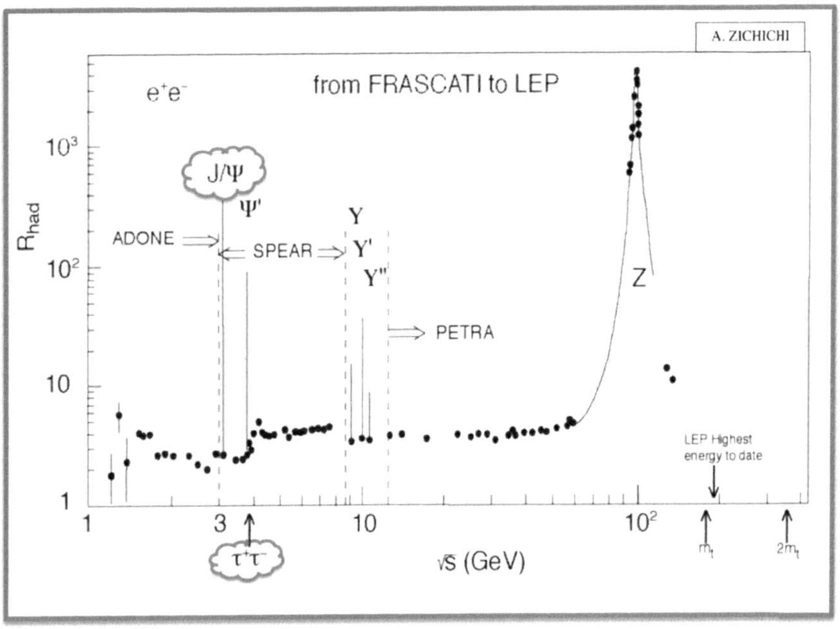

Something similar did happen with PETRA whose lowest energy was not immediately after SPEAR but with the incredible gap where the Lederman's (Υ, Υ', Υ") could have been discovered.

Let me go back to the origin of this new venture in elementary particle physics based on the search for new physics using non Bubble Chamber detectors. Needless to say that, in addition to the search for "narrow resonances" the search for the 3rd lepton was the reason for a stronger energy increase in ADONE.

3

In another occasion it was necessary "memory" in order to recall people the origin of the "God Particle", as reported on pages 625 and 626 of my Closing Lecture at the Pontifical Academy of Sciences.

NOTE ADDED IN JUNE 2012

• **The "God particle"** treated by Professor Englert in his Lecture has been discovered during the time interval between the 1st Subnuclear Physics Seminar of our Pontifical Academy of Science and the printing of our Proceedings. Let me express our warmest wishes to Professors Englert and Higgs for their theoretical work, to the CERN experimental physicists, to the CERN Research Director, Professor Sergio Bertolucci, and to the CERN Director General, Professor Rolf-Dieter Heuer.

As said in my Opening Lecture this work allow us to have our theoretical description of the Logic of Nature with non-zero Masses. This problem goes back to 1964 when Brout, Englert, Guralnik, Hagen, Higgs and Kibble (see References quoted below) gave to the physics community their proposal to solve the problem of introducing the "Mass" in the theoretical description of Nature.

Article	Reception date	Publication date
F. Englert and R. Brout *Phys. Rev. Letters* ***13*** *(1964) 321*	26/06/1964	31/08/1964
P.W. Higgs *Phys. Letters* ***12*** *(1964) 132*	27/07/1964	15/09/1964
P.W. Higgs *Phys. Rev. Letters* ***13*** *(1964) 508*	31/08/1964	19/10/1964
G.S. Guralnik, C.R. Hagen and T.W.B. Kibble *Phys. Rev. Letters* ***13*** *(1964) 585*	12/10/1964	16/11/1964

I was many decades younger that today when the discovery of the S^0 (S ≡ Scalar; zero indicates the value of the electric charge) was announced in 1967 [12]. It was the first time that a particle with all zero quantum numbers (zero electric charge, zero Lorentz spin, zero spin in all internal symmetries) was discovered. From 1897 when the first particle (electron) was discovered to 1947 only six particles were known to exist (e, γ, p, n, π, ν). From 1947 to 1967, during two decades, a large number of particles were discovered: all with at least a non-zero intrinsic quantum number.

The S^0 was the first candidate for what became later to be known as the "God particle". It was the era of bubble-chamber dominance and the NBC (non-bubble-chamber) physicists were invited to corroborate this first great discovery using NBC technology. The S^0 discovery attracted everybody's attention.

An exact repetition of the experiment was needed: the same Energy, the same solid angle and all other essential details. We had at CERN a powerful NBC set-up and the results found were not showing any evidence for the existence of the S^0–meson [13]: in disagreement with those of [12]. The group leader came to visit my experimental set-up and, after three days of analysis, declared our experimental result to be correct. During the next 45 years no-one was able to find the scalar meson with all zero properties. Leon Lederman devoted all efforts of Fermi Lab and enormous amount of time to prove its existence. The enormous difficulty for its discovery is the only justification for the name given "God particle" to the boson of Brout-Englert-Guralnik-Hagen-Higgs and Kibble.

References

[1] INTERNATIONAL SCHOOL OF COSMOLOGY AND GRAVITATION
J. Wheeler, 1972–*High Energy Astrophysics and its Relation to Elementary Particle Physics*; and 1992–*String Quantum Gravity and Physics at the Planck Energy Scale*.

[2] INTERNATIONAL SCHOOL OF COSMOLOGY AND GRAVITATION
N. Rosen, 1977–*Theories of Gravitation*.

[3] INTERNATIONAL SCHOOL OF COSMOLOGY AND GRAVITATION
P.G. Bergmann, 1979–*Spin, Torsion, Rotation and Supergravity*; 1982–*Unified Field Theories of More Than 4 Dimensions Including Exact Solutions*; 1985–*Topological Properties and Global Structure of Space-Time*; 1987–*Gravitation Measurements, Fundamental Metrology and Constants*;
1990–*Symposium on the Problem of the Cosmological Constant in Honor of Peter Gabriel Bergmann's 75th Birthday*;
P.G. Bergmann and Zheniju Zhang, 1991–*Black Hole Physics*;

«ETTORE MAJORANA» FOUNDATION AND CENTRE FOR SCIENTIFIC CULTURE
25TH ANNIVERSARY OF THE ERICE STATEMENT
1982 – 2007
ANTONINO ZICHICHI
NEITHER SCIENCE NOR CIVILIZATION COULD EXIST WITHOUT MEMORY
THE SCIENCE FOR PEACE ERICE PRIZE
WFS

On the occasion of the twenty-fifth anniversary of the Ettore Majorana Foundation and Centre for Scientific Culture (EMFCSC), in order to promote the values of scientific culture worldwide and following a proposal by the World Federation of Scientists (WFS), a special law was voted unanimously by the Sicilian Parliament to establish the

"Ettore Majorana Prize – Erice – Science for Peace".

The Prize has been awarded to fellows who played a leading role in promoting and implementing the goals outlined in the "Erice Statement" for a Science without secrets and without borders.

P.A.M. Dirac, P.L. Kapitza, A.D. Sakharov, E. Teller, V.F. Weisskopf, J.B.G. Dausset, S.D. Drell, M. Gell-Mann, H.W. Kendall, L.C. Pauling, A. Salam, C. Villi, R. Doll, J.C. Eccles, T.D. Lee, L. Montagnier, Qian Jaidong, J.S. Schwinger, U. Veronesi, G.M.C. Duby, R.L. Garwin, S.L. Glashow, D.C. Hodgkin, R.Z. Sagdeev, K.M.B. Siegbahn, Y.P. Velikhov, J. Karle, J.M.P. Lehn, A. Magnéli, N.F. Ramsey, H. Rieben, J.J. van Rood, C.S. Wu, R.L. Mössbauer, A. Müller, H. Kohl, M.S. Gorbachev, H.H. John Paul II, R. Clark, M. Cosandey, A. Peterman, R. Wilson, J. Alderdice, J.J. Friedman, M. Koshiba, S. Coleman, A.N. Chilingarov, P.C.W. Chu, L. Esaki, W.N. Lipscomb Jr., J. Szysko, M.-K. Wu, H.A. Hauptman, D.H. Hubel, R. Huber, B.I. Samuelsson, H. Sun, A.E. Yonath, G. 't Hooft, Y.T. Lee, W. Arber, S.C.C. Ting.

In 1985 at the famous Geneva meeting when the two most powerful world leaders met (Reagan and Gorbachev) they both agreed that the biggest enemy for Peace in the world is the existence of secret Labs. They both declared that they would have opened their secret Labs, thus following the Erice Statement signed by more than 100.000 scientists the world over.

See the next page for the Erice Statements written in 1982 by Paul A.M. Dirac, Piotr Kapitza (the fellow near to Blackett in the 1929 photo, page 22) and A.Z.

ETTORE MAJORANA FOUNDATION AND CENTRE FOR SCIENTIFIC CULTURE

THE ERICE STATEMENT

- *It is unprecedented* in human history that mankind has accumulated such a military power to destroy, at once, all centres of civilization in the world and to affect some vital properties of the planet.

  The *danger* of a nuclear holocaust is not the unavoidable consequence of the great development of pure Science.

  In fact, *Science* is the study of the Fundamental Laws of Nature.

  Technology is the study of how the power of mankind can be increased.

  Technology can be for peace and for war. The choice between peace and war is not a scientific choice. It is a cultural one: the *culture of love* produces peaceful technology. The *culture of hatred* produces instruments of war. Love and hatred have existed forever. In the bronze and iron ages, notoriously pre-scientific, mankind invented and built tools for peace and instruments of war. In the so called "modern era" it is imperative that *culture of love* wins.

  An enormous number of scientists share their time between pure Science research and military applications. This is a fundamental source for the arms race.

  It is necessary that a *new trend* develops, inside the scientific community and on an international basis.

It is of vital importance to identify the basic factors needed to start an effective process to protect human life and culture from a third and unprecedented catastrophic war. To accomplish this it is necessary to change the peace movement from a unilateral action to a truly international one involving proposals based on mutual and true understanding.

- **Here are our proposals:**

  1. Scientists who wish to devote all of their time, fully, to study theoretically or experimentally the basic laws of Nature, should in no case suffer for this free choice, to do only pure Science.
  2. All Governments should make every effort to reduce or eliminate restrictions on the free flow of information, ideas and people. Such restrictions add to suspicion and animosity in the world.
  3. All Governments should make every effort to reduce secrecy in the technology of defense. The practice of secrecy generates hatred and distrust. To start a ban for military secrecy will create greater stability than offered by deterrence alone.
  4. All Governments should continue their action to prevent the acquisition of nuclear weapons by additional nations or non-national groups.
  5. All Governments should make every effort to reduce their nuclear weapons stockpiles.
  6. All Governments should make every effort to reduce the causes of insecurity of non-nuclear powers.
  7. All Governments should make every effort to ban all types of nuclear tests in war technology.

- **Conclusion**

  Those scientists — in the East and in the West — who agree with this «Erice Statement», engage themselves morally to do everything possible in order to make the *new trend*, outlined in the present document, become effective all the world over and as soon as possible.

- This Statement was written in ERICE, August 1982, by Paul A.M. DIRAC, Piotr KAPITZA and Antonino ZICHICHI. By now the number of signatories of the Erice Statement has exceeded 100'000, the world over.

- The 'Erice Statement' has attracted, in the eighties, the attention of World Leaders such as Deng Xiao Ping (China), Mikhail Gorbachev (USSR), Olof Palme (Sweden), Sandro Pertini (Italy), Ronald Reagan (USA), Pierre Trudeau (Canada) and stimulated various actions on their part for a Science without secrecy and without frontiers.

VII – THE FUTURE

The Future

- Complexity exists at all scales, including the Fundamental level
- Past ≡ From Galilei to strange particles→**SM&B**

  ok
- Future ≡ Why not?

The Future

COMPLEXITY EXISTS AT ALL SCALES, INCLUDING THE SUBNUCLEAR UNIVERSE

The Logic of Nature allows the existence of a large variety of structures with their regularities and laws which appear to be independent from the basic constituents and fundamental laws of Nature which govern their interactions.

But, without these laws it would be impossible to have the real world which is in front of us and of which we are part of. A series of complex systems is shown in Figure 9.

As you can see, we go from traffic flux, to the internet network, to earthquakes and seismicity, to social and economic systems, to the behaviour of financial markets, to the study of cosmological structures, and so on.

The experimental evidences for the **existence** of **Complexity are two**:

1) The Anderson-Feynman-Beethoven-type phenomena (**AFB**) i.e. phenomena whose laws and regularities ignore the existence of the Fundamental Laws of Nature from which they originate.

2) The Sarajevo-type effects, i.e. Unexpected Events of quasi irrelevant magnitude which produce Enormous Consequences (**UEEC**).

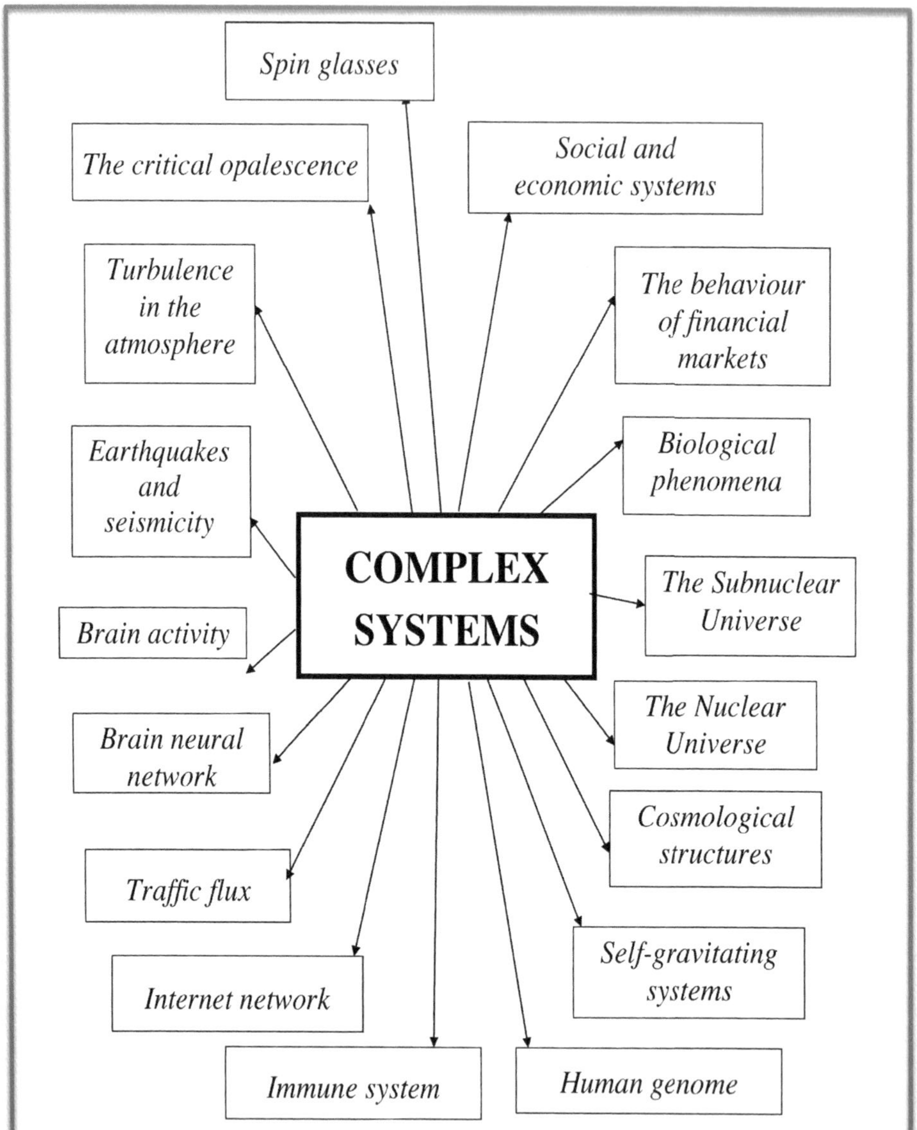

Figure 9

The only certainty about Complexity is the existence of these experimentally observable effects. Nature tell us that AFB and UEEC exist at all scales, and therefore Complexity exists at all scales, as illustrated in Figure 10.

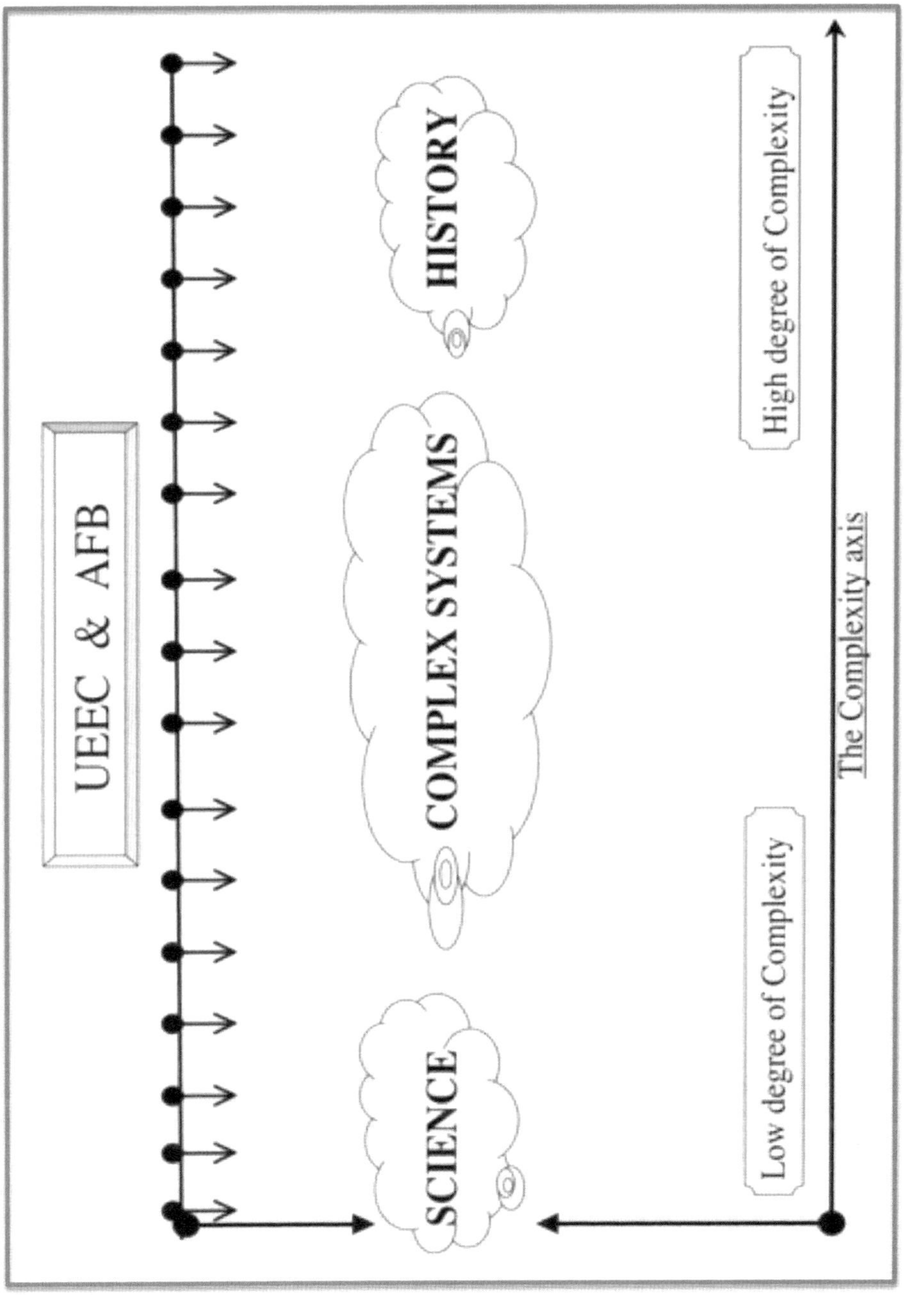
UEEC & AFB
SCIENCE
COMPLEX SYSTEMS
HISTORY
Low degree of Complexity
High degree of Complexity
The Complexity axis

Figure 10

AFB Phenomena from Beethoven to the Superworld

Beethoven and the laws of acoustics.

Beethoven could compose superb masterpieces of music without any knowledge of the laws governing acoustic phenomena.

But these masterpieces could not exist if the laws of acoustics were not there.

The living cell and QED.

To study the mechanisms governing a living cell, we do not need to know the laws of electromagnetic phenomena whose advanced formulation is QED (Quantum ElectroDynamics).

All mechanisms needed for life are, to a great extent, examples of electromagnetic processes. If QED was not there, Life could not exist.

Nuclear physics and QCD.

Proton and neutron interactions appear as if a fundamental force of nature is at work: the nuclear force, with its rules and its regularities.

These interactions ignore that protons and neutrons are made with quarks and gluons.

Nuclear physics does not appear to care about the existence of Quantum ChromoDynamics (QCD), the fundamental force acting between quarks and gluons at the heart of the Subnuclear world.

Nuclear physics ignores QCD but all phenomena occurring in nuclear physics have their roots in the interactions of quarks and gluons.

In other words, protons and neutrons behave like Beethoven: they interact and build up nuclear physics without 'knowing' the laws governing QCD.

The most recent example of AFB type phenomenon comes from the frontier of our scientific knowledge: the Superworld. What we call the **World could apparently not care less about the existence of the Superworld**, whose foundation is not in the four dimensional Space-Time but in a Superspace with 43 dimensions.

UEEC EVENTS, FROM GALILEI UP TO SM&B

In Figure 11 there is a sequence of **UEEC** events from Galilei to Fermi-Dirac and the 'strange particles'.

Details can be found in my book whose front page is reproduced below.

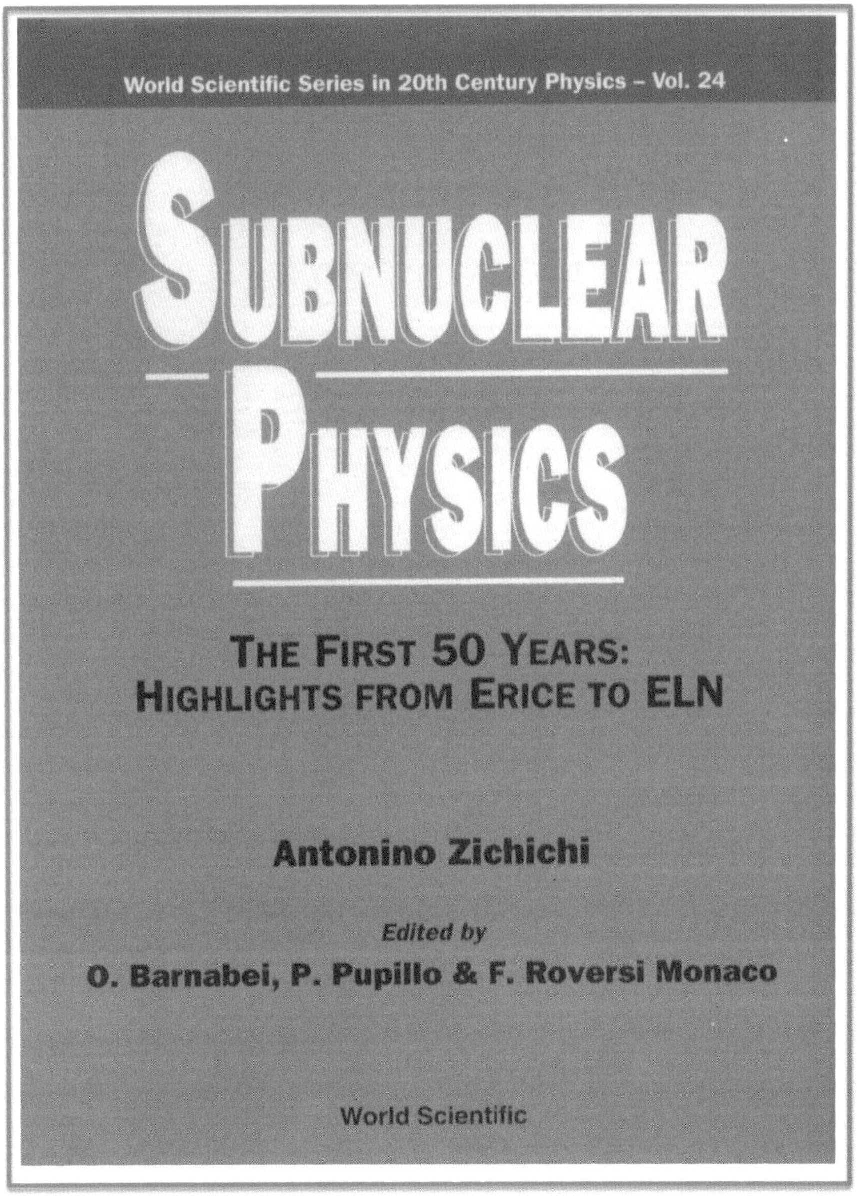

"UEEC"
TOTALLY UNEXPECTED DISCOVERIES
FROM GALILEI TO THE "STRANGE" PARTICLES

I	Galileo Galilei discovery of F = mg.
II	Newton discovery of $F = G\frac{m_1 \cdot m_2}{R_{12}^2}$.
III	Maxwell discovers the unification of electricity, magnetism and optical phenomena, which allows him to conclude that light is a vibration of the EM field.
IV	Planck discovery of $h \neq 0$.
V	Lorentz discovers that Space and Time cannot be both real.
VI	Rutherford discovers the nucleus.
VII	Hess discovers the cosmic rays.
VIII	Dirac discovers his equation, which opens new horizons, including the existence of the antiworld.
IX	Blackett discovers the simultaneous production of (e^+e^-) thus giving life to the vacuum polarization effect.
X	Fermi–Dirac and Bose–Einstein discover two completely different statistical laws.
XI	The "strange particles" are discovered in the Blackett Lab.
XII	The existence of mesons with positive and negative strangeness is discovered in the Blackett Lab.

Figure 11

Let me show few examples (Figure 12) where I have been involved in the construction of the Standard Model with Totally Unexpected Events.

① **The 3rd lepton, HL** (now called τ) with its own neutrino, ν_{HL} (now called ν_τ),
despite *the abundance of neutrinos:* ν_e *and* ν_μ.

② **Antimatter**
despite *S-matrix and C, P, CP, T breakings.*

③ **Nucleon Time-like** EM structure
despite *S-matrix.*

④ **No quarks** in violent (pp) collisions
despite *scaling.*

⑤ **Meson mixings**
$\theta_V \neq \theta_{PS}$: $(51^o) \neq (10^o) \neq 0$ ***despite*** $SU(3)_{uds}$.

⑥ **Effective energy:** the Gribov QCD-light
despite *QCD-confinement.*

⑦ **The running** of α_1 α_2 α_3 versus **energy**:
the EGM effect, the GAP between E_{GUT} *and* E_{SU}, *and the absence of the Platonic straight line convergence.*

Figure 12

I will only show in Figure 13 the incredible consequences due to the "Effective Energy": a series of different initial states produce the same final states.

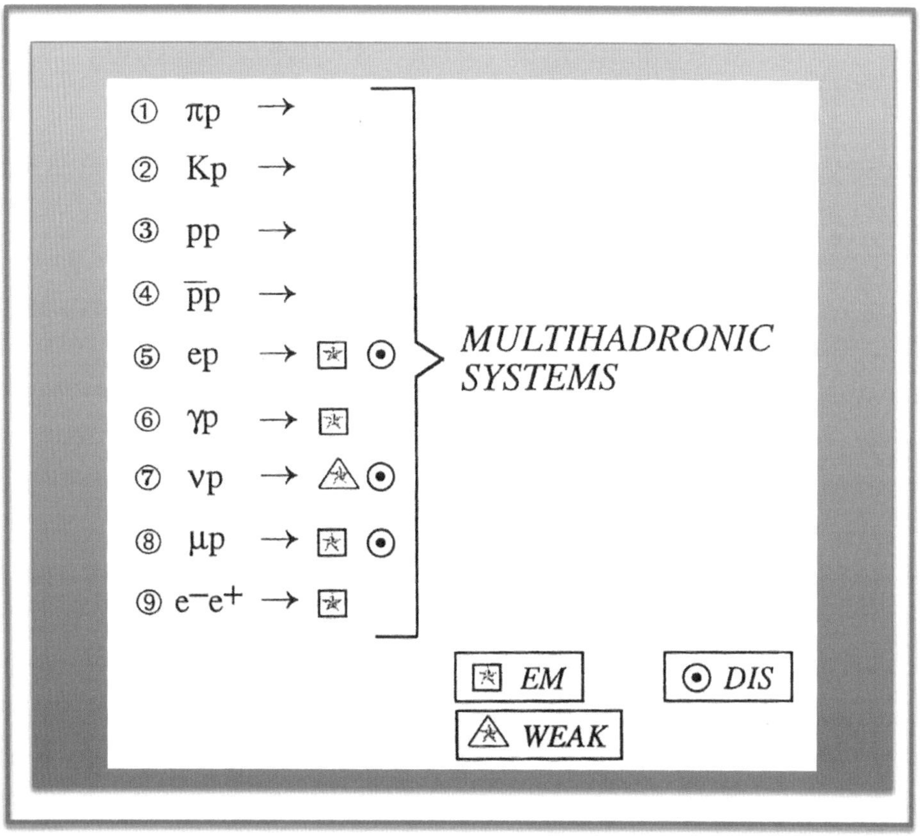

Figure 13: A list of initial states yielding multihadronic final states with the same properties, provided the "effective energy" is the same.

The Lesson Needed for the Future

For the Future what is to be Expected?

We have proved that AFB and UEEC – which are at the origin of Complexity, with its consequences permeating all our existence, from molecular biology to life in all its innumerable forms up to our own, including History – do exist at the fundamental level [19–22] and [23].

It turns out that Complexity in the real world exists, no matter the mass-energy and Space–Time scales considered.

Therefore the only possible prediction is that:

- **Totally Unexpected Effects** should **show up**.
- **Effects**, which are impossible to be predicted on the basis of **present knowledge**.

We should be prepared with **powerful experimental instruments, technologically at the frontier of our knowledge**, to discover Totally Unexpected Events in all laboratories, the world over (including CERN in Europe, Gran Sasso in Italy, and other facilities in Japan, USA, China and Russia). All the pieces of our Physics could not have been discovered if the experimental technology was not at the frontier of our knowledge.

Example: the cloud-chambers (**Blackett**, Anderson, Neddermeyer), the photographic emulsions (Lattes, Muirhead, Occhialini, Powell), the high power magnetic fields (Conversi, Pancini, Piccioni) and the powerful particle accelerators and associated detectors for the discovery – the world over – of the **SM&B**.

This means that we must be prepared with the most advanced technology for the discovery of totally unexpected events like the ones found in the **Blackett group**, from (e^+e^-) production to strange particles.

The mathematical descriptions, and therefore the predictions – for new phenomena to be discovered in the field opened by the given UEEC event – come after the UEEC event, never before.

<u>**Recall**</u>:

- The **discoveries in Electricity**, **Magnetism** and **Optics** (UEEC).
- **Radioactivity** (UEEC).
- The **Cosmic Rays** (UEEC).
- The **Weak Forces** (UEEC).
- The **Nuclear Physics** (UEEC).
- The **Strange Particles** (UEEC).
- The **3 Columns** (UEEC).
- The **origin of the Fundamental Forces** (UEEC).

VIII – CONCLUSIONS – FROM BLACKETT TO PRESENT DAY PHYSICS

THE THREE FUNDAMENTAL FORCES

• **QED ≡ Quantum ElectroDynamics** ≡ Telephones, TV, Computers, Satellites, Internet … .	① ≡ F_{EW}
• **QFD ≡ Quantum Flavour Dynamics** ≡ Stability of matter; Safety valves of Stars; Billion years life for nuclear candles.	
• **QCD ≡ Quantum ChromoDynamics** ≡ Laws governing the Subnuclear Universe ($\lesseqqgtr 10^{-13}$ cm).	② ≡ F_{QCD}
• **Gravitational Forces** ≡ The Solar System and the Universe.	③ ≡ F_G

Figure 14

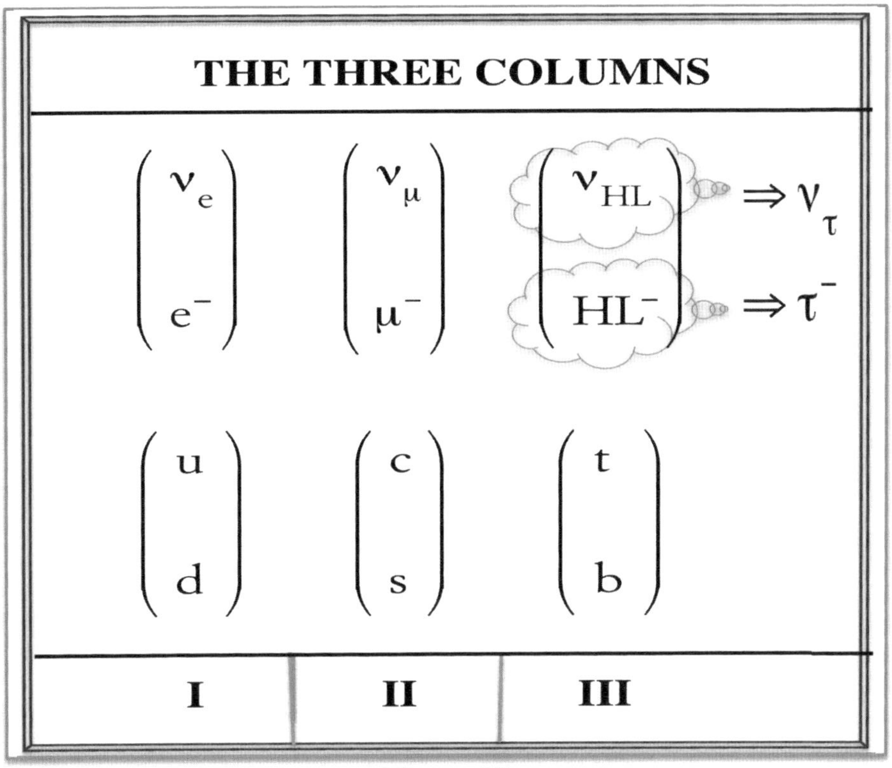

Figure 15

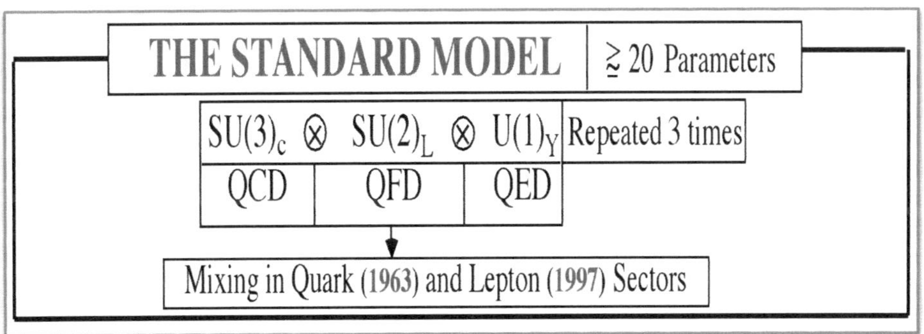

Figure 16

We are sure that it is necessary to go beyond the Standard Model whose five basic steps are reported in Figure 17. **Where can SM&B come from?** A possible answer is in Figure 18 which illustrates the consequences of introducing α_G of the gravitational forces. The note below is needed for the symbols used in Figures 17 and 18.

NOTE

q	≡	quark and squark;
l	≡	lepton and slepton;
G	≡	Gauge boson and Gaugino;
H	≡	Higgs and Shiggs;
m_F	≡	Fermi mass scale;
m_P	≡	Planck mass scale;
k	≡	quadrimomentum;
C	≡	Charge Conjugation;
P	≡	Parity;
T	≡	Time Reversal;
≠	≡	Breakdown of Symmetry Operators;
RGEs	≡	Renormalization Group Equations;
GUT	≡	Grand Unified Theory;
SUSY	≡	Supersymmetry;
RQST	≡	Relativistic Quantum String Theory;
SSB	≡	Spontaneous Symmetry Breaking.

SM&B
THE STANDARD MODEL AND BEYOND

THE FIVE BASIC STEPS NEEDED FOR SM&B

① The renormalization group equations (RGEs) imply that the gauge couplings (α_i) and the masses (m_j) all run with k^2. It is this running which allows GUT, suggests SUSY and produces the need for a non point-like description (RQST) of physics processes, thus opening the way to quantize gravity.

RGEs $(\alpha_i\ (i \equiv 1, 2, 3);\ m_j\ (j \equiv q,\ l,\ G,\ H)) :\ f(k^2)$.

- GUT ($\alpha_{GUT} \cong 1/24$) & GAP ($10^{16} - 10^{18}$) GeV.
- SUSY (to stabilize $m_F/m_P \cong 10^{-17}$).
- RQST (to quantize Gravity).

② All forces originate in the same way: the gauge principle.

Gauge Principle (hidden and expanded dimensions).

— How a Fundamental Force is generated: SU(3); SU(2); U(1) and Gravity.

③ Imaginary masses play a central role in describing nature: SSB & Confinement.

The Physics of Imaginary Masses: SSB.

— The Imaginary Mass in SU(2)×U(1) produces masses thanks to SSB: $(m_{W^\pm};\ m_{Z^0};\ m_q;\ m_l)$, including $m_\gamma = 0$.

— The Imaginary Mass in SU(5)⇒SU(3)×SU(2)×U(1) or in any higher (not containing U(1)) Symmetry Group ⇒ SU(3)× SU(2)×U(1) produces Monopoles.

— The Imaginary Mass in $SU(3)_c$ generates Confinement.

④ The mass-eigenstates are mixed when the Fermi forces come in: the matrix describing the mixing is the product of two fundamental matrices.

Flavour Mixings & CP ≠ , T ≠ (direct ≠ , not via SSB).

— Why the mixing is there? No need for it but it is there.

⑤ The Abelian force QED has lost its role of being the guide for all fundamental forces. The non-Abelian gauge forces dominate and have features which are not present in QED.

Anomalies & Instantons.

— Basic Features of all Non-Abelian Forces.

Figure 17

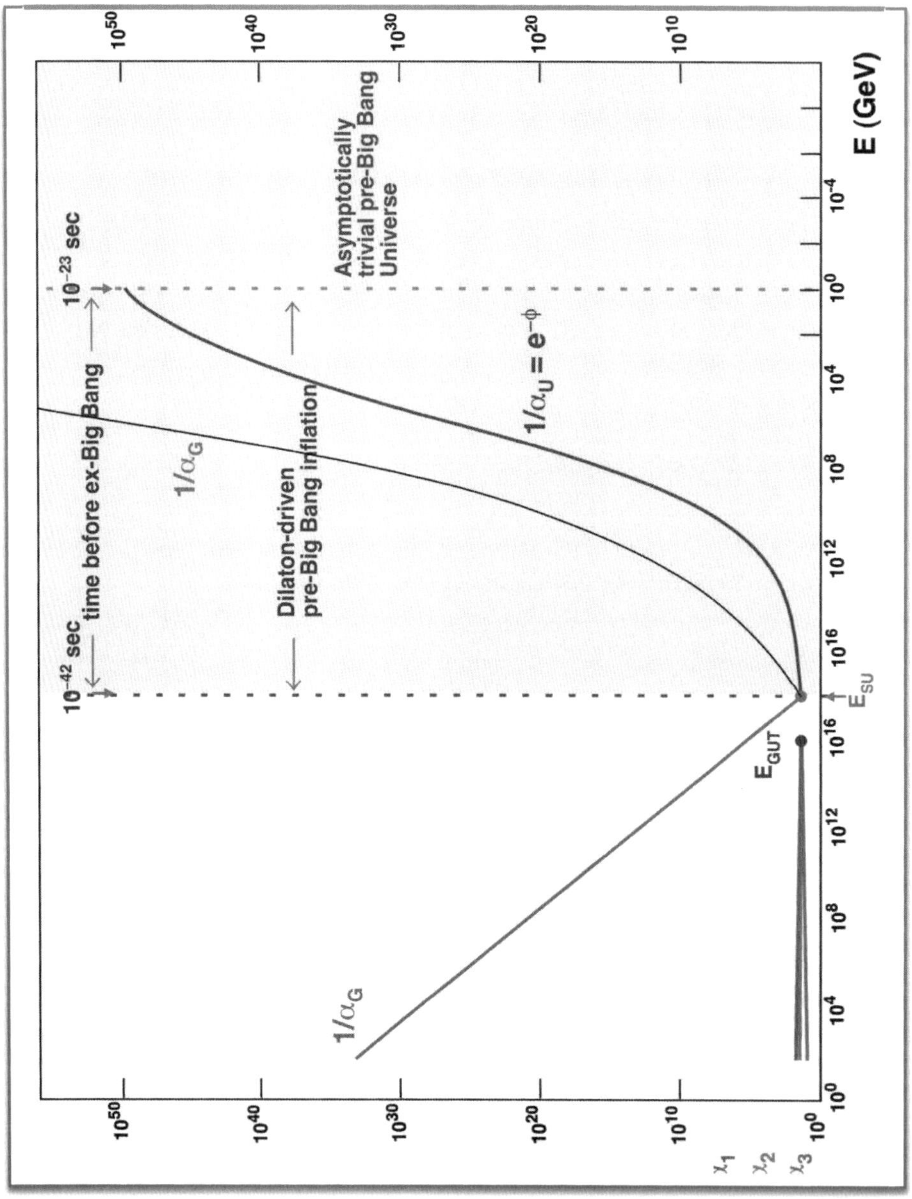

Figure 18: This Figure illustrates the energy dependence of all couplings, including the gravitational one. In the pre-Big-Bang region all couplings are controlled by the dilaton (α_U). The gravitational coupling α_G has the extra factor E^2/E_{SU}^2 [24]. It should be pointed out that the introduction of α_G makes the three α_1 α_2 α_3 looking like a single line in Figure 18 (left bottom). See Figure 4 to compare α_1 α_2 α_3 with Figure 18.

SM&B TELLS US THAT ELN IS NEEDED

A view of the 300 km ring.

ELN is the Euroasiatic Long Intersecting Storage Accelerator, a (pp) collider (300 Km ring) able to reach the highest energy and luminosity [25]. ELN could be built with simple extrapolation of the presently known technologies.

The ELN project is very ambitious but we should be encouraged by our previous experiences.

In fact, the path leading to the ELN project had to go through the Gran Sasso project, now the largest and most powerful underground Laboratory in the world, the LEP-white-book, which allowed this great European venture to overcome the many difficulties that had blocked its implementation during many years, the HERA collider, now successfully completed, the roots of LHC as, for example, the 27 Km long (rather than the 13 Km initially planned) LEP tunnel whose dimensions are big enough for two colliders (not only for the (e^+e^-) initially planned) and the LAA-R&D project, implemented to study the detector technologies needed for the LHC.

These past achievements in project realization are mentioned in order to corroborate my optimism and enthusiasm in encouraging new actions and new ideas for the **future of Subnuclear Physics in Europe and in the world**, all having as focus CERN, the greatest Subnuclear Physics Lab in the world.

The present status of Science is reported in Figure 19.

It could be that Science will be mathematically proved to be 'NP-complete'. This is the big question for the immediate future [26].

It is therefore instructive to recall what I learned when I joined the Blackett group: how Science fits in the whole of our knowledge as reported in Figure 8 (page 28) and Table 3 (page 26) where the *Blackett interest in understanding the Logic of Nature* was given in terms of *Science and History*. All this started in the middle of last century. Nowadays is one of the biggest problems to understand how and why **the Logic of Nature** allows to go **from the lowest limit of Complexity, Science, to the highest limit of Complexity, History**.

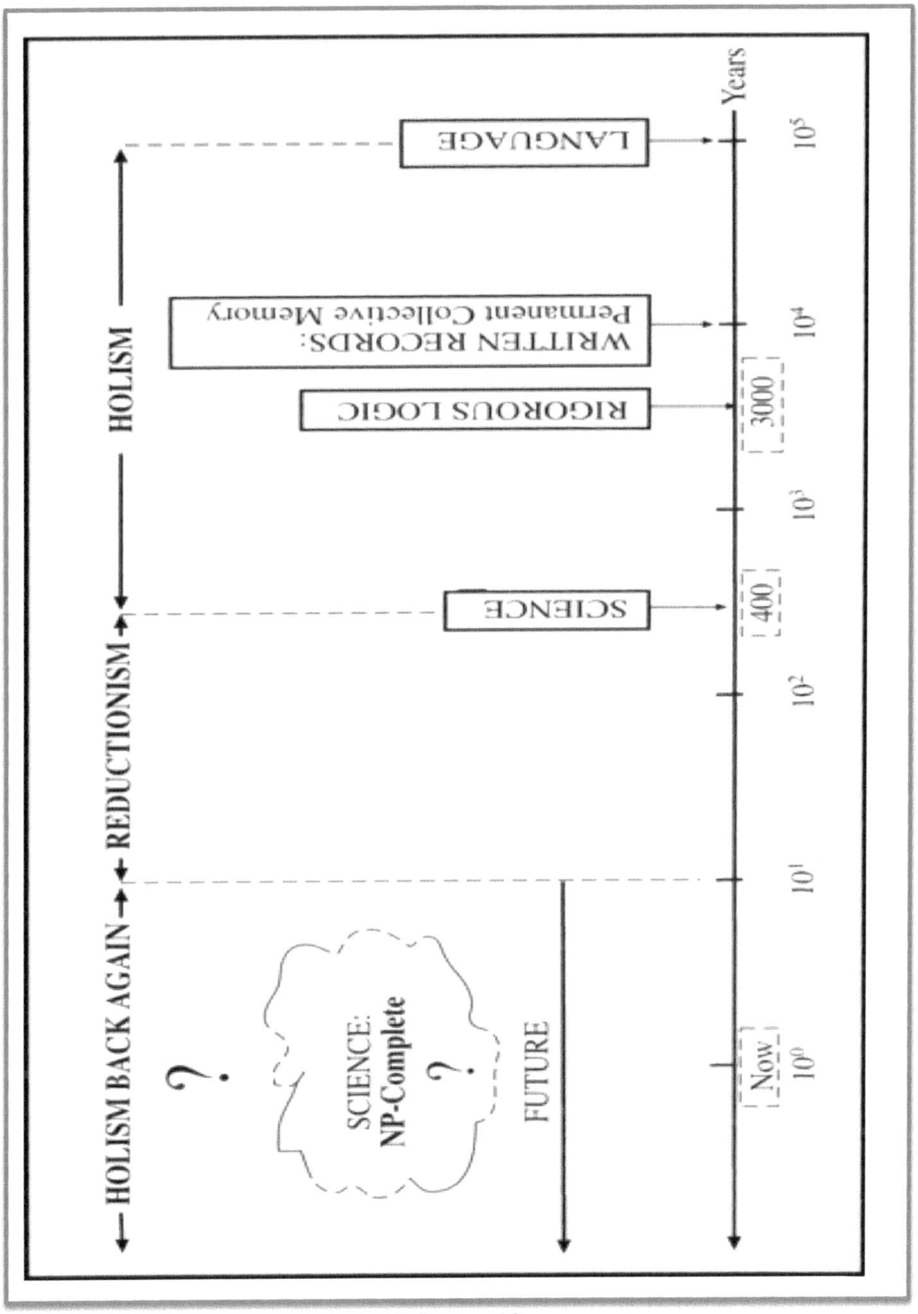
HOLISM BACK AGAIN
REDUCTIONISM
HOLISM
LANGUAGE
WRITTEN RECORDS:
Permanent Collective Memory
RIGOROUS LOGIC
SCIENCE
SCIENCE:
NP-Complete
?
?
FUTURE
Now
10^0
10^1
10^2
400
10^3
3000
10^4
10^5
Years

Figure 19

«ETTORE MAJORANA» FOUNDATION AND CENTRE FOR SCIENTIFIC CULTURE

INTERNATIONAL SCHOOL OF SUBNUCLEAR PHYSICS

Patrick M.S. Blackett
Scholarship and Diplomas from 1974

1974 **Jonathan Schonfeld**, *Princeton University, Princeton, NJ, USA*
1975 **Michael E. Peskin**, *Cornell University, Ithaca, NY, USA*
1976 **William J. Marciano**, *Rockefeller University, New York, USA*
1977 **William Celmaster**, *Harvard University, Cambridge, MA, USA*
1978 **Anthony D. Kennedy**, *University of Sussex, Brighton, Sussex, UK*
1979 **Paul Ginsparg**, *Cornell University, Ithaca, NY, USA*
1980 **Enore Guadagnini**, *University of Pisa, Italy*
1981 **Jay L. Banks**, *Ohio State University, Columbus, OH, USA*
1982 **Leonardo Castellani**, *State University of New York, Stony Brook, NY, USA*
1983 **Robert H. Bernstein**, *University of Chicago, IL, USA*
1984 **Rosario Nania**, *CERN, Geneva, Switzerland*
1985 **Gerard Gilbert**, *The University of Texas, TX, USA*
1986 **Andrea Pasquinucci**, *University of Milano, Italy*
1987 **Jun Liu**, *University of Texas at Austin, TX, USA*
1988 **Gerjan Bobbink**, *NIKHEF, Amsterdam, The Netherlands*
1989 **Bindu A. Bambah**, *Panjab University, Chandigarh, India*
1990 **Mark Wexler**, *Princeton University, Princeton, NJ, USA*
1991 **Matthias Neubert**, *Universität Heidelberg, Heidelberg, Germany*
1992 **Rongzhi Liu**, *Caltech, Pasadena, CA, USA*
1993 **Konstantinos Skenderis**, *State University of New York, Stony Brook, NY, USA*
1994 **Nicola Fabiano**, *INFN-Frascati, Italy*
1995 **Vladimir Anferov**, *University of Michigan, Ann Arbor, MI, USA*
1996 **Stefan Schönert**, *Technische Universität, München, Germany*
1997 **William E. Brown**, *University of Oxford, UK*
1998 **Salvatore Mele**, *CERN, Geneva, Switzerland*
1999 **Alan Marcus**, *Tel Aviv University, Israel*
2000 **Arno Straessner**, *CERN, Geneva, Switzerland*
2001 **Kristian Harder**, *DESY, Hamburg, Germany*
2002 **Mark Ramtohul**, *University of Edinburgh, UK*
2003 **Wolfgang Menges**, *DESY, Hamburg, Germany*
2004 **Sevil Salur**, *Yale University, New Haven, CT, USA*
2005 **Kazunori Hanagaki**, *Fermilab, Batavia, IL, USA*
2006 **Paolo Aschieri**, *University of Piemonte Orientale, Alessandria, Italy*
2007 **Mira Krämer**, *DESY, Hamburg, Germany*
2008 **Zachary Marshall**, *Columbia University, New York, NY, USA*
2009 **Mickey Guotai Chiu**, *Brookhaven National Laboratory (BNL), Upton, NY, USA*
2010 **Mohd Danish Azmi,** *Aligarh Muslim University, Aligarh, India*
2011 **Magnus Mager,** *CERN, Geneva, Switzerland*
2012 **Max Zoller,** *Karlsruhe Institute für Tecnologie, Karlsruhe, DE*
2013 **Silvia Nagy,** *Imperial College of London, UK*

IX – REFERENCES

[1] *Search for Supersymmetric Particles using Acoplanar Charged Particle Pairs from Z^0 decays*
ALEPH Collab., D. Decamp et al., *Phys. Lett.* B236, 86 (1990).

[2] *Search for Neutral Higgs Bosons from Supersymmetry in Z decays*
ALEPH Collab., D. Decamp et al., *Phys. Lett.* B237, 291 (1990).

[3] *Search for Neutralino Production in Z decays*
ALEPH Collab., D. Decamp et al., *Phys. Lett.* B244, 541 (1990).

[4] *Search for the Neutral Higgs Bosons of the MSSM and other two Doublet Models*
ALEPH Collab., D. Decamp et al., *Phys. Lett.* B265, 475 (1991).

[5] *Search for Heavy Charged Scalars in Z^0 decays*
DELPHI Collab., P. Abreu et al., *Phys. Lett.* B241, 449 (1990).

[6] *Search for Pair Production of Neutral Higgs Bosons in Z^0 decays*
DELPHI Collab., P. Abreu et al., *Phys. Lett.* B245, 276 (1990).

[7] *Search for Scalar Quarks in Z^0 decays*
DELPHI Collab., P. Abreu et al., *Phys. Lett.* B247, 148 (1990).

[8] *A Search for Sleptons and Gauginos in Z^0 Decays*
DELPHI Collab., P. Abreu et al., *Phys. Lett.* B247, 157 (1990).

[9] *Mass Limits for Scalar Muons, Scalar Electrons and Winos from e^+e^- Collisions near S**(1/2) = 91–GeV*
L3 Collab., B. Adeva et al., *Phys. Lett.* B233, 530 (1989).

[10] *Search for the Neutral Higgs Bosons of the Minimal Supersymmetric Standard Model from Z^0 Decays*
L3 Collab., B. Adeva et al., *Phys. Lett.* B251, 311 (1990).

[11] *Search for the Charged Higgs Boson in Z^0 decay*
L3 Collab., B. Adeva et al., *Phys. Lett.* B252, 511 (1990).

[12] *A Search for Acoplanar Pairs of Leptons or Jets in Z^0 decays: Mass Limits on Supersymmetric Particles*
OPAL Collab., M.Z. Akrawy et al., *Phys. Lett.* B240, 261 (1990).

[13] *A Search for Technipions and Charged Higgs Bosons at LEP*
OPAL Collab., M.Z. Akrawy et al., *Phys. Lett.* B242, 299 (1990).

[14] *A Direct Search for Neutralino Production at LEP*
OPAL Collab., M.Z. Akrawy et al., *Phys. Lett.* B248, 211 (1990); P.D. Acton et al., preprint CERN-PPE/91-115, 22 July 1991.

[15] *Searches for Supersymmetric Particles Produced in Z Boson decay*
MARK II Collab., T. Barklow et al., *Phys. Rev. Lett.* 64, 2984 (1990).

[16] *Searches for New Particles at LEP*
M. Davier, LP-HEP 91 Conference, Geneva, CH, Preprint LAL 91-48, December 1991.

[17] *The Evolution of Gaugino Masses and the SUSY Threshold*
F. Anselmo, L. Cifarelli, A. Peterman and A. Zichichi, *Nuovo Cimento* 105A, 581 (1992).

[18] *A Detailed Comparison of LEP Data with the Predictions of the Minimal Supersymmetric SU(5) GUT*
J.R. Ellis, S. Kelley, D.V. Nanopoulos, preprint CERN-TH/6140-91, *Nucl. Phys.* B373, 55 (1992).

The References [19–22] refer to the various occasions where I have presented papers on highly specialized topics and discussed the connection of these topics with Complexity. Each title on the upper part refers to the connection with Complexity while the specialized topic is reported in the detailed references.

[19] *Complexity at the Fundamental Level*
A. Zichichi
presented at:

- International Conference on 'Quantum [un]speakables' in Commemoration of John S. Bell, International Erwin Schrödinger Institut (ESI), Universität Wien (Austria), November 2000, '*John Bell and the Ten Challenges of Subnuclear Physics*'.
- 40th Course of the International School of Subnuclear Physics, Erice (Italy), September 2002, '*Language Logic and Science*'.
- 31st, 32nd and 33th Course of the International School of Solid State Physics, Erice (Italy), July 2004, '*Complexity at the Elementary Level*'.
- 42nd International School of Subnuclear Physics, Erice (Italy), August - September 2004, '*Complexity at the Elementary Level*'.
- Trinity College, Dublin (Ireland), February 2005, '*Complexity at the Elementary Level*'.
- Department of Physics, University of Padova (Italy), March 2005, '*Complexity at the Elementary Level*'.
- 43th Course of the International School of Subnuclear Physics, Erice (Italy), September 2005, '*Complexity at the Elementary Level*'.
- Italian Physics Society (SIF) XCI Annual National Congress, University of Catania (Italy), September 2005, '*Complexity at the Elementary Level*'.
- Desy, Hamburg, November 2005, '*Complexity at the Fundamental Level*'.
- 44th Course of the International School of Subnuclear Physics, Erice (Italy), September 2006, '*Complexity at the Fundamental Level*'.

[20] *The Logic of Nature and Complexity*
A. Zichichi
presented at:

- Pontificia Academia Scientiarum, The Vatican, Rome (Italy), November 2002, '*Scientific Culture and the Ten Statements of John Paul II*'; '*Elements of Rigour in the Theory of Evolution*'.
- The joint Session of:
6th Course of the International School of Biological Magnetic Resonance; Erice (Italy), July 2003, '*Language Logic and Science*'.
- 2nd Workshop on Science and Religion of the Advanced School of History of Physics; Erice (Italy), July 2003, '*Language Logic and Science*'.
- 10th Workshop of the International School of Liquid Crystals; Erice (Italy), July 2003, '*Language Logic and Science*'.
- International School on Complexity, 1st Workshop on Minimal Life, Erice (Italy), December 2004, '*Evolution and Complexity at the Elementary Level*'.

[21] *Complexity and New Physics*
A. Zichichi
presented at:

- INFN-Alice Meeting, University of Catania (Italy), January 2005, '*Complexity at the Elementary Level*'.
- INFN Eloisatron Project 'The 1st Physics ALICE Week', Erice (Italy), December 2005, '*Complexity and New Physics with ALICE*'.
- 50th Anniversary of INFN Bologna - ALICE Week, Bologna (Italy), June 2006, '*Complexity at the Fundamental Level*'.

[22] *Complexity and Planetary Emergencies*
A. Zichichi
presented at:

- 27th Sessions of the International Seminars on Planetary Emergencies, Erice (Italy), August 2002, '*Language, Logic and Science*'.
- 28th Sessions of the International Seminars on Planetary Emergencies, Erice (Italy), August 2003, '*Language Logic and Science, Evolution and Planetary Emergencies*'.
- 36th Sessions of the International Seminars on Planetary Emergencies, Erice (Italy), August 2006, '*Complexity and Planetary Emergencies*'.

[23] *Complexity exists at the Fundamental Level*
A. Zichichi in Proceedings of the 2004–Erice Subnuclear Physics School '*How and Where to Go Beyond the Standard Model*', pages 251-333, World Scientific (2007).

[24] *Relic Gravitational Waves from String Cosmology*
R. Brustein, M. Gasperini, M. Giovannini and G. Veneziano, *Phys. Lett.* B361, 45 (1995).

Primordial Magnetic Fields from String Cosmology
M. Gasperini, M. Giovannini and G. Veneziano, *Phys. Rev. Lett.* 75, 3796 (1995).

Quantization of Point Particles in (2+1)-Dimensional Gravity and Spacetime Discreteness
G. 't Hooft, *Class. Quantum Grav.* 13, 1023 (1996); and

The Scattering Matrix Approach for the Quantum Black Hole: an Overview
G. 't Hooft, *Int. Journal Mod. Phys.* A11, 4623 (1996).

For a review see G. Veneziano '*An Amusing Cosmology from the String Effective Action*' in '*Effective Theories and Fundamental Interactions*', Erice 1996, World Scientific, 300 (1997).

Quantised Singularities in the Electromagnetic Field
P.A.M. Dirac, *Proc. Roy. Soc.* A133, 60 (1931).

[25] *Fifty Years of Subnuclear Physics: From Past to Future and the ELN Project*
A. Zichichi, *Rivista del Nuovo Cimento* 21 N12, 1-98 (1998); see also

ELOISATRON (The European LOng Intersecting Storage Accelerator)
A. Zichichi et al, INFN/AE-83/7, June 1983. INFN/AE-84/2, January 1984 and 1985. Presented at the Galileo Galilei and Alfred B. Nobel Celebrations '*Science for Peace*', San Remo and Rome, Italy (1-11 May 1983).

[26] *The Logic of Nature, Complexity and New Physics: From Quark-Gluon Plasma to Superstrings, Quantum Gravity and Beyond*
Proceedings of the 2006–Erice Subnuclear Physics School, World Scientific (2008).

Details on heavy mesons with positive and negative strangeness discovered by the Blackett group can be found in the following paper.

W. A. Cooper, etc.
1957, Giugno
Il Nuovo Cimento
Serie X, Vol. 5, pag. 1386-1397

Examples of the Production of $(K^0, \bar{K}^0)$ and $(K^+, \bar{K}^0)$ Pairs of Heavy Mesons.

W. A. Cooper, H. Filthuth, J. A. Newth, G. Petrucci,
R. A. Salmeron and A. Zichichi
C.E.R.N. - Geneva

(ricevuto il 14 Gennaio 1957)

Summary. — Two simple nuclear interactions that produce pairs of K-mesons are described and discussed. They are interpreted as examples of the processes $n + p \to K^0 + \bar{K}^0 + n + p$ and $n + p \to K^+ + \bar{K}^0 + n + n$ where $\bar{K}^0$ is the anti-particle of the K^0-meson.

$np \to K^0 \bar{K}^0 + \ldots$

$np \to K^+ \bar{K}^0 + \ldots$

1. – Introduction.

The production, in elementary reactions, of pairs of K-mesons without other strange particles has been predicted theoretically by Gell-Mann and Pais (1). Two examples of (K^+, K^-) pairs, interpreted in this way, have been observed in emulsion experiments (2,3). One event seen in a cloud chamber attached to the Berkeley accelerator has been interpreted as a $(K^0, \bar{K}^0)$ pair in which only one of the K^0-mesons was seen to decay (4).

During a systematic study of the associated production of heavy mesons and hyperons in a cosmic-ray cloud chamber we have found two examples

(1) M. Gell-Mann and A. Pais: *Proceedings of the Glasgow Conference on Meson and Nuclear Physics* (1954), p. 342.

(2) M. W. Friedlander, D. Keefe and M. G. K. Menon: *Nuovo Cimento*, 2, 660 (1955).

(3) M. Ceccarelli, M. Grilli, M. Merlin, G. Salandin and B. Sechi: *Nuovo Cimento*, 2, 828 (1955).

(4) W. B. Fowler, G. Maenchen, W. M. Powell, G. Saphir and R. W. Wright: *Phys. Rev.*, 103, 208 (1956).

6. – Conclusions.

The two events reported in this paper extend our knowledge of the production processes of heavy mesons. To the best of our knowledge, they are the first examples that have been observed of $(K^0, \bar{K}^0)$ and $(K^+, \bar{K}^0)$ pairs produced in simple interactions where the K-mesons have been identified directly.

The events are evidence that K^0-mesons with both positive and negative strangeness exist. That this should be so was predicted by GELL-MANN and PAIS who suggested that the K^+- and K^--mesons were particle and anti-particle with $S = +1$ and $S = -1$ and that, likewise, the K^0-mesons should exist in two states as particle and anti-particle with $S = +1$ and $S = -1$.

We interpret our observations, therefore, as being examples of $(K^0, \bar{K}^0)$ and $(K^+, \bar{K}^0)$ pairs produced in elementary neutron-proton interactions.

There seems to be good evidence that the production of K-meson pairs of this type is an important process at cosmic-ray energies.

* * *

We are grateful for the facilities provided for us at the Jungfraujoch Research Station where the photographs discussed in this paper were taken. In particular, we have received great help from the manager of the Station, Mr. HANS WIEDERKEHR.

Among our colleagues we should like to thank Mr. S. O. LARSON for the construction of apparatus and assistance in running it, Mlle. E. JOVET for making measurements of the photographs and Mr. A. H. CHAPMAN, Mr. G. D. JAMES and Mr. H. STEINER who all helped us with our work at the Jungfraujoch.

In writing this paper we have had the good fortune to be able to discuss it with Dr. B. D'ESPAGNAT and Dr. J. PRENTKI.

APPENDIX A

Event VB 536.

The measurements made on this event are given in Table II. None of the four tracks from the two neutral V-events shows a measurable curvature and the lower limits to the momenta of the four particles are given by the maximum detectable momentum. This is assumed to be proportional to the square of the track length for tracks less than 22 cm long and is equal to 4 GeV/c for tracks of this length or longer.

CELEBRATION LECTURE

Concrete Quarks

George Zweig
MIT, Research Laboratory of Electronics,
26-169, 77 Massachusetts Ave, Cambridge, MA 02139, USA
zweig@mit.edu

A short history of the physics of strongly interacting particles is presented. Events leading to the discovery, and eventual acceptance, of concrete quarks are described.*

1. Introduction

This year is the fiftieth anniversary of the discovery of quarks, and last year was the fortieth anniversary of the birth of QCD (Quantum Chromodynamics). QCD developed in two phases, the first involving the discovery of quarks, the second specifying the nature of their interactions. These phases arose from two very different traditions, that of Rutherford and Bohr, and that of Einstein. The first was grounded in observation, and the startling interpretation of what was observed, the second in a triumph of the imagination made possible by the power of quantum field theory. Here the first phase, culminating in the discovery and acceptance of concrete quarks, is described.

2. How it Started

The story of QCD begins with the discovery of spontaneous radioactivity by Henri Becquerel in 1896 who thought that X-rays, discovered just a few months earlier, might be emitted by phosphorescent substances. He noticed, quite by accident, that crystalline crusts of uranium salts created silhouettes of great intensity when left in a drawer next to a photographic plate wrapped in black paper, even though the salt had not been exposed to sunlight (Fig. 1).

The nature of radioactivity was elucidated three years later by Ernest Rutherford who found that two types of particles, distinguished by their penetration power, were present in uranium radiation, particles he called α

*Other aspects of quark history are given in Refs. 1 and 2.

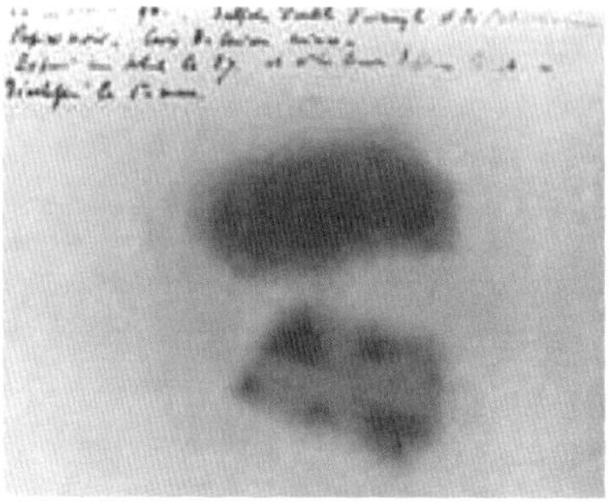

Fig. 1. One of Becquerel's photographic plates fogged by exposure to radiation from potassium uranyl sulfate. The shadow of a metal Maltese Cross placed between the plate and the uranium salt is visible.

and β. After three more years, Rutherford, with his young collaborator Frederick Soddy, interpreted the phenomena of radioactivity as "the spontaneous disintegration of [a] radio-element, whereby a part of the original atom was violently ejected as a radiant particle, and the remainder formed a totally new kind of atom with distinct chemical and physical character." Rutherford received the Nobel Prize in Chemistry in 1908. Soddy also received a Nobel Prize, but in 1921 for other work.[a] A photo of Rutherford with his group at Manchester University, taken two years after his prize, is shown in Fig. 2.

QCD speaks of protons, neutrons, and more remotely, the nuclei of atoms. Thompson's discovery of the electron in 1897 indicated that the atom was divisible. This meant that the charge on an electron in a neutral atom must be cancelled by a positive charge. The first indication that this positive charge is point-like was reported by Hans Geiger and Ernest Marsden in 1909.[3] Geiger was a postdoctoral fellow from Germany who came to Rutherford's lab to study the deflection of α-particles as they passed through thin metal plates. Marsden was a 20-year-old undergraduate from New Zealand, Rutherford's own country of origin. Geiger and Marsden reported that

> "conclusive evidence was found of the existence of a diffuse reflection of the α-particles. A small fraction of the α-particles falling upon a metal plate have their directions changed to such an extent that they emerge again at the side of incidence."

This innocuous sounding statement, when properly interpreted, was revolutionary, completely contradicting Newton's laws of mechanics, and Maxwell's

[a]All told, eleven of Rutherford's students, collaborators, and members of his laboratory went on to receive the Nobel Prize in Physics or Chemistry. Many more made remarkable but less recognized contributions. Rafi Muhammad Chaudhry went on to pioneer experimental nuclear physics in Pakistan, and with his student Mustafa Yar Khan, founded Pakistan's successful nuclear weapons program.

equations of electricity and magnetism as they had been formulated some 50 years earlier. Two years later in 1911 Rutherford published the proper interpretation,[4] which negated the common view that an atom consisted of negatively charged electrons embedded within a ball of positive charge. He showed that a large point-like concentration of charge was essential to account for the dramatic reflection of the positively charged α-particles back towards their source.[b] This meant that positive and negative charges were separated in an atom, but how? Not statically, for they would fall together. Not circling one another, for they would radiate energy like tiny antennae spiraling together. Rutherford had discovered the nucleus, an object that could not exist according to the existing laws of physics.

Fig. 2. Rutherford's group at Manchester University in 1912. Rutherford is seated second row, center. Also present: C. G. Darwin, J. M. Nuttall, J. Chadwick; H. Geiger, H. G. J. Moseley, and E. Marsden.

When Marsden in 1914 and 1915 presented the first experimental evidence that nuclei contain protons, a second contradiction appeared. Why didn't the positively charged nucleus explode? What was binding protons together?

Shortly after receiving his doctorate in 1911, Niels Bohr visited Rutherford's lab for several months, eventually settling there from 1914 to 1916. During his stay Bohr correctly combined two incomprehensible ideas, that of charge separation in the atom, and that of Planck's quantization of

[b]Rutherford concluded the nuclear charge could have either sign, was roughly proportional to the atomic number, and approximately $\pm 100e$ for gold.

radiation, into one incomprehensible idea: electrons exist in "stationary states" within the atom, emitting quanta of light as they jumped from one state to another. This view of the atom solved none of the contradictions with classical physics, but provided a conceptual framework within which the frequencies of spectral lines could be fruitfully organized, and their patterns contemplated.

The contradictions present in the Bohr atom, but not the nucleus,[c] were resolved with the introduction of quantum mechanics a decade later, which was formulated for atoms by Heisenberg solely in terms of the possible frequencies of light emitted as electrons changed their stationary states. His formulation consisted both of an equation involving a matrix of these possible frequencies, and *a philosophy of how theoretical physics should be practiced.* He emphasized that the contradictions inherent in Bohr's formulation of atomic physics would have been avoided if the equations of physics had been formulated entirely in terms of observables.

When experimental particle physics came to consist primarily of two-particle collisions, the observables were scattering amplitudes, like the observables in the Geiger–Marsden experiment. In three papers published during World War II Heisenberg argued that scattering amplitudes should also be organized into a matrix he called the S-matrix, and that the fundamental laws governing particles and their strong interactions should be formulated solely in terms of this matrix. Field theory, whose constructs and interactions were difficult or impossible to observe, was suspect. S-matrix theory in the form of the bootstrap[5] became the dominant school of thought in particle physics in the early 1960s.

My first exposure to nuclear physics came at the age of 10 in 1947, two years after the atomic bomb had been dropped on Hiroshima. One of my favorite after-school radio programs was the "Lone Ranger," sponsored by Kix breakfast cereal. Quite unexpectedly during a commercial break, the announcer asked his little listeners to send away for an "Atomic Bomb Ring."

[c]Rutherford and colleagues kept scattering α-particles off ever lighter nuclei. Rutherford and Chadwick, who later discovered the neutron, write:[7] "The study of the collisions of α-particles with hydrogen nuclei has shown that the force between the α-particle and the hydrogen nucleus obeys Columb's law for large distances of collision, but that it diverges very markedly from this law at close distances. The experiments of Chadwick and Bieler showed that for distances less than about 4×10^{-13} cm, the force between the two particles increased much more rapidly with decrease of distance than could be accounted for on an inverse square law of force. ... Possible explanations of the origin of these additional forces are discussed, and it is suggested tentatively that they may be due to magnetic fields in the nuclei." Here is the first direct observation of the nuclear force, the force holding protons together in the nucleus, but no new force is postulated to make the connection! The correct interpretation of truly novel phenomena can be difficult, even for giants like Rutherford and Chadwick.

I jumped at the chance. After mailing in my name, address, 15 cents, and a Kix box top, I received the ring (Fig. 3), took it into a dark closet full of winter coats, and waited till my eyes had adapted to the dark. Removing the red cap and peering down the long axis of the "bomb," I was rewarded with brilliant punctate flashes of light as one α-particle after another, emitted from a tiny piece of radioactive polonium, barreled into a zinc sulfide screen (a spinthariscope invented by William Crookes in 1903).[d] This was the same kind of screen used by Geiger and Marsden in their 1909 α-particle scattering experiment.[e]

In that same year of 1947, C. F. Powell discovered the pion in emulsions. Earlier in 1932 J. Chadwick had discovered the neutron. With these discoveries the cast of characters within the nucleus was complete. Both Chadwick and Powell had been Rutherford's students.

But also in 1947 a strange form of matter was discovered. G. D. Rochester and C. C. Butler published two cloud chamber photographs of cosmic-ray events, providing the first evidence of the existence of the K meson. Unlike the pion, whose existence had been predicted by Yukawa in 1935 to provide the force necessary to bind protons and neutrons together, the appearance of the K meson was entirely unexpected. As I. I. Rabi famously quipped when the muon was discovered, "Who ordered that?"[f].

This proliferation of elementary particles led E. Fermi and C. N. Yang to publish a paper in 1949 titled "Are Mesons Elementary Particles?" with the abstract:[6]

> "The hypothesis that π-mesons may be composite particles formed by the association of a nucleon with an antinucleon is discussed. From an extremely crude discussion of the model it appears that such a meson in most respects would have properties similar to those of the meson of the Yukawa theory."

However, in the body of the paper they caution,

> "Unfortunately we have not succeeded in working out a satisfactory relativistically invariant theory of nucleons among which such attractive forces act."

[d]Crookes was remarkable. He also discovered thallium in 1861, invented the radiometer, and developed the Crookes tube that was later used by W. C. Roentgen and J. J. Thomson in their discoveries of the X-ray and electron. As a young teenager wandering the stacks of the Detroit Public Library, I was fascinated by a book written by Crookes chronicling seances held in his house in the 1870s, not realizing his seminal contribution to the development of nuclear physics and the creation of the atomic bomb ring.

[e]I was so moved by the magic of the ring that I got a second one, not knowing that the half-life of polonium was only 138 days.

[f]We still don't know who ordered the muon.

Lone Ranger Atom Bomb Ring Spinthariscope (1947 - early 1950s)

This ring spinthariscope was known as the Lone Ranger Atom Bomb Ring and advertised as a "seething scientific creation." The Lone Ranger was more closely associated with silver bullets than atomic bombs but that's what it was called. When the red base (which served as a "secret message compartment") was taken off, and after a suitable period of time for dark adaptation, you could look through a small plastic lens at scintillations caused by polonium alpha particles striking a zinc sulfide screen.

Distributed by Kix Cereals (15 cents plus a boxtop), the instructions stated: "You'll see brilliant flashes of light in the inky darkness inside the atom chamber. These frenzied vivid flashes are caused by the released energy of atoms. PERFECTLY SAFE - We guarantee you can wear the KIX Atomic "Bomb" Ring with complete safety. The atomic materials inside the ring are harmless."

The following advertisement was appearing in newspapers in early 1947.

Fig. 3. The Atom Bomb Ring based on Crookes's 1903 spinthariscope.

This unsurmountable problem was to haunt Sakata's 1956 extension[8] of their work. He augmented the proton and neutron with the newly discovered Λ to construct K mesons out of pairs like the Λ and antineutron ($\Lambda\bar{n}$), but the binding mechanism was still obscure. Creating the light π meson from the other hadrons, all much heavier, would also be problematic for the bootstrap,[5] the fulfillment of Heisenberg's S-matrix theory. Advocates of the bootstrap ignored this problem.

By 1957 the list of "elementary particles" had grown to 19.[g] M. Gell-Mann and A. Rosenfeld summarized the situation in a paper for the Annual Review of Nuclear Sciences.[9] Working with E. P. Rosenbaum, an editor for Scientific American, Gell-Mann reworked this material for the general public in an article titled "Elementary Particles."[10] The 19 elementary particles they listed are shown in Fig. 4.

In the same issue of the Annual Review of Nuclear Science, S. Lindenbaum discussed a relatively new phenomenon, *resonances* in pion nucleon scattering, of which there were at least two (Fig. 5). Elementary particles and resonances coexisted side by side in journals and theoretician's minds, unconnected. Resonances, the "elephants in the room," were about to explode in number.

What distinguished resonances from the so-called elementary particles was their lifetimes. Resonances were created and decayed via the strong interactions. They lived for only 10^{-23} seconds, about the time it takes light to travel across a proton. Elementary particles lived much longer, either being stable, or decaying slowly through the electromagnetic or weak interactions. Elementary particles were grouped into two classes, point-like and extended in size. Only later was it realized that the long-lived extended elementary particles and the short-lived resonances were cut from the same cloth.

In the summer of 1957 I had just finished my sophomore year as a math major at the University of Michigan. I remember reading the Gell-Mann–Rosenbaum Scientific American article. My reaction can be summarized by lines from a Bob Dylan song of later years,

> "But something is happening here
> And you don't know what it is
> Do you, Mister Jones?"[h]

---

[g] Willis Lamb, in the first paragraph of his 1955 Nobel Prize Lecture, joked that he had "heard it said that 'the finder of a new elementary particle used to be rewarded by a Nobel Prize, but such a discovery now ought to be punished by a $10,000 fine.'"

[h] Bob Dylan, Ballad of a Thin Man, final track on Side One of Highway 61 Revisited, 1965.

Point particles

Spin 1/2 leptons	
Particle	Mass
e^-	1
μ^-	206.7
ν	0

Spin 1 photon	
Particle	Mass
γ	0

Extended particles (strongly interacting)

Spin 1/2 baryons		
Multiplet	Particle	Mass (m_e)
Ξ	Ξ^0	?
	Ξ^{-1}	2585
Σ	Σ^{-1}	2341
	$\Sigma+$	2325
	Σ^0	2324
Λ	Λ	2182
N	n	1838.6
	p	1836.1

Spin 0 mesons		
Multiplet	Particle	Mass
π	π^+	273.2
	π^{-1}	273.2
	π^0	264.2
K	K^+	966.5
	K^-	966.5
	K_1^0	965
	K_2^0	965

Fig. 4. The elementary (long-lived) particles in 1957.[10]

3. Caltech

Two years later in 1959, 50 years after the Geiger–Marsden experiment, I started graduate school in physics at Caltech,[i] and shortly thereafter started my PhD thesis as an experimentalist in Alvin Tollestrup's group, piggy-

[i]The Caltech Physics Department was remarkable. Carl Anderson, chairman, discovered both the positron and muon. His department included six soon-to-be Nobel Prize winners: R. Feynman, W. Fowler, M. Gell-Mann, Shelly Glashow, R. Mössbauer, and K. Wilson; Sidney Coleman and R. Dashen were students, and Y. Ne'eman and J.J. Sakurai visitors. If that wasn't enough, I could always go across campus and talk with ex-particle-physicist Max Delbrück, who had invented molecular biology, or Linus Pauling, a phenomenologist par excellence. I got the impression that it was possible to do things. Discoveries were happening, just down the hall!

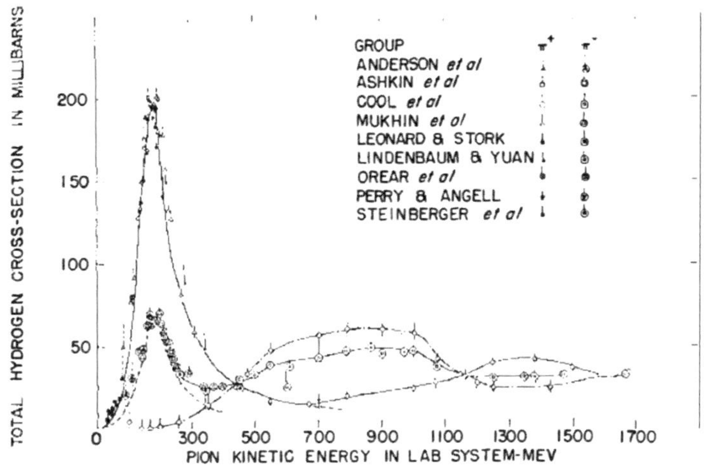

Fig. 5. $\pi^{\pm} + p$ total cross-sections as of 1957 showing evidence for the first pion–nucleon resonances.[11]

backing on their proposed study of $K^+ \to \pi^+ + \pi^0 + \gamma$, to measure the polarization of the μ, out of the plane of decay, in $K^+ \to \mu^+ + \pi^0 + \nu$. A non-zero value would imply a violation of time-reversal symmetry. This was the first "user's group" experiment, involving one faculty member, two research fellows, and two graduate students. The experimental equipment was built and tested at Caltech over a two-year period, but run at the Bevatron in Berkeley. After 21 half-days of running in the early spring of 1962, an adventure that should be chronicled elsewhere, I returned to Caltech with several hundred thousand spark-chamber photographs. A preliminary scan found no effect. I then faced an additional two years of tedious analysis determining the magnitude of errors, and establishing an accurate upper bound. With this unpleasant prospect, I embraced denial and went camping in the Yucatán.

On returning at the end of summer, I switched to theory. Screwing up my courage, I asked Murray Gell-Mann if he would be my thesis advisor. He seemed the natural choice: he supervised many graduate students, all of whom graduated quickly. I had spoken with him on several occasions after Alvin suggested that I talk to him about Alvin's K-decay experiment, meant to elucidate the mysterious $\Delta I = 1/2$ rule for weak nonleptonic decays. When asked, Murray immediately said "No," paused, and then said he was going to the East Coast on sabbatical, but would "talk to Dick." It wasn't until I returned from CERN two years later that I would speak with Murray once again.

"Dick" was Richard Feynman. I would never have had the courage to approach him. I had just watched him, in the second lecture of his gravity course, write down the Feynman rules for gravity for the first time, and compute the scattering of Mercury off the sun, getting the advance of the perihelion in 45 minutes, a result that I had seen H. P. Robertson, the grand old man of general relativity, obtain only after seven months into his course on general relativity.[j]

About a week later I asked Feynman if he would be my thesis advisor. He replied in a hoarse low-pitched voice, "Murray says you're okay, so you must be okay." Then he laid down the ground rules. I was to see him every Thursday afternoon from 1:30 to 4:15 when we would adjourn for tea, and then proceed to the physics colloquium. Each week I prepared frantically for our Thursday sessions, trying to pick some subject that would interest him, never touching any subject more than once, not even my thesis. My job was to make sure that Feynman was never bored. During the 1962–63 academic year we covered essentially all of particle physics. Here is some of what we talked about:

- Theory in the abstract:
  - (a) Axiomatic field theory was championed by Arthur Wightman at Princeton. Feynman hated it, and I didn't see how it helped explain the wealth of observations being made.
  - (b) Theory related to belief was championed by Geoffrey Chew at Berkeley. At a La Jolla conference in June 1961 in support of the bootstrap Chew said:

    > "I believe the conventional association of fields with strongly interacting particles to be empty. ... field theory..., like an old soldier, is destined not to die but just fade away."

  Feynman liked the idea of "nuclear democracy," but never tried to develop it.
- Theory related to experiment:
  - (a) Particle classification (no dynamics): G_2 was first championed by Lee and Yang, and by Gell-Mann, with Gell-Mann eventually switching to SU_3, also Ne'eman's choice. Feynman, on the sidelines, had little

[j]Robertson, of the "Robertson–Walker metric," had the distinction of rejecting a paper submitted to the Physical Review by Einstein. Einstein claimed that an accelerating mass would *not* emit gravitational radiation. Robertson's review was longer than Einstein's paper. Einstein was furious, but Robertson's evaluation was correct (`http://scitation.aip.org/content/aip/magazine/physicstoday/article/58/9/10.1063/1.2117822`). Einstein never again submitted a paper to the Physical Review.

doubt that SU_3 was the correct symmetry group, and that the 1, 8, and 10 dimensional representations were useful for organizing the low-lying mesons and baryons. Glashow and Sakurai were using the 27-dimensional SU_3 representation to classify hadrons.

In the fall of 1962 I organized the weak and electromagnetic currents into representations of SU_3, an octet of currents for the familiar weak and electromagnetic interactions, but also included currents in a 27-dimensional representation to accommodate what was then thought to be experimental evidence for weak interactions with $\Delta S = -\Delta Q$ (the Barkas event, $\Sigma^+ \to n + \mu^+ + \nu$). Feynman said "Forget the experiment! Throw away the 27, and keep the 8," but I couldn't because the experiment was clean, and the chance that the Barkas event was background was only 2 or 3×10^{-5}. Feynman was right. The event turned out to be an unlikely statistical fluctuation.[k]

(b) Dynamics (no particle classification): Although the bootstrap had its problems, Fred Zachariasen was able to bootstrap the ρ meson from two pions (Fig. 6). Forces between particles are created by the exchange of particles. If the ρ exists, it will be exchanged between two pions, creating an attractive force strong enough to bind them into a state with the quantum numbers of the ρ. Equating the particle exchanged (ρ) with the state it creates leads to a determination of the ρ mass and $\rho\pi\pi$ coupling constant that have the right order of magnitude.[12] As simply described by Chew, Gell-Mann, and Rosenfeld, the bootstrap was the future.[13]

Feynman was sympathetic to the goals of S-matrix theory and the bootstrap, but never used it himself. I didn't see how to bootstrap the π.

- Experimental physics:

  (a) More elementary particles and many more resonances were discovered:

    i. Point particles: The 4th lepton (ν_μ),
    ii. Extended elementary particles: An 8th spin 0 meson (η) and an 8th spin 1/2 baryon (Ξ^0),
    iii. Resonances: 26 meson resonances (ρ, ω, $K^*, \ldots$), the first of which had been discovered only two years earlier in 1961.

[k] More than 30 years later I ran into Nicola Cabibbo and asked him if he hadn't worried about the Barkas event before he published his famous 1963 paper on the weak interactions postulating the existent of an octet of currents. "Yes," he said, "but I had an important advantage over you. I was at CERN and there they had observed an additional 10^4 Σ^+ decays, none with $\Delta S = -\Delta Q$, so I decided to publish."

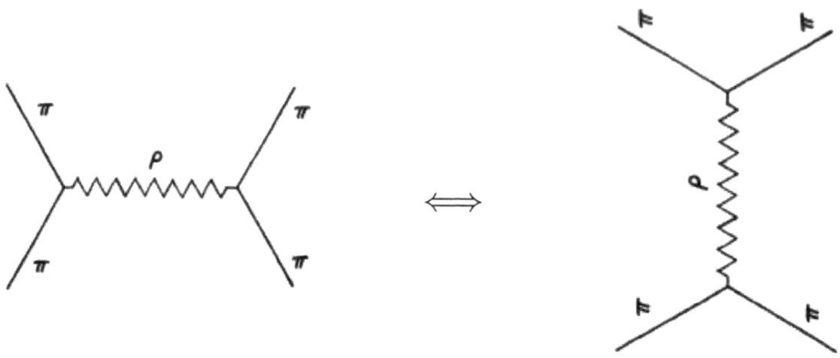

Fig. 6. The ρ bootstrap. Exchanging a ρ binds two pions into a ρ.[12]

4. Something Unbelievable

One Thursday afternoon I showed Feynman a paper in Physical Review Letters titled "Existence and Properties of the ϕ Meson."[14] Other papers had presented evidence for the existence of the ϕ, but this paper found no evidence where evidence was primarily expected. It was the nonexistence of a decay mode that should have broadened the width of the ϕ that fascinated me. Figure 7 shows the Dalitz plot for $K^- + p \to \Lambda + K + \bar{K}$. Note the long thin vertical pencil of $K\bar{K}$ events at the left edge of phase space clearly indicating the existence of the ϕ. So far, so good.

But now look at Fig. 8 where the 3-π mass distribution is displayed. There is no statistically significant evidence for a peak, so there is no evidence for the decay $\phi \to \pi^+ + \pi^- + \pi^0$, which I thought should be dominant. This decay channel is expected to proceed through the chain of reactions $\phi \to \rho + \pi$, followed by $\rho \to \pi + \pi$. The authors comment on this unexpected absence:

> "The observed rate [for $\phi \to \rho + \pi$] is lower than ... predicted values by one order of magnitude; however the above estimates are uncertain by at least this amount so that this discrepancy need not be disconcerting."

But I was very disconcerted. My calculations indicated that the $\rho\pi$ mode was suppressed by at least *two* orders of magnitude,[1] an *unprecedented* suppression for the strong interactions. I had learned from Feynman that in the

[1]The simple estimate was:

$$\begin{aligned} \frac{\Gamma_{K\bar{K}}}{\Gamma_{\rho\pi}} &\sim \left(\frac{p_{K\bar{K}}}{p_{\rho\pi}}\right)^3, \\ &= 1/4 \text{ (expected)}, \\ &\geq 35 \quad \text{(observed)}. \end{aligned}$$

Here p_{ij} is the momentum of either particle i or j in the rest frame of the ϕ.

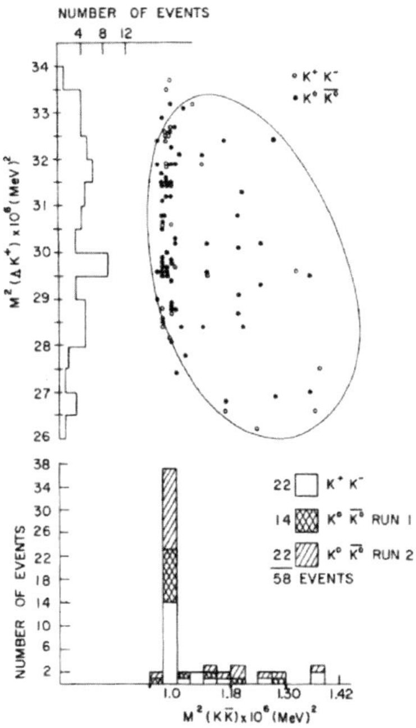

Fig. 7. Figure 1 from Ref. 14. "Dalitz plot for the reaction $K^- + p \to \Lambda + K + \bar{K}$. The effective-mass distributions for $K\bar{K}$ and ΛK^+ are projected on the abscissa and ordinate." The cluster of events at the edge of phase space indicate the existence of a meson called the ϕ decaying into $K + \bar{K}$.

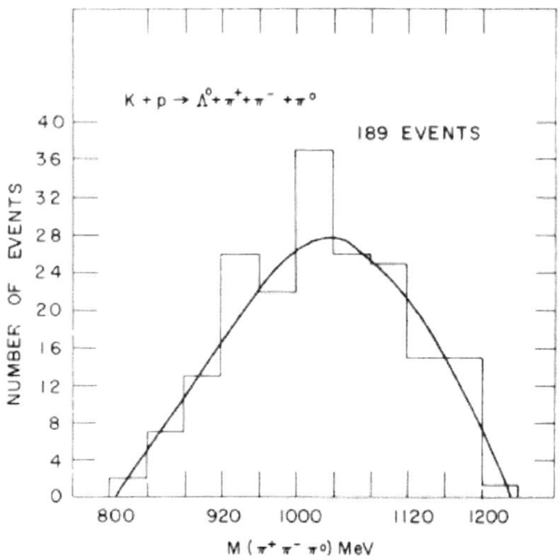

Fig. 8. Figure 4 from Ref. 14. "The $M(\pi^+\pi^-\pi^0)$ distribution from the reaction $K^- + p \to \Lambda + \pi^+ + \pi^- + \pi^0$." The absence of a large peak above phase space (solid line) indicates the suppression of the decay $\phi \to \rho + \pi$.

strong interactions everything that can possibly happen does, and with the maximum strength allowed by unitarity. Suppression implied symmetry, but no symmetry was present. Therefore the suppression must be dynamical, but no dynamical mechanism was available.

Feynman was not disconcerted. He launched into a tirade about how unreliable experiments were, and explained that at the time he proposed the V−A theory for the weak interactions all experiments were against him, and those experiments were wrong.[m]

5. The Explanation

I couldn't stop thinking about the suppression of ϕ decay, and finally realized that this paradox could be resolved by assuming that hadrons had constituents that obeyed a simple *dynamical* rule when they decayed.[15,16] Like Fermi and Yang, and later Sakata, I assumed that mesons are constructed out of fermion-antifermion pairs, but the fundamental fermions are not nucleons, but new fields called "aces," for reasons that will become apparent (they are now associated with constituent quarks, but unlike constituent quarks, aces are also the fundamental fields in electromagnetic and weak currents). With this assumption, the problems Fermi and Yang had in making the pion a nucleon-antinucleon bound state disappeared. Some unknown force, not the nuclear force, bound aces together. As in the Sakata model, there are three building blocks: a doublet $N_0 \equiv \{p_0, n_0\}$ with isotopic spin $I = 1/2$ and strangeness $S' = 0$, analogous to the nucleons $\{p, n\}$, and a singlet Λ_0 with isotopic spin $I = 0$ and strangeness $S' = -1$, analogous to the Λ. Following Dirac, aces are paired with antiaces, $\{\bar{p}_0, \bar{n}_0, \bar{\Lambda}_0\}$. The vector mesons with their constituents are shown in Fig. 12(d). The sawtooth line connecting an ace a (shaded circle) with an antiace $\bar{a}$ (open circle) represents the spring holding them together. The $a\bar{a}$ pair has angular momentum $L = 0$, and total spin $S = \frac{1}{2} + \frac{1}{2} = 1$. The pion, a pseudoscalar meson, has the same constituents as the ρ, but the spins of its constituents cancel, making $S = 0$. A meson decays when its $a\ \bar{a}$ constituents fly off in different directions. Their separation induces a polarization of the vacuum creating an $\bar{a}'a'$ pair that also splits, forming $a\bar{a}'$ and $a'\bar{a}$, the decay products (Fig. 9).

[m] The V−A theory (1957) was initially at variance with angular correlations measured in He^6 decay and the absence of the decay $\pi^- \to e^- + \bar{\nu}$, which was not seen in two independent experiments by Jack Steinberger in 1955, and Herb Anderson in 1957. In 1958, in CERN's first major discovery, Fazzini, Fidecaro, Merrison, Paul and Tollestrup observed this decay at the predicted rate, confirming V−A. As a testimony to the difficulty of measurement, both Steinberger and Anderson were outstanding experimentalists, students of Fermi. Steinberger later shared the Nobel Prize for demonstrating that the electron and muon each have their own neutrinos.

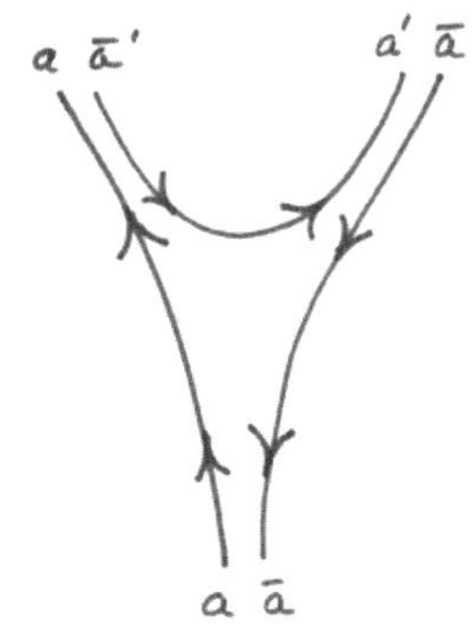

Fig. 9. The decay of meson $a\bar{a}$, a is an ace, $\bar{a}$ an antiace.

With this rule of ace conservation (a component of Zweig's rule), aces in the initial state must appear in the final state, i.e., "aces don't eat each other." Since the ϕ consists only of Λ_0 and $\bar{\Lambda}_0$, constituents not found in ρ or π, the decay ϕ to $\rho + \pi$ cannot take place, and empirically is strongly suppressed.[n] Remarkably, the constituent assignment for the mesons required to suppress $\phi \to \rho+\pi$ gave mass formulae for the vector mesons that worked exceptionally well (see below).[o]

In order to obtain the observed baryon spectrum it is necessary to build baryons out of three aces, giving them each baryon number $B = 1/3$. This distinguishes them from the $\{p, n, \Lambda\}$ of the Sakata model, which fails to group the low-lying baryons into the observed group of 8 (Fig. 4). The charge Q of an ace is then given by the NNG charge formula $Q = e[I_z + (B+S')/2]$, where I_z is the $z-$projection of the isotopic spin. With the I_z values of $\{1/2, -1/2, 0\}$ for $\{p_0, n_0, \Lambda_0\}$, their charges are $\{2/3, -1/3, -1/3\}$.

These ideas were extended to characterize additional properties of hadrons and their strong, electromagnetic, and weak interactions. The essence of the model as it appeared in two CERN preprints at the start of 1964, (Fig. 10[15,16]), is as follows:[p]

---

[n]Originally a graphically different, but functionally equivalent, visual representation of the decay was given, involving a "tinker toy" construction (see Fig. 13 of Ref. 16).

[o]Okubo obtained the suppression of ϕ decay, and mass formulae, with a different argument that didn't involve constituents.[17]

[p]Work on aces was almost finished by Thanksgiving 1963 when Ricardo Gomez, a Caltech Research Fellow I had worked with on the K-decay experiment at Berkeley, came to visit. When I told him what I was doing, he smiled and said I was "crazy." He wanted to go to the Bataclan, the one and only strip club in town (Calvin's Geneva). Much to my surprise, we skipped a very long line and were seated at a small table up front with a free bottle of champagne. The manager thought Ricardo was *Ricardo Gomez*, the famous bicycle racer, and Ricardo did nothing to disabuse him of this idea.

AN SU_3 MODEL FOR STRONG INTERACTION SYMMETRY AND ITS BREAKING

G. Zweig *)
CERN - Geneva

A B S T R A C T

Both mesons and baryons are constructed from a set of three fundamental particles called aces. The aces break up into an isospin doublet and singlet. Each ace carries baryon number $\frac{1}{3}$ and is consequently fractionally charged. SU_3 (but not the Eightfold Way) is adopted as a higher symmetry for the strong interactions. The breaking of this symmetry is assumed to be universal, being due to mass differences among the aces. Extensive space-time and group theoretic structure is then predicted for both mesons and baryons, in agreement with existing experimental information. An experimental search for the aces is suggested.

*) This work was supported by the Air Force Office of Scientific Research and the National Academy of Sciences - National Research Council, U.S.A.

8182/TH.401
17 January 1964

Fig. 10. Title page of Ref. 15. Only certain SU_3 representations, quantum numbers, and decays are allowed, constraints not found in the Eightfold Way.

(1) Hadrons have *point* fermion constituents with baryon number 1/3.
(2) There is a correspondence between leptons, the point particles of the weak interactions, and aces, the point particles of the strong interactions. In 1963 four leptons were known, hence the name aces.[q] This correspondence was just a hunch, in the Einstein tradition.
(3) The ultimate number of constituents was unknown. Representing them by ♣, ♡, ♠ and ♢ was not general enough. To allow for the possibility of an expanding set, each constituent was represented by a regular polygon of ever increasing size, corresponding to increasing mass: a circle represented p_0, a triangle n_0, a square Λ_0, a pentagon the fourth ace, and so on.
(4) Each hadron is represented by a linear combination of $a\bar{a}'$ pairs (deuces) for mesons, and $aa'a''$ triplets (treys) for baryons. The deuces or treys are weighted by SU_3 coefficients to form meson and baryon wave functions, e.g., $\rho^0 = \frac{1}{\sqrt{2}}(p_0\bar{p}_0 - n_0\bar{n}_0)$, indicating that the ρ^0 is equally likely to be a $p_0\bar{p}_0$ or $n_0\bar{n}_0$, neglecting electromagnetic interactions.

Aces provide a rationale for the existence of SU_3 symmetry, but more restrictively, lead to unique predictions for the existence of only certain SU_3 representations and quantum numbers, and provide relations among masses, and coupling constants. Baryons only come in groups of 1, 8, and 10 ($3 \times 3 \times 3 = 1 + 8 + 8 + 10$), mesons only in groups of 1 and 8 ($3 \times \bar{3} = 1 + 8$), with these two SU_3 representations mixing strongly for vector mesons, leading to a group of 9. Keeping track of ace spins, and allowing aces to have angular momentum, connects the spin-parity of baryons with the dimension of their SU_3 representations, and creates higher mass excited states. Assuming aces in the lightest mesons and baryons have zero angular momentum leads to the observed baryon octet ($J^P = \frac{1}{2}^+$), decuplet ($\frac{3}{2}^+$), and singlet ($\frac{1}{2}^-$), as well as a pseudoscalar meson octet ($J^{PC} = 0^{-+}$) and vector meson nonet (1^{+-}). Mesons with certain quantum numbers are forbidden, i.e., those with $J^{PC} = 0^{--}$, and those in the series $0^{+-}, 1^{-+}, 2^{+-}, \ldots$. Such "exotic" states may still be absent, but weakly bound deuteron-like bound states with exotic quantum numbers may now exist.

[q]A more obvious reason for choosing the word ace is that it is derived from the frequently used Latin word *as*, designating a *unit*, *whole* or *one*, and also designating a *small copper coin*. In English, it meant the side of the die with only one mark before it meant the playing card. In High Energy Physics, an ace might now be taken to mean any one of the six faces of the die. Who says that "God doesn't play dice with the Universe?"

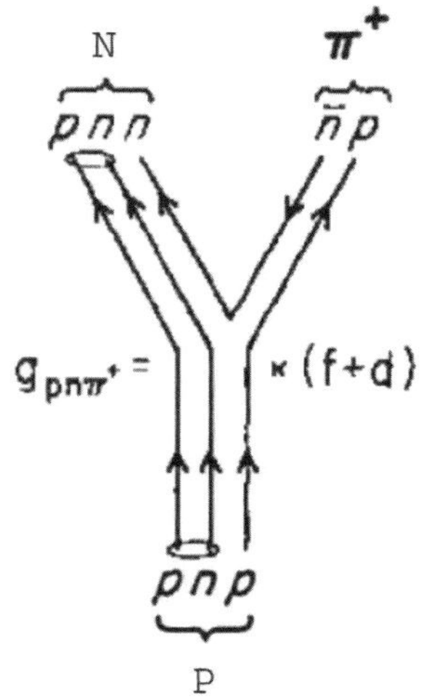

Fig. 11. Graphical representation of the "$f + d$" contribution to meson-baryon coupling. The "little loop" encloses anti-symmetrized aces. A different contribution to the coupling arises if the anti-symmetrized aces are separated in the decay, one ace terminating in the baryon, the other in the meson. The subscript "0" on aces is suppressed. Adapted from Ref. 18.

(5) *Aces, not hadrons, interact:*

(a) For the strong interactions, hadrons decay when their constituents separate and initiate the formation of decay products (Zweig's rule,[r] Figs. 9 and 11).

Graphically, when a meson decays into other mesons, it decays in all ways the deuces present can be connected, each connection having the same amplitude. The decay amplitude is proportional to the sum of these amplitudes, each amplitude weighted by the amplitude of the deuces in the meson wave functions (examples given in Figs. 10 and 11 of Ref. 16). This is a simple application of Feynman's "sum over all paths" formulation of quantum mechanics.

(b) The electromagnetic or weak interactions of hadrons occur through the interaction of their constituents with the photon or intermediate vector boson. For example, when the neutron n undergoes β-decay

[r] Gell-Mann enjoyed using Rosner's term, the "Twig rule" ("twig" is derived from the German word "zweig," which is conventionally — and more flatteringly! — translated as "branch"). Zweig's rule differs from that of Okubo's or Iizuki's because it not only says what is forbidden, but also what is allowed, *and by how much.* It involves enumerating, weighting, and summing all graphs available to a hadron for decay. An application of Zweig's rule that is not covered by the OZI rule is found in Ref. 16, and illustrated in Fig. 11.

$(n \to p + e^- + \nu)$, it is really its n_0 constituent that β-decays $(n_0 \to p_0 + e^- + \nu)$, the lighter p_0 becoming a constituent of the final-state proton. In this respect aces behave like fields in a field theory, i.e., like "current quarks."[19]

(6) The mass of a hadron is the weighted average of the masses of its deuces or treys, the weights given by the relative probabilities of the hadron existing in its possible deuce or trey configurations. The mass of a deuce $D_a^{\bar{a}'}$ or trey $T_{a\,a'a''}$ is the sum of their constituent masses minus their *pairwise* binding energies. Subscripts refer to aces, superscripts to antiaces. Three-body forces are negligible. Ace mass differences Δm are assumed to be significantly greater than their binding energy differences ΔE, i.e., $|\Delta m| \gg |\Delta E|$, so symmetry breaking arises primarily from ace mass splittings:

(a) For mesons: $m(D_a^{\bar{a}'}) = m(a) + m(\bar{a}') - E_a^{\bar{a}'}$. The binding energy $E_a^{\bar{a}'}$ depends on the SU_3 representation of the meson containing the deuce, and on the total spin $\vec{S}$ and angular momentum $\vec{L}$ of the $a\,\bar{a}$ system.

(b) For baryons: $m(T_{a\,a'a''}) = m(a) + m(a') + m(a'') - E_{a\,a'\,*} - E_{a\,*\,a''} - E_{*\,a'\,a''}$. A trey $T_{a\,a'a''}$ in a baryon is represented by a triangle with the aces a, a' and a'' at its vertices, with $E_{a\,a'\,*} = E_{a'\,a\,*}$. The binding energies depend on the SU_3 representation of the baryon containing the trey, and on the spins and angular momentum of the aces.

The strong interaction symmetry SU_3 is broken by distinguishing the mass and binding energies of Λ_0 from N_0, (Fig. 12).

The electromagnetic symmetry SU_2 is broken by distinguishing the mass and binding energies of n_0 from p_0 (Fig. 13).

No potential function is assumed. This is not the naive quark model! The naive quark model missed the point. It made detailed assumptions about the potential that were surely incorrect, with very little to show for it. I had no idea how the potential varied with distance, just that it gave rise to a very strong two-body force.[s]

[s]When Finn Ravndal first came to Caltech as a graduate student in 1968 he was eager to work on the quark model, and was very surprised to find that Murray and I were not thinking about quarks. I had done what I could five years earlier, and didn't know how to find the force binding aces. Even if I guessed correctly, I thought the theory would be so nonlinear that I couldn't prove the guess correct. Murray wasn't trying to find the force because he didn't think quarks existed. Ken Young, one of Murray's brightest graduate students, laments that Murray didn't ask him to compute the force between two quarks when he asked Murray for a thesis problem in 1970, a problem that was assigned to Politzer and Wilczek by their thesis advisors two years later, after Gerard 't Hooft had found the answer, but didn't fully publish his result.

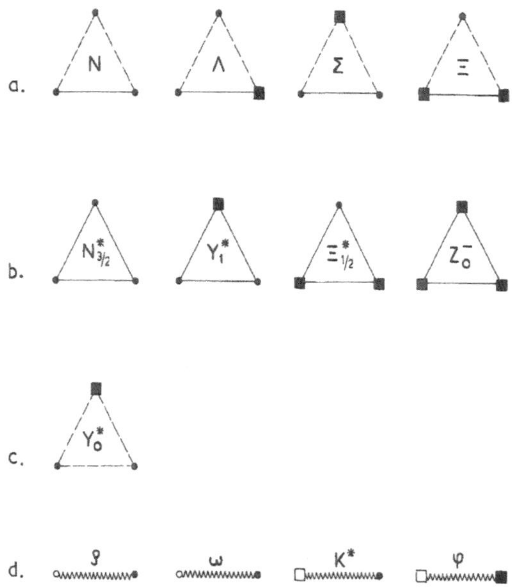

Fig. 12. Figure 2 of Ref. 15. "We view the particle representations with unitary symmetry broken. One of the three aces has now become distinguishable from the other two. It is pictured as a shaded square. ... The mass splittings within representations are induced by making the squares heavier than the circles. Since the same set of aces are used to construct all hadrons, mass relations connecting mesons and baryons may be obtained." The assignment of constituents to the vector mesons was suggested by the observation that the ρ and ω had the same mass, and the square of the K^* mass was the average of the mass squares of the ρ and ϕ.

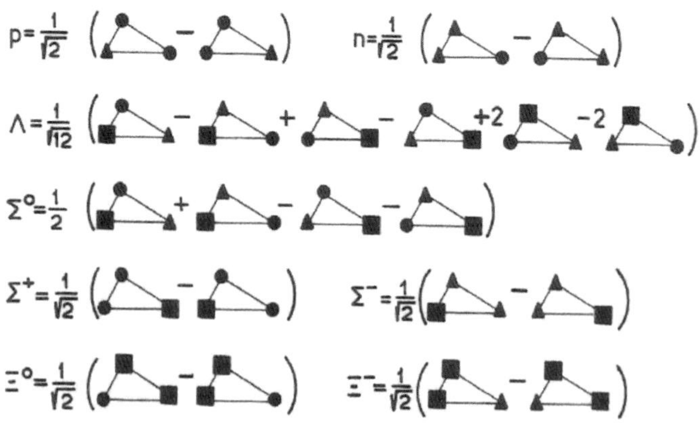

Fig. 13. Figure 3 of Ref. 16. "After SU_3 has been broken by the strong and electromagnetic interactions the baryon octet looks like this. The three aces are now completely distinguishable from one another." A baryon wave function is a linear combination of treys. The trey coefficients come from SU_3. Although aces are point-like, the mass of an ace is represented by the size of its symbol.

6. The Question

Would you, the reader, have adopted aces as constituents? To help answer this question, some results beyond SU_3 symmetry from the 80 page CERN report are outlined below. Except when in curly brackets, the masses in MeV are those known at the time.

(1) SU_3 symmetry breaking: $m(p_0) = m(n_0) < m(\Lambda_0)$, leading to:
A hierarchy of baryon relations, e.g., for the $J^P = \frac{1}{2}^+$ octet:

(a) If all binding energies are equal, then the stranger a baryon in a multiplet, the heavier it is:

i. $m(N) < m(\Lambda) \approx m(\Sigma) < m(\Xi)$,
939 1115 1193 1315

ii. $\frac{1}{2}[m(\Xi) + m(N)] \approx \frac{1}{2}[m(\Lambda) + m(\Sigma)]$.
1127 1154

(b) If averaging relations for the binding energies hold, i.e., $\frac{1}{2}(E_{\Lambda_0\Lambda_0.} + E_{\alpha\beta.}) \approx E_{\Lambda_0\alpha.} \approx E_{\Lambda_0\beta.}, \alpha, \beta = p_0, n_0$, then:

i. $\frac{1}{2}[m(\Xi) + m(N)] \approx \frac{1}{4}[3m(\Lambda) + m(\Sigma)]$.
1127 1134
(Gell-Mann–Okubo mass formula)

A hierarchy of meson relations,[t] e.g., for the $J^{PC} = 1^{--}$ nonet:

(a) If all binding energies are equal, then:

i. $m^2(\rho) \approx m^2(\omega) < m^2(K^*) < m^2(\phi)$.
750^2 784^2 888^2 1018^2

(b) If averaging relations for the binding energies hold, i.e., $\frac{1}{2}(E^{\bar{\Lambda}_0}_{\Lambda_0} + E^{\bar{\beta}}_{\alpha}) \approx E^{\bar{\beta}}_{\Lambda_0} \approx E^{\bar{\Lambda}_0}_{\alpha}, \quad \alpha, \beta = p_0, n_0$, then:

i. $m^2(\phi) \approx 2m^2(K^*) - m^2(\rho)$.
1018^2 1007^2

(2) SU(2) symmetry breaking: $m(p_0) < m(n_0)$, leading to:
A hierarchy of baryon relations, e.g., for the $J^P = \frac{1}{2}^+$ octet:

(a) If all binding energies are equal, then the more negative a baryon in a multiplet, the heavier it is:

[t]Mass relations were linear for baryons, and quadratic for mesons, in analogy with the linear Dirac equation for fermions, and the quadratic Klein-Gordon equation for mesons.

i. $m(p) < m(n)$,
938 939

ii. $m(\Sigma^+) < m(\Sigma^0) < m(\Sigma^-)$,
1190 1193 1196

iii. $m(\Xi^0) < m(\Xi^-)$.
1315 1321

(b) If averaging relations for the binding energies hold, i.e., $\frac{1}{2}(E_{p_0 p_0 .} + E_{n_0 n_0 .}) \approx E_{p_0 n_0 .}, \cdots$, then:

i. $\frac{1}{2}[m(\Sigma^+) + m(\Sigma^-)] \approx m(\Sigma^0)$.
1193.4 ± 0.3 1193.2 ± 0.7

(c) Even if all binding energies are different,

i. $m(n) - m(p) + m(\Xi^-) - m(\Xi^0) = m(\Sigma^-) - m(\Sigma^+)$.
7.3 ± 1.3 8.3 ± 0.5

2013 masses: $\{8.14 \pm 0.21\}$ $\{8.08 \pm 0.08\}$

2013 mass differences: $\{0.06 \pm 0.22\}$

This remarkable relation currently still holds to the known accuracy of the mass differences.

(3) Relations for the baryon decuplet and pseudoscalar meson octet.

(4) Additional mesons with orbital excitations ($L = 1, 2, \cdots$) and $\vec{L} \cdot \vec{S}$ mass splittings.

(5) Cross multiplet relations:

(a) $m(N) < m(\Lambda) \Longrightarrow m^2(\rho) < m^2(K^*)$,

(b) $m(\Xi^*) - m(\Sigma^*) \approx m(\Xi) - m(\Sigma)$,
145 122

(c) $m^2(K^*) - m^2(\rho) \approx m^2(K) - m^2(\pi)$.
0.22 Gev^2 0.22 Gev^2

I thought the empirical evidence for the existence of aces was overwhelming. Were these many successful relations a result of over-fitting?[u] Probably not, because over the past 50 years the accuracy of these relations has either remained essentially the same, or improved.

7. The Difficulties

All this was not as easy as it looks. Although aces resolved the problem of ϕ decay, explained many regularities, and made predictions that could easily be tested, some predictions contradicted experiment, and even violated basic theoretical principles.

Matt Roos's "definitive" compilation of resonances, published in the 1963 Reviews of Modern Physics, lists 26 meson resonances.[20] *We now know that 19 of these resonances do not exist, and of those 19, 7 are exotic, i.e., cannot be made from deuces.* As recounted previously:[1]

> "Matt Roos's ... compilation of particles and their properties referred to several hundred experimental papers. I read essentially all of them, taking care to understand how each measurement was made. Then an accurate appraisal of the results of each experiment was possible. Rational choices between conflicting experiments usually could be made."

My training both in experimental and theoretical physics made it possible to separate "wheat from chaff." Others did not have the training, interest, or patience.

A major theoretical problem was the existence of the celebrated Ω^- with spin $S = 3/2$ constructed from three identical aces all in the same state, thereby *violating Pauli's spin-statistics theorem.* I didn't have a resolution to this problem, but thought it eventually would be resolved, much as the contradictions with Rutherford's atom and Bohr's orbits were eventually resolved by quantum mechanics.[v]

Equally problematic, aces had no place in current dogma. They were incompatible with nuclear democracy, the basis of the bootstrap, and much prevailing thinking. In addition, because aces were not observed as free particles, formulating a theory of strong interactions in terms of them was

[u]Certain simple relations like $m(\Xi) = [3m(\Sigma) - m(N)]/2$, which were correct to electromagnetic mass splittings, were considered accidental.

[v]Feynman's 1970 solution to the spin-statistics problem was to make quarks bosons! The abstract to a paper written by Feynman, Kislinger, and Ravndal titled "The $\Delta I = 1/2$ Rule from the Symmetric Quark Model" reads "The $\Delta I = 1/2$ rule for the weak non-leptonic hyperon decays will result from quark currents interacting at a point, if the quarks obey Bose statistics." The paper was withdrawn before publication when Feynman learned that this idea had been previously proposed (Ravndal, private communication).

incompatible with Heisenberg's requirement that theory be based solely on observables. Like Copernicus's view of the solar system, simplicity was no excuse for challenging what was known to be true.[w]

Finally, working with aces made me uncomfortable. I knew what real theories looked like. As a student I had read and understood Schwinger's papers on Quantum Elecrodynamics. I had written "professional" papers.[21] Nothing I now did looked like that. Talking about aces in public was embarrassing, because I understood how theorists thought, and what many thought of me.[x] In spite of this, a good part of me believed that aces were real particles, and that the beginnings of a solution to the proliferation problem of hadrons had finally been found. Aces were real particles because they had *dynamics*. They moved from one hadron to another, avoiding annihilation, either forwards or backwards in time. They spun and rotated around one another giving rise to $\vec{L} \cdot \vec{S}$ mass splittings. In addition, they interacted with light and underwent β decay. I applied the "duck" test: If it walks like a duck, swims like a duck, and quacks like a duck, it's probably a duck. Others disagreed.

8. The Reaction

After returning from CERN in the early fall of 1964, I went into Murray's office and told him all about aces. Sometimes Murray would close his eyes when someone talked to him, but this time his eyes stayed open. After I finished at the blackboard he exclaimed from behind his desk "Oh, the *concrete* quark model. That's for blockheads!" As late as five years after the deep inelastic scattering experiments at the Stanford Linear Accelerator provided conclusive evidence for electrons scattering off point particles in nucleons, Murray still did not accept the existence of constituent quarks inside of hadrons. He writes:[22]

> "In these lectures I want to speak about at least two interpretations of the concept of quarks for hadrons and, the possible relations between them.
>
> First I want to talk about quarks as 'constituent quarks'. *These were used especially by G. Zweig (1964)* [italics added] who referred to them as aces. ..."

[w]Wegener's theory of continental drift provides another example of discovery contradicting dogma. Evidence from geology and paleontology clearly showed that Africa and South America were connected sometime in the past, but this idea was not accepted because the Earth's crust was thought to be "frozen." A dynamical mechanism for driving continents apart was not yet known.

[x]Having worked at the Bevatron in the early 1960s, I appreciated the excitement and beauty of Berkeley and the San Francisco area. I wanted a job at UC Berkeley. Gerson Goldhaber presented my application package at a physics faculty meeting, but a senior theorist blocked the appointment, passionately arguing that the ace model was the work of a "charlatan" (Goldhaber, private communication).

It is more precise to say: "*These were introduced by G. Zweig* (1964)..." After all, they did not exist at the time as a tool in a toolbox for anyone to use.

Murray then goes on to say:

> The whole idea is that hadrons act as if they are made up of quarks, but the quarks do not have to be real. ..."

That's a mischaracterization of my idea. I believed that they *were* made up of quarks.

> "There is a second use of quarks, as so-called 'current quarks' which is quite different from their use as constituent quarks ...
>
> If quarks are only fictitious there are certain defects and virtues. The main defect would be that we never experimentally discover real ones and thus will never have a quarkonics industry. The virtue is that then there are no basic constituents for hadrons — hadrons act as if they were made up of quarks but no quarks exist — and, therefore, *there is no reason for a distinction between the quark and bootstrap picture: they can be just two different descriptions of the same system, like wave mechanics and matrix mechanics.* [italics added]"

This was Murray's vision. Aces, which had aspects of both current and constituent quarks, are not mentioned.

Although Feynman had no quarrel with "current quarks,"[19] or the current-quark aspect of aces, he disliked Zweig's rule. Every time I would "explain" it to him he became angry, and said that it didn't make sense. Unitarity mixed all states with the same quantum numbers, so the suppression of the ϕ to $\rho\pi$ was not possible. He also believed that the correct theory of the strong interactions should not allow one to say which particles are elementary, a key element of the bootstrap.

And what about Heisenberg? In an interview in the early 1970s for broadcast as part of a CBC radio documentary series entitled Physics and Beyond he said:[24]

> "Even if quarks should be found (and I do not believe that they will be), they will not be more elementary than other particles, since a quark could be considered as consisting of two quarks and one anti-quark, and so on. I think we have learned from experiments that by getting to smaller and smaller units, we do not come to fundamental units, or indivisible units, but we do come to a point where division has no meaning. This is a result of the experiments of the last twenty years, and I am afraid that some physicists simply ignore this experimental fact."

No, I'm afraid not. Quarks certainly are "more elementary than other particles," and a quark is not "considered as consisting of two quarks and one anti-quark." Heisenberg's variant of the Sakata model, but at a smaller scale, simply doesn't work.

9. Acceptance

Late in May of 1968 I bumped into Feynman as we both were walking to the Greasy (the Caltech cafeteria) for lunch. He was very excited about the High Energy Physics course he had just finished teaching, and was back in full swing after a lull in research following his unsuccessful attempt to renormalize gravity. After listing the many areas he had covered in his course he stopped, turned to me and asked, "Did I miss anything Zweig?" Patiently, once again, I told him about aces. This time he was quiet, intent, and listened. After I finished, he hitched up his pants with both thumbs, looked straight into my eyes, and in his most official voice replied: "All right, I'll look into it!" Shortly thereafter he created the parton model, and the following fall Feynman, with Finn Ravndal and Mark Kislinger, two graduate students, looked for evidence of concrete quarks in pion photoproduction, a place where I, and others, had never looked.

Three years later, as I was walking down the corridor on the fourth floor of Lauritsen where we both had offices, I noticed Feynman in the distance with an enormous grin, swaggering like a sailor, thumbs hooked inside his belt, fingers splayed apart. When he was almost in my face, no more than a foot away, he extended his right hand and said "Congratulations Zweig! You got it right." By that time I had switched to neurobiology and didn't know that his work with Finn and Mark had finally been completed.

9.1. *Bayes' theorem*

In retrospect, Bayes' theorem could have been used to elucidate some of the issues involved in deciding if aces should have been accepted as constituents of hadrons. With some massaging, Bayes' theorem says that the probability that hadrons have ace constituents, given the experimental evidence ($\phi \not\to \rho + \pi$, mass relations, etc.), is

$$P(A|E) = \frac{1}{1+\lambda}, \text{ where } \lambda = \frac{P(E|\bar{A})}{P(E|A)}\frac{P(\bar{A})}{P(A)}.$$

Here $P(E|\bar{A})$ is the probability of the evidence given that aces do not exist, $P(A)$ is the a priori probability that they do exist, and $P(\bar{A})$ that they don't.

Since

$$P(\bar{A}) \approx P(E|A) \approx 1,$$

the expression for $P(A|E)$ simplifies,

$$P(A|E) \approx \frac{1}{1 + P(E|\bar{A})/P(A)}.$$

Whether or not a rational person would have believed in the existence of aces depends only on the ratio λ of two small numbers. *They would have believed if and only if the likelihood of obtaining the evidence, assuming that aces didn't exist, was much lower than the a prior probability of aces existing* i.e., if and only if

$$P(E|\bar{A}) \ll P(A).$$

When this inequality holds, even though initially $P(\bar{A}) \approx 1$, when the experimental evidence is taken into account, just the opposite holds, i.e., $P(A|E) \approx 1$.

Because it is not possible to empirically determine these probabilities, how should this argument be interpreted? It suggests that in deciding whether hadrons have ace constituents, consider how easy it would be to find an alternative explanation for all the relations that follow assuming their existence, and weigh that against how much you like the idea of ace constituents in the abstract, without any experimental evidence for or against their existence. I suspect that those working in the Rutherford–Bohr tradition found acceptance easier than those trying to follow in Einstein's footsteps.[y]

9.2. *Different priors, different times*

The early adopters were Linus Pauling and Dick Dalitz, with Pauling coming first. Shortly after the CERN preprints were circulated, I received galley proofs of the third edition of Linus Pauling's "College Chemistry,"[23] where Pauling presented aces to undergraduates. He asked for my comments and corrections. Essentially none were necessary. Pauling recognized a good thing when he saw it. Aces didn't have anything to do with college chemistry, but I suspect he thought aces were so beautiful, and so right, that he couldn't resist ($P(A)$ was not *that* small for him).

Dalitz's conversion is less surprising. He came from nuclear physics where nucleon constituents, and the forces between them, were his daily bread and butter. He just changed constituents.

[y] A Baysian inspired evaluation of current speculative theories is also possible, even if $P(E|Theory)$ is unknown. If a theory has not yet advanced to the stage where it can compute observables, the evidence E should be replaced by the Standard Model.

For Feynman acceptance of concrete quarks came when he discovered them for himself in photoproduction amplitudes, where no one else had looked. Others were convinced by deep inelastic scattering experiments at SLAC. The holdouts finally folded with the discovery of the ψ(J), whose forbidden decay modes and narrow width were just like ϕ with the fourth ace replacing Λ_0. It was "déjà vu all over again."

10. Invention or Discovery

Last summer Murray and I were present at a talk given by Chris Llewellyn Smith at the Santa Fe Institute. In his introduction, Geoffrey West said "Murray and George invented quarks, which were later discovered at SLAC." I interrupted to suggest a different ending to Geoffrey's sentence: "quarks, whose existence was later confirmed at SLAC." But what was invented, and what discovered? According to Merriam-Webster, discovery is "the act of finding or learning something for the first time," and invention is "a product of the imagination." Current quarks were invented in the tradition of Einstein. Aces were discovered in the Rutherford–Bohr tradition, buried in the data, obscured by the contradictions they implied. Bohr would have loved them.

11. Final Thoughts

The CERN report concludes:[15]

> "There are, however, many unanswered questions. Are aces particles? If so, what are their interactions? Do aces bind to form only deuces and treys? What is the particle (or particles) that is responsible for binding the aces? Why must one work with masses for the baryons and mass squares for the mesons? And more generally, why does so simple a model yield such a good approximation to nature?"

After dutifully listing possible interpretations of the results, the report concludes with:

> "there is also the outside chance that the model is a closer approximation to nature than we may think, and that fractionally charged aces abound within us."

This then was the beginning of QCD, the theory that will eventually give us a real understanding of how it came to be that Becquerel's photographic plate clouded in his drawer.

12. Epilogue

In modern terms, an ace, or equivalently a concrete quark, was used as a point particle in a field theory to construct the electromagnetic and weak currents, and as a much heavier particle (the point particle dressed in glue) to construct the hadrons.[z] I thought observation required a dual description.

Although Bohr was able to calculate the spectral lines of hydrogen, his model could not accurately account for the spectrum of helium, and was conceptually incomplete. For example, the wave nature of particles had not yet been discovered by de Broglie, and was absent from Bohr's thinking. There was no Lagrangian. There was no theory!

The concrete quark model provided predictions for the particle spectrum, and enabled calculations of hadronic masses and couplings that were accurate to varying degree, but it was conceptually incomplete, lacking a specification of the interaction between quarks. Like the wave–particle duality of quantum mechanics, concrete quarks were chimeric: sometimes acting as fields in a field theory for weak decay, sometimes as convenient objects for the calculation of masses and coupling constants. And like the Bohr atom, the Ω^-, made from three quarks in the same state, could not exist according to the laws that were known. There was no Lagrangian. There was no theory!

Details aside, the important observation was that hadrons had fermion constituents of baryon number 1/3, with dynamics that suggested they were real, and therefore should have corresponding fields. It was generally believed that hadrons had constituents, but those constituents were other hadrons. Some even thought that field theory was irrelevant. No! Concrete quarks said that there is a deeper level of reality to be described with field theory, and channelled thinking into more productive directions, eventually leading to QCD. It was this deeper level of reality, with fractional charges, that made the acceptance of concrete quarks so difficult, and truly revolutionary.

The regularities in the spectral lines of the hydrogen atom responsible for the creation of the Bohr model, that eventually led to the invention of quantum mechanics, were immediately derived from quantum mechanics. Although ever more accurate numerical calculations of masses of the low-lying hadrons are now possible with QCD, the *regularities* among these masses that led to the discovery of the concrete quark model have not been derived. Their derivation would provide a satisfying test of that fledgling theory.

[z]By constructing the weak currents from aces I hoped that it would be possible to eliminate the unintelligible Cabibbo angle. The suppression of strangeness-changing decays would then be dynamical.

How can such a complicated nonlinear theory as QCD give rise to the simple low-energy relations of the concrete quark model? There should be some way to approximate QCD so that the existence and hierarchy of relations among masses become apparent. It's not like the physics of water where the greater the nonlinearity, the more complex the flow. Quite the contrary. In QCD, when energies are small and nonlinearities large, the particle spectrum and its couplings are simple.

Recall the history of classical mechanics. Newton's laws were reformulated again and again over centuries, each time with a different purpose. In the following centuries QCD will be reformulated again and again, and with one of those reformulations the concrete quark model may appear, a fortiori.

Acknowledgments

I thank Erica Jen and Jeffrey Mandula for their insightful comments, and David Donoho, Stanislaw Mrowczynski, Finn Ravndal, and Harvey Shepard for improving the accuracy and content of the text.

References

1. G. Zweig, Origins of the Quark Model, in *Proc. Fourth Int. Conf. Baryon Resonances*, ed. N. Isgur, University of Toronto, Canada, pp. 439–479 (1980), `http://www-hep2.fzu.cz/~chyla/talks/others/zweig80.pdf`.
2. G. Zweig, Memories of Murray and the Quark Model, *Proc. Conf. in Honor of Murray Gell-Mann's 80th Birthday*, eds. H. Fritzsch, K. K. Phua and B. E. Baaquie, World Scientific, Singapore, pp. 7–20 (2010), `http://arxiv.org/abs/1007.0494`.
3. H. Geiger and E. Marsden, *Proc. R. Soc. Lond. A* **82**, No. 557, 495 (1909).
4. E. Rutherford, *Phil. Mag.*, Series 6, **21**, 669 (1911).
5. G. F. Chew and S. Frautschi, *Phys. Rev. Lett.* **7**, 394 (1961).
6. E. Fermi and C. N. Yang, *Phys. Rev.* **76**, 1739 (1949).
7. E. Rutherford and J. Chadwick, *Phil. Mag.* Series 7 **4:22**, 605 (1927).
8. S. Sakata, *Progr. Theor. Phys.* **16**, 686 (1956).
9. M. Gell-Mann, and A. H. Rosenfeld, *Annu. Rev. Nucl. Sci.* **7**, 407 (1957).
10. M. Gell-Mann and E. P. Rosenbaum, *Scientific American*, 72 (July 1957).
11. S. J. Lindenbaum, *Annu. Rev. Nucl. Sci.* **7**, 317 (1957).
12. F. Zachariasen, *Phys. Rev. Lett.* **7**, 112 (1961); *Erratum* **7**, 268 (1961).
13. G. F. Chew, M. Gell-Mann, and A. H. Rosenfeld, *Scientific American*, **74** (February 1964).
14. P.L. Connolly *et al.*, *Phys. Rev. Lett.* **10**, 371 (1963).
15. G. Zweig, An SU_3 Model for Strong Interaction Symmetry and its Breaking, *CERN Report 8419/TH.401* (January 17, 1964), `http://cdsweb.cern.ch/record/352337?ln=en`.

16. G. Zweig, An SU_3 Model for Strong Interaction Symmetry and its Breaking II, *CERN Report 8419/TH.412* (February 21, 1964), in *Developments in the Quark Theory of Hadrons, A Reprint Collection*, Vol. I: 1964–1978, eds. D. B. Lichtenberg and S. P. Rosen, Hadronic Press, Inc., Nonantum, MA, pp. 22–101 (1980), `http://cdsweb.cern.ch/record/570209?ln=en`.
17. S. Okubo, *Phys. Lett.* **5**(2), 165 (1963).
18. J. Mandula, J. Weyers, and G. Zweig, *Annu. Rev. Nucl. Sci.* **20**, 289 (1970).
19. M. Gell-Mann, *Phys. Lett.* **8**, 214 (1964).
20. M. Roos, *Rev. Mod. Phys.* **35**, 314 (1963).
21. G. Zweig, *Il Nuovo Cimento* **XXXII**, No. 5, 689 (1964).
22. M. Gell-Mann, *Acta Physica Austriaca, Suppl. IX*, 733 (1972).
23. L. Pauling, *College Chemistry*, 3rd edn. (W. H. Freeman and Co., San Francisco, 1964).
24. W. Heisenberg, *Glimpsing Reality: Ideas in Physics and the Link to Biology*, P. Buckley and F. D. Peat eds., University of Toronto Press, Canada, 15 (1996).

LECTURER: G. ZWEIG

Concrete Quarks

Scientific Secretaries: J.Jejelava, J. Krog

DISCUSSION

- *Haidt:*

Since you have 4 aces, but used only three, did you worry about the fourth (charmed) one?

- *Zweig:*

I thought there had to be a fourth ace because there were four leptons, and I had a hunch that there was a symmetry between the two different types of fundamental fields. I didn't think that the particle spectrum, as we knew it then, had evidence for the fourth one. Ace, by the way, derives from the old Latin word "*as*" meaning a "unit" or "one." I wanted a word for the fundamental fields from which hadrons are built, that meant one, but also came in groups of four.

- *Haidt:*

The only evidence, which existed at the time, was from a CERN neutrino experiment; the observation of di-lepton events. This would be most easily explained by charm production; by the decay of charm particles.

- *Zweig:*

The experiment to which you refer was only published in October 1964 (Block, M.M. et al. 1964. Phys. Lett. 12: 281), long after the CERN Ace preprints had been distributed.

- *t' Hooft:*

Still it was very insightful, that you had some premonition that there should be something like a fourth quark.

- *Zweig:*

If you have basic units out of which you want to create your field theory, leptons and aces, then why should you have three of these and four of those?

- *t' Hooft:*

But didn't it cross your mind that perhaps quarks are leptons?

- *Zweig:*

It did, but leptons don't have any connection with the strong interactions, so I didn't know how to proceed.

- *t' Hooft:*

In the early days this wasn't all so clear and obvious.

- *Nanopoulos:*

What George says is right. In Gell-Mann's paper on quarks, he has, at the end, the comment: Because there are four leptons, exactly with same logic, there should be a fourth quark. He calls it 'p-prime'.

- *Zweig:*

In enumerating the possibilities for constructing toy field theories, Gell-Mann mentions the possibility of having 4 fundamental fields, in analogy with the leptons, but his four fields have integer charges. He then dismisses this possibility and goes on to say: "A simpler and more elegant scheme can be constructed if we allow non integral values for the charges....," and describes the quark model with three quarks.

- *t' Hooft:*

Today we know that it has to be like that, the number of quarks has to equal the number of leptons because of anomalies, but that's our present argument. It could not have been otherwise.

- *Inguglia:*

This is not a technical question, but I am a bit curious about the title for your talk. Why did you call it the ``Beginning of he end?''

- *Zweig:*

The talk is titled "Concrete quarks: The beginning of the end." The "end" is QCD, and "concrete quarks" were its starting point.

- *Minkowski:*

Now we have QCD, but at the time I concluded that these quarks or aces must be fields. Local fields, to maintain locality.

- *Zweig:*

I too believed in field theory for aces, but didn't think I could figure out which field theory it was. I was familiar with other areas of physics that contained much simpler nonlinearities, like the physics of fluids. Turbulence experiments were simple and cheap, yet no one could figure out what was really going on. So, I asked myself: can you figure out what's going on when two particles collide and all hell breaks loose? Even if I could guess the right field theory, *I wouldn't be able to prove it.* Proof demands prediction, and the required computations seemed much too difficult, even if the field theory was known. That thought, the impossibility of prediction, was very demoralizing. The problem was just too difficult.

- *Korthals-Altes:*

Well, this is not a question but a remark, and maybe the director of the school (t'Hooft), would agree with it. During the school we should systematically talk about aces, and not about quarks.

- *Zweig:*

Chris, that's very kind, but asking too much. Call them "concrete quarks." That name comes from the comment Gell-Mann made in the fall of 1964 after I told him all about aces: "Oh, the concrete quark model," he said when I finished. "That's for block heads." Murray had a way with words. Here we have a multi-layered pun, the kindest layer referring to the blocks of different size and shape with which I represented aces.

- *t' Hooft:*

Talking about concrete quarks, did you encourage the search for quarks based on their fractional charge?

- *Zweig:*

Absolutely! When I returned to Caltech I began working with Ricardo Gomez and others looking for fractionally charged particles in cosmic rays. The negative result was published in in '67 in the Physical Review Letters. The fundamental unit of charge is e/3, so it's natural to ask if free particles exist with the fundamental unit of charge. Nothing in theory forbids their existence, if they don't have color. QCD confines color, not fractional charge, by polarizing the vacuum *and creating quark-antiquark pairs*. QED can't do this. I realized in 1978, long after I had switched to neuro-biology, that free fractionally charged particles could catalyze fusion at room temperature. If free fractionally charged particles exist, at least one must be stable, because charge is conserved. Either that particle, or its anti-particle, has negative charge. That negatively charged particle would be heavy, because it hasn't been seen. So a stable "heavy electron" would exist. The size of a hydrogen atom is set by the reduced mass of the electron and proton, essentially the lighter mass. If the electron is replaced by a heavy fractionally charged particle, the size of the atom is set by the proton mass, about 10^-12 cm. If you take a molecule consisting of two deuterons and replace their electrons with heavy negative fractionally charged particles, the deuterons will be close enough to tunnel together and fuse, freeing the heavy fractionally charged particles to catalyze the next fusion reaction (G. Zweig, Science 201, 973-979 (1978)). So where in nature should you look for fractionally charged particles? If you're looking for gold, you look in gold mines where gold is concentrated, not, for example in pyrolytic carbon, one of the prime compounds used in early fractionally charged particle searches (pyrolytic carbon is an extremely pure form of carbon, a condensate of carbon vapor). Together with Klaus Lackner, and later Joe Smyth, we worked out the chemistry (Phys. Rev. D, {\bf 28}, 1671-1691 (1983)) and geochemistry of fractionally charged particles (Joe and I couldn't get our geochemistry paper published in a Journal because the geologists said that fractionally charged particles don't exist! However, the paper was issued as a Los Alamos National Laboratory Report: LA-UR 87-2219). I also started a project in 1982 with Chuck Hendricks at Livermore National Laboratory to use ink jet technology to measure the charge on droplets containing tiny particles of ground-up material. These droplets were slurries of carbonaceous chondrites, iron meteriorites, and other substances. Carbonaceous chondrites contain material that the solar system condensed from, and have a little bit of everything. Looking for fractionally charged particles in iron meteorites is a good idea. Feynman thought that aces were in the center of the earth, because they were heavy. You can't get to the center of the earth, but you can look at the center of asteroids that fragmented during collisions, i.e., iron meteorites. It was a difficult experiment because the jet kept plugging up. We lost our funding after 3 years.

- *t'Hooft :*

I would have expected that a negatively charged particle would be attracted by a heavy nucleus, forming a fractionally charged nucleus.

- *Zweig:*

Yes it would. To liberate this negative charge, the heavy nucleus would have to be fractured at an accelerator. Heavy elements would be excluded from the reaction chamber.

- *Tannenbaum:*

What do you mean by "Concrete Quarks?" Steve Weinberg distinguishes constituent quarks from partons or current quarks, which are massless. The constituent quarks have masses of a third of the mass of the proton, but the current quarks are massless. Which ones are your concrete quarks?

- *Zweig:*

In today's language, concrete quarks, or equivalently aces, were used in the two CEN preprints as point particles in a field theory to construct the electromagnetic and weak currents (like current quarks), and as much heavier particles (the point particles dressed in glue and possibly ace-antiace pairs) to construct the hadrons (like constituent quarks). I thought observation required a dual description.

- *t'Hooft :*

I read an interview by you, where you mentioned difficulties you had getting your paper on aces published. I wonder if you can say a few words about that.

- *Zweig:*

It's a strange story. While a visitor at CERN, I wrote up my work on aces in two preprints, and scheduled a seminar titled "Dealers choice: Aces are wild." I wanted to publish the content of these preprints as one paper in the Physical Review. At the time all papers coming from my Division had to be approved for publication by the head of the Division, Leon van Hove. When he heard that I wanted to publish in the Physical Review he came into my office and said that I was not allowed to publish in American journals, only European ones. He then instructed the Division secretary, Madam Fabergé, not to type any of my papers. When I found him taking down the announcements for my seminar, I asked him what he was doing, and he replied that I was not allowed to talk about aces at CERN. The first public lectures were given about six months later at Erice. That's why I'm here today.

- *Korthals-Altes:*

Did you get it published in the end?

- *Zweig:*

Eventually the longer preprint was published (*Developments in the Quark Theory of Hadrons, A Reprint Collection, Volume I: 1964-1978*, D. B. Lichtenberg and S. P. Rosen eds., Hadronic Press, Inc., Nonantum, Mass., pp. 22-101 (1980)), but it was widely circulated in 1964, even reissued to meet demand. Van Hoove and one of his collaborators, Kokkedee, edited a book, published by Benjamin, called 'The Quark Model.' They reprinted many papers about quarks, including Murray's, but without my CERN preprints, not even the first and shorter one.

- *Korthals-Altes:*

Kokkedee and Van Hove cooperated on many papers on the quark model.

- *Zweig:*

That's right. And they put their papers in the book, but not mine.

STATUS OF THEORETICAL UNDERSTANDING

CERN-PH-TH/2014-134

Higher Curvature Supergravity, Supersymmetry Breaking and Inflation*

Sergio Ferrara[a,b,c] and Alex Kehagias[d,e]

[a] *Physics Department, Theory Unit, CERN, CH 1211, Geneva 23, Switzerland*

[b] *INFN - Laboratori Nazionali di Frascati*
Via Enrico Fermi 40, I-00044 Frascati, Italy

[c] *Department of Physics and Astronomy, University of California*
Los Angeles, CA 90095-1547, USA

[d] *Physics Division, National Technical University of Athens, 15780 Zografou, Athens, Greece*

[e] *Department of Theoretical Physics 24 quai E. Ansermet, CH-1211 Geneva 4, Switzerland*

dedicated to the memory of Bruno Zumino

Abstract

In these lectures, after a short introduction to cosmology, we discuss the supergravity embedding of higher curvature models of inflation. The supergravity description of such models is presented for the two different formulations of minimal supergravity.

* Lectures given by S.Ferrara at the Erice ISSSP 2014, 52th Course, 24 June-3 July, 2014

1 Introduction

These lectures are devoted to the application of higher curvature supergravity to a particular class of cosmological models for inflation in which the "inflaton" field is identified with the "scalaron". The latter is a purely gravitational mode which arises when we add to the Einstein-Hilbert action a term quadratic in the (scalar) curvature

$$\mathcal{L}_{modified} = \mathcal{L}_{EH} + \alpha R^2 = \frac{1}{2\kappa^2} R + \alpha R^2, \tag{1}$$

where $\kappa^2 = 8\pi G = M_P^{-2}$. This theory is "dual" to standard Einstein gravity coupled to a scalar field. The revival of these models was motivated by the fact that the recent experiments Planck [1] and BICEP2 [2] seem to favor simple one-field cosmological models for inflation even if there is a tension between the two experiments. In fact, while for the slow roll parameter

$$n_S = 1 - 6\epsilon + 2\eta \approx 1 - \frac{2}{N} \approx 0.96 \tag{2}$$

i.e., the spectral index of scalar perturbations, the same formula agrees, for the other slow-roll parameter r, the tensor-to-scalar ratio, a value of

$$r = \frac{12}{N^2} \tag{3}$$

as in Starobinsky inflation [3] and Higgs Inflation [4], seems to be favored by the Planck collaboration. The latter reported

$$r < 0.08, \tag{4}$$

whereas chaotic models [5] like the quadratic one, predicts typically

$$r = \frac{8}{N} \approx 0.2, \tag{5}$$

and are favored by BICEP2. The slow-roll parameters ϵ, η and the number of e-folds $N(\sim 50-60)$ during inflation [6] are defined in terms of the scalar potential $V(\phi)$ of a canonically normalized inflaton field with Lagrangian

$$\mathcal{L} = \mathcal{L}_E - \frac{1}{2} g^{\mu\nu} \partial_\mu \phi \partial_\nu \phi - V(\phi) \tag{6}$$

as

$$\epsilon = \frac{M_P^2}{2} \left(\frac{V'}{V}\right)^2, \quad \eta = \frac{M_P^2 V''}{V}, \quad N = \frac{1}{M_P^2} \int_{\phi_{end}}^{\phi_{init}} \frac{V}{V'} d\phi. \tag{7}$$

In the next section, we will recall some well known matter from Cosmology, and, in particular, the description of the inflaton as a component of a cosmological perfect fluid. In section 3. we describe the Starobinsky model. In section 4. we present the supergravity embedding of the Starobinsky and chaotic inflation in the two different formulations of the $\mathcal{N}=1$ supergravity. In addition, we show how integrating out the sgoldstino multiplet, the Volkov-Akulov-Starobinsky supergravity emerges. Finally, section 5 contains our conclusions.

2 A (short) Introduction to Cosmology

The Standard Model for Cosmology describes the Universe as made of different forms of energy densities acting as sources of the gravitational field. The latter in turn is described by a Friedmann-Lemaître-Robertson-Walker geometry (FLRW) with metric

$$ds^2 = -dt^2 + a(t)^2 \left(\frac{dr^2}{1-kr^2} + r^2 d\Omega^2 \right) , \tag{8}$$

where the three-dimensional constant-time slice is a maximally symmetric space, the Riemann tensor of which satisfies (for constant t)

$$R_{ijkl} = \frac{k}{a^2}(g_{ik}g_{jl} - g_{il}g_{jk}) , \quad k = 0, \pm 1 . \tag{9}$$

For $k>0, k<0$ and $k=0$ we refer to closed, open or flat Universe, respectively. $a(t)$ is the "scale factor" which tells us how big is the 3D slice at (comoving) time t. The above assumptions are motivated by the "Copernican Principle", namely that our Universe looks isotropic and homogeneous. Isotropy says that space looks the same in any direction and homogeneity that the metric looks the same everywhere. If a space is isotropic everywhere, then it is homogeneous as well. If we have isotropy and homogeneity, then the 3D slice is a maximally symmetric space, i.e.,

$$a)\ \frac{SO(4)}{SO(3)}\ (k>0), \quad b)\ \frac{SO(3,1)}{SO(3)}\ (k<0), \quad c)\ E_3\ (k=0) . \tag{10}$$

The Einstein equations are written as

$$G_{\mu\nu} = R_{\mu\nu} - \frac{1}{2} g_{\mu\nu} R = \kappa^2 T_{\mu\nu} , \quad (R = R_\mu{}^\mu) , \tag{11}$$

where $T_{\mu\nu}$ is the energy-momentum (stress energy) tensor, or, equivalently,

$$R_{\mu\nu} = \kappa^2 \left(T_{\mu\nu} - \frac{1}{2} g_{\mu\nu} T \right) , \tag{12}$$

which implies

$$T = T_\mu{}^\mu = -\frac{1}{\kappa^2} R . \tag{13}$$

Note that if the energy-momentum tensor is that of a vacuum energy,

$$T_{\mu\nu} = -\Lambda g_{\mu\nu} \tag{14}$$

where $\Lambda = const.$, by the energy-momentum conservation

$$\nabla_\mu T^{\mu\nu} = 0, \tag{15}$$

we get that

$$R = 4\kappa^2\Lambda = const. \tag{16}$$

and we have a maximally symmetric space-time, which is de Sitter (DS) ($\Lambda > 0$), anti-de Sitter (AdS) ($\Lambda < 0$) or Minkowski ($\Lambda = 0$). But what is the general form of the energy-momentum tensor $T_{\mu\nu}$? We can answer to this question if we make the picture of the Universe as being made by perfect fluids, described by an energy density ρ and a pressure p. In this case, the energy-momentum tensor can be expressed as

$$T_{\mu\nu} = (\rho + p)U_\mu U_\nu + p g_{\mu\nu} = (\rho + p)\left(U_\mu U_\nu + \frac{1}{4}g_{\mu\nu}\right) + \frac{1}{4}(3p - \rho)g_{\mu\nu}\,, \tag{17}$$

where U^μ is the relativistic four-velocity vector. We may impose an equation of state

$$p = w\,\rho\,, \tag{18}$$

where $p = p(t)$, $\rho = \rho(t)$ and $w = const.$, with $|w| \leq 1$ by the dominant energy condition. Moreover, the stress tensor conservation (15) implies

$$\frac{\dot{\rho}}{\rho} = -3(1+w)\frac{\dot{a}}{a}\,, \tag{19}$$

which gives that

$$\rho \propto a^{-3(1+w)}\,. \tag{20}$$

Note in particular that a vacuum energy $T_{\mu\nu} \propto g_{\mu\nu}$ as in Eq. (14) leads to

$$\rho + p = 0, \quad (w = -1)\,, \tag{21}$$

while a traceless stress tensor $T_\mu{}^\mu = 0$ implies

$$3p - \rho = 0, \quad (w = \frac{1}{3}, \text{ radiation})\,. \tag{22}$$

The value of the scalar curvature is

$$R = -\kappa^2 T_\mu{}^\mu = \kappa^2(1-3w)\rho \tag{23}$$

for a single component perfect fluid, or

$$R = \kappa^2 \sum_i (1-3w_i)\rho_i \tag{24}$$

in the case of a multi-component perfect fluid. For $w = -1$, we have de Sitter or anti-de Sitter space-time depending on the value of ρ (DS for $\rho > 0$ and AdS for $\rho < 0$) whereas, $R \geq 0$ for $w \leq \frac{1}{3}$ and $\rho_i \geq 0$.

Because of the symmetries of the FLRW geometry, the Einstein equations now read

$$G_{00} = \kappa^2 T_{00}\,, \tag{25}$$

$$G_{ij} = \kappa^2 T_{ij}\,, \tag{26}$$

which give the equations for the scale factor

$$\left(\frac{\dot{a}}{a}\right)^2 = \frac{8\pi G}{3}\rho - \frac{k}{a^2}\,, \tag{27}$$

$$\frac{\ddot{a}}{a} = -\frac{4\pi G}{3}(\rho + 3p)\,, \quad (k = \pm 1,\ 0)\,. \tag{28}$$

A dot and a double dot denote first and second derivatives with respect to comoving time t. By introducing the Hubble parameter

$$H = \frac{\dot{a}}{a} \tag{29}$$

and noticing that

$$\frac{\ddot{a}}{a} = \dot{H} + H^2\,, \tag{30}$$

we can rewrite the Einstein equations as (Friedmann equations)

$$H^2 = \frac{8\pi G}{3}\rho - \frac{k}{a^2}\,, \tag{31}$$

$$\dot{H} = -4\pi G(\rho + p) + \frac{k}{a^2}\,. \tag{32}$$

By defining

$$\rho_k = -\frac{3k}{8\pi G}\frac{1}{a^2}\,, \quad w_k = -\frac{1}{3}\,, \tag{33}$$

Friedmann equations (31,32) are written as

$$H^2 = \frac{8\pi G}{3}(\rho + \rho_k)\,, \tag{34}$$

$$\dot{H} = -4\pi G(1+w)\rho - \frac{8\pi G}{3}\rho_k\,, \tag{35}$$

or, in the case of a multicomponent perfect fluid

$$H^2 = \frac{8\pi G}{3}\sum_i \rho_i\,, \tag{36}$$

$$\dot{H} = -4\pi G\sum_i (1+w_i)\rho_i\,, \tag{37}$$

(where ρ_i includes ρ_k). By dividing Eq. (31) by H^2 and defining the density parameter

$$\Omega = \frac{8\pi G}{3H^2}\rho\,, \tag{38}$$

we get

$$\Omega - 1 = \frac{k}{H^2 a^2}\,. \tag{39}$$

In addition, if we also define the critical energy density

$$\rho_{crit} = \frac{8\pi G}{3H^2}\,, \tag{40}$$

we have that

$$\Omega = \frac{\rho}{\rho_{crit}}\,. \tag{41}$$

Similarly, Eq. (36) can be compactly written as

$$\sum_i \Omega_i = 1\,, \tag{42}$$

where

$$\Omega_i = \frac{\rho_i}{\rho_{crit}}\,, \tag{43}$$

including ρ_k. The second Friedmann equation (32) can be written in terms of the "deceleration parameter"

$$q = -a\frac{\ddot{a}}{\dot{a}^2} = -\frac{1}{H^2}(\dot{H} + H^2)\,, \tag{44}$$

as

$$q = \frac{4\pi G}{3H^2}(\rho + 3p) = \frac{4\pi G}{3H^2}(1 + 3w)\rho\,, \tag{45}$$

or for several components of energy densities

$$q = \frac{4\pi G}{3H^2}\sum_i(1 + 3w_i)\rho_i = \frac{1}{2}\sum_i(1 + 3w_i)\Omega_i\,. \tag{46}$$

Note that Ω_k does not contribute to q (since $w_k = -1/3$). As a result, Einstein equations for an FLRW geometry may be written as

$$\sum_i \Omega_i = 1, \quad q = \frac{1}{2}\sum_i(1 + 3w_i)\Omega_i\,. \tag{47}$$

Values for w include

$$w = 0 \qquad \text{(dust-baryonic matter)}\,, \tag{48}$$

$$w = -\frac{1}{3} \qquad \text{(curvature)}\,, \tag{49}$$

$$w = 1 \qquad \text{(fast roll scalar field)}\,. \tag{50}$$

Whenever a single term dominates, we say that the Universe is dominated by that component of (p_i, ρ_i) or $(w_i = p_i/\rho_i,\ \rho_i)$. For example, for $w = 0$ we have a matter dominated Universe, for $w = \frac{1}{3}$ a radiation dominated, for $w = -1$ a vacuum-energy dominated and for $w = -\frac{1}{3}$ a curvature dominated Universe. When only one component dominates, the Friedmann equation

$$\left(\frac{\dot{a}}{a}\right)^2 \propto \frac{8\pi G}{3}a^{-3(1+w)} \tag{51}$$

can easily be integrated to give

$$a \sim \begin{cases} t^{\frac{2}{3(1+w)}}, & w \neq -1 \\ e^{Ht}, & w = -1, \quad (H = const.)\,. \end{cases} \tag{52}$$

A scalar field (inflaton) can be viewed as a perfect fluid component of the Universe with Lagrangian

$$\mathcal{L}_\phi = -\frac{1}{2}g^{\mu\nu}\partial_\mu\phi\partial_\nu\phi - V(\phi)\,, \tag{53}$$

and stress energy tensor

$$T^\phi_{\mu\nu} = -\frac{\partial\mathcal{L}_\phi}{\partial\partial_\mu\phi}\partial_\nu\phi + g_{\mu\nu}\mathcal{L}_\phi\,. \tag{54}$$

In components, we find for the latter

$$T^{\phi}_{00} = \frac{1}{2}\dot{\phi}^2 + V(\phi), \tag{55}$$

$$T^{\phi}_{ij} = \left(\frac{1}{2}\dot{\phi}^2 - V(\phi)\right) g_{ij}\,, \tag{56}$$

which gives, after comparing with the standard form of the stress tensor of a perfect fluid

$$T_{00} = \rho, \quad T_{ij} = p g_{ij}\,, \tag{57}$$

that the energy density and the pressure of the inflaton are

$$\rho = \frac{1}{2}\dot{\phi}^2 + V(\phi), \quad p = \frac{1}{2}\dot{\phi}^2 - V(\phi)\,. \tag{58}$$

Therefore, we get that

$$\rho + p = \dot{\phi}^2, \quad \rho + 3p = 2(\dot{\phi}^2 - V)\,, \tag{59}$$

$$\rho - p = 2V, \quad T_{\mu}{}^{\mu} = \dot{\phi}^2 - 4V\,, \tag{60}$$

so that we have

$$p = -\rho, \quad (w = -1) \ \text{ for } \ \dot{\phi}^2 \ll V(\phi)\,, \tag{61}$$

$$p = \rho, \quad (w = 1) \ \text{ for } \ \dot{\phi}^2 \gg V(\phi)\,, \tag{62}$$

wheras, in general

$$\frac{1}{2}\dot{\phi}^2(1 - w) = (1 + w)V\,. \tag{63}$$

The matter equations

$$\frac{\delta \mathcal{L}_\phi}{\delta \phi} - \partial_\mu \frac{\delta \mathcal{L}_\phi}{\delta \partial_\mu \phi} = 0\,, \tag{64}$$

together with the Einstein equations are explicitly written as

$$H^2 + \frac{k}{a^2} = \frac{8\pi G}{3}\left(\frac{1}{2}\dot{\phi}^2 + V(\phi)\right), \tag{65}$$

$$\dot{H} - \frac{k}{a^2} = -4\pi G \dot{\phi}^2\,, \tag{66}$$

$$\ddot{\phi} + 3H\dot{\phi} + V_\phi = 0\,. \tag{67}$$

Inflation claims to solve the flatness and horizon problems. The flatness problem is the explanation for having today $\Omega \sim 1$ without fine-tuning of $|\Omega - 1|$ close to zero at early times, whereas, the horizon problem is the problem of homogeneity of the observed Universe arising form seemingly non-causally connected initial regions. Inflation occurs as long as

$$\dot{\phi}^2 \ll V(\phi), \quad |\ddot{\phi}| \ll |3H\dot{\phi}|, |V_\phi|, \tag{68}$$

where the potential energy dominates the kinetic energy for sufficient period. The inflationary regime is usually parametrized by the slow-roll parameters $(\epsilon, \eta) \ll 1$ defined in Eq. (7).

Let us also note at this point that the number of e-foldings N is defined as

$$N = \int_{t_i}^{t_f} H(t)dt = \ln \frac{a_f}{a_i}, \tag{69}$$

where $t_i(t_f)$ are some initial(final) time and $a_{i,f} = a(t_{i,f})$. The definition (69) reduces to that of eq.(7) once the field equations (65-67) are used in the slow-roll approximation. Therefore we have that

$$a_f = a_i e^N = a_i \prod_m e^{N_m}, \quad \text{for} \quad N = \sum_m N_m. \tag{70}$$

In addition, for a de Sitter background with

$$H(t) = H_0 = const. \tag{71}$$

we have

$$H_0 \Delta t = N. \tag{72}$$

3 The Starobinksy Model

The Starobinsky model is the $R + R^2$ theory. It is dual (conformally equivalent) to standard gravity coupled to a scalar field [7] with a potential giving rise to inflation. Indeed, the Starobinsky Lagrangian is (in $M_P^{-2} = 8\pi G = 1$ units)

$$\mathcal{L} = \frac{1}{2}R + \alpha R^2. \tag{73}$$

By introducing new fields σ and Λ, the Lagrangian (73) can be written equivalently as

$$\mathcal{L} = \frac{1}{2}R + \frac{1}{2}\sigma(R - \Lambda) + \alpha\Lambda^2, \tag{74}$$

where the field σ is a Langrange multiplier which enforces the constraint $\Lambda = R$. We may write (74) as

$$\mathcal{L} = \frac{1}{2}R(1+\sigma) - \frac{1}{2}\sigma\Lambda + \alpha\Lambda^2\,, \tag{75}$$

where we observe that we have a Jordan frame function $(1+\sigma)$. Going to the Einstein frame through the change of variables

$$g'_{\mu\nu} = g_{\mu\nu}(1+\sigma)^{-1}\,, \tag{76}$$

we get that the Lagrangian (75) is written as

$$\mathcal{L} = \frac{1}{2}R - \frac{1}{2}\partial_\mu\phi\partial^\mu\phi - \frac{1}{16\alpha}(1-e^{-\sqrt{\frac{2}{3}}\phi})^2\,, \tag{77}$$

where the field ϕ, the "scalaron" is defined as

$$1+\sigma = e^{\sqrt{\frac{2}{3}}\phi}\,. \tag{78}$$

In Fig.1 below, the scalaron potential

$$V = \frac{1}{16\alpha}(1-e^{-\sqrt{\frac{2}{3}}\phi})^2\,, \tag{79}$$

has been plotted. At $V_\phi = \partial V/\partial\phi = 0$, supersymmetry is unbroken whereas during the inflationary phase ("de Sitter plateau"), we have $\partial V/\partial\phi \neq 0$ and supersymmetry is broken.

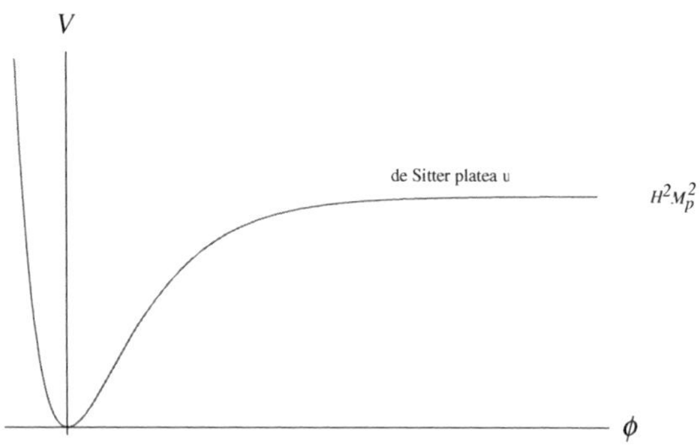

Figure 1: The "scalaron" potential in $R+R^2$ theory.

In trying to describe the Starobinsky model in supergravity, let us note that the off-shell components of the gravity field give rise to extra massive modes in higher curvature supergravity. For example let us consider a Lagrangian quadratic to curvature of the form

$$\mathcal{L} = \mathcal{L}_E + \mathcal{L}_{R^2} + \mathcal{L}_{Weyl^2}$$
$$= \frac{1}{2\kappa^2}R + \alpha R^2 + \beta W^2_{\mu\nu\rho\sigma}.$$

The graviton $g_{\mu\nu}$ has a total of six degrees of freedom, since the total number of the ten independent components of $g_{\mu\nu}$ is reduced by the number four, which the number of diffeomorphisms

$$10 - \underbrace{4}_{\text{diffeomorphisms}} = 6. \tag{80}$$

These six degrees of freedom of $g_{\mu\nu}$ give rise to a scalar degree of freedom with mass m_0 and a massive spin-2 state with mass m_2, according to the splitting

$$6 = \underbrace{1}_{\text{scalar}} + \underbrace{5}_{\text{spin 2}} \tag{81}$$

where

$$m_0^2 \sim \frac{1}{\kappa^2\alpha}, \quad m_2^2 \sim -\frac{1}{\kappa^2\beta}. \tag{82}$$

Therefore, the massive spin-2 state is a tachyon ($\beta > 0$) or a ghost ($\beta < 0$). This state decouples in the $\beta \to 0$ limit, leading to Starobinsky $R + R^2$ theory with dynamical degrees of freedom a massless graviton and a scalar field.

The higher curvature theory should be compared with the chaotic model [5]. The Lagrangian of the latter is

$$\mathcal{L} = \frac{1}{2\kappa^2}R - \frac{1}{2}(\partial_\mu\phi)^2 - \frac{1}{2}m^2\phi^2, \tag{83}$$

where the mass of the inflaton is $m \sim H$. It can be seen that during inflation, the de Sitter cosmological constant Λ is

$$\Lambda \sim H^2 M_P^2, \tag{84}$$

where $H/M_p \sim 10^{-5}$ for both the higher curvature and the chaotic model.

4 Supergravity Embedding

Here we present the supergravity embedding of these two models which is minimal in two respects: It uses the minimal set of multiplets needed to describe these models. It also uses the minimal off-shell representations of the underlying local supersymmetry algebra. The latter introduces new fields which are "auxiliary" (not propagating) in the standard Einstein supergravity but become propagating when higher curvature terms are introduced. The minimal supergravity extension of such a model was derived in the late eighties [8,9] in two different forms depending of two different off-shell completion of the supergravity multiplet

$$a) \quad V_\mu^a, \ \psi_\mu, \ A_\mu, \ S, \ P, \tag{85}$$

$$b) \quad V_\mu^a, \ \psi_\mu, \ A_\mu, \ b_{\mu\nu}, \qquad A_\mu \to A_\mu + \partial_\mu a, \ \ b_{\mu\nu} \to b_{\mu\nu} + \partial_\mu \xi_\nu - \partial_\nu \xi_\mu \,. \tag{86}$$

The first case (a) is the old-minimal supergravity and the second case (b) is the new-minimal one. The six bosonic degrees of freedom which make the gravity multiplet to have the same number of bosons and fermions ($12b + 12f$) give two different supergravity extensions of the Starobinsky model.

The off-shell components of ($\mathcal{N} = 1$) supergravity fields give extra massive supermultiplets in higher curvature supergravity. In the old-minimal supergravity we have

$$\text{graviton} \quad g_{\mu\nu}: \quad 6 = 1_0 + 5_2, \quad A_\mu, \ S, \ P: \quad 6 = 3_1 + 1_0 + 1_0 + 1_0 \,, \tag{87}$$

$$\text{gravitino} \quad \psi_\mu: \quad 16 - 4 = 12 = 2 \times \frac{3}{2} + 2 \times \frac{1}{2} \,, \tag{88}$$

which describes two massive chiral multiplet $2(\frac{1}{2}, 2(0))$ with $(4b+4f)$ and a ghost spin-2 multiplet $(2, 2(\frac{3}{2}), 1)$. In the new-minimal supergravity, the $3 + 3$ degrees of freedom of the gauge fields $A_\mu, b_{\mu\nu}$ fill the bosonic part of a physical massive vector multiplet $(1, 2(\frac{1}{2}), 0)$ [10].

The "dual" standard supergravity action contains, in the a) formulation two "matter" chiral (massive) multiplets T, S $(4b + 4f)$ while in the b) formulation contains a "massive" vector (or tensor) multiplet $\mathcal{V}$ $(4b + 4f)$. The main difference is that in the a) theory we are in a presence of a "four-field" model whereas in the b) theory we have a "single-field" inflaton model since the other three bosonic degrees of freedom combine in a massive vector.

Standard supergravity formulae allow to describe the a) theory in terms of a Kähler potential K and a superpotential W. It turns out that their form is

$$K = -3 \log \left(1 + T + \bar{T} - h(S, \bar{S})\right), \quad W = \lambda T S \,, \tag{89}$$

where λ is a constant related to the α parameter and $h(S, \bar{S})$ is an arbitrary real function which starts with

$$h(S, \bar{S}) = S\bar{S} + \mathcal{O}(S^3) \,. \tag{90}$$

It is possible to choose the function $h(S,\bar{S})$ appropriately in order to make the inflationary trajectory stable. Here the "inflaton" is identified with the ReT, the real part of the scalar T, while the other three scalars are "extremized". The potential for

$$ReT = e^{-\sqrt{\frac{2}{3}}\phi} \tag{91}$$

is the Starobinsky potential. It can be shown that this theory, for any $h(S,\bar{S})$ is "dual" to a higher curvature supergravity theory. The scalar supercurvature $\mathcal{R}$ is a chiral superfield [11]

$$\bar{D}_{\dot{a}}\mathcal{R} = 0\,, \tag{92}$$

and $h(S,\bar{S})$ corresponds to terms of the form $h(\mathcal{R},\bar{\mathcal{R}})$ in the supergravity side. It is important to notice that the inflaton potential is an "F-term" potential, which means that it comes from the standard expression

$$V(T,S) = e^K\left(D_iW\,D_{\bar{j}}\bar{W}K^{i\bar{j}} - 3|W|^2\right), \quad (i,j = S,T)\,. \tag{93}$$

The inflaton potential is then

$$V(\phi) = V(T,S)\Big|_{\frac{\partial V}{\partial S}=0,\ \frac{\partial V}{\partial ImT}=0}\,. \tag{94}$$

It happens that all supersymmetric models for the inflaton potential considered in the literature [12–24] are mostly deformations of the previous model with modification of $K(T,\bar{T},S,\bar{S})$ and of $W(T,S)$ but still keeping the same (S,T) chiral multiplet content.

It is possible to show that at least two multiplets are needed to get an inflationary potential. In fact former theories with higher supercurvature terms of F-term type with chiral function $f(\mathcal{R})$ ($\bar{D}f = 0$) were considered in the past [25] but were shown [13,26] not to produce an inflationary potential.

An important deformation of the (S,T) model from which the concept of "attractor" [18] came from, is a superpotential of the type

$$W(S,T) = Sf(T)\,, \tag{95}$$

which allows bosonic potentials containing arbitrary functions of the inflaton $f(\tanh\frac{\phi}{\sqrt{6}})$. These theories are no longer equivalent to pure higher curvature supergravity but in certain cases, to higher curvature coupled to a (single) chiral multiplet. For instance, taking K as in (89) but W now as (95), the dual higher derivative supergravity is a matter coupled theory with [27]

$$\Phi = e^{-\frac{1}{3}K} = 1 + T - \frac{f(T)}{f'(T)} + \bar{T} - \frac{\bar{f}(\bar{T})}{\bar{f}'(\bar{T})} \tag{96}$$

and a term

$$\frac{1}{|f'(T)|^2}\mathcal{R}\bar{\mathcal{R}}\,. \tag{97}$$

Both terms become T-independed if $f(T) = a\,T$.

The $b)$ formulation gives directly a single-field inflation model where a "D-term" potential for the massive superfield is generated. The most general self-interaction of such massive vector multiplet with spin content $(1, 2(\frac{1}{2}), 0)$ resides on a real function J of a real variable C: $J(C)$ [28] The bosonic part of the supergravity action is

$$\mathcal{L} = \frac{1}{2}R - \frac{1}{4}F_{\mu\nu}(B)F^{\mu\nu}(B) + \frac{g^2}{2}J''(C)B_\mu B^\mu + \frac{1}{2}J''(C)\partial_\mu C\partial^\mu C - \frac{g^2}{2}J'(C)^2\,, \tag{98}$$

so that the potential is

$$V(C) = \frac{g^2}{2}J'(C)^2\,. \tag{99}$$

The equation (98) actually coincides with the self-interaction of a massive vector multiplet in global supersymmetry [29]. Note also that (98) depends only on J', J'', so a linear term in J shifts J' by a constant but leaves J'' invariant. This constant is the so-called Fayet-Iliopoulos (FI) term. By using the Stückelberg trick one writes (98) as a gauge theory by shifting

$$A_\mu = B_\mu - \frac{1}{g}\partial_\mu a\,, \tag{100}$$

so that

$$\frac{g^2}{2}J''(C)B_\mu B^\mu = \frac{g^2}{2}J''(C)\Big(A_\mu + \frac{1}{g}\partial_\mu a\Big)^2\,. \tag{101}$$

In the limit $g \to 0$ the theory becomes

$$\mathcal{L} = \frac{1}{2}R - \frac{1}{4}F_{\mu\nu}(A)F^{\mu\nu}(A) + \frac{1}{2}J''(C)\Big(\partial_\mu a\partial^\mu a + \partial_\mu C\partial^\mu C\Big)^2\,. \tag{102}$$

The (a, C) variables can be complexified to $T = -C + i\sqrt{\frac{2}{3}}\,a$ and the J-function can be interpreted as a Kähler potential

$$J = -\frac{1}{2}K(ReT)\,. \tag{103}$$

The higher curvature supergravity in the b) formulation is "dual" to a self-interacting massive vector multiplet with a very precise choice [14] of

$$J(C) = \frac{3}{2}\Big(\log(-C) + C\Big). \tag{104}$$

Computation of the potential, for a canonically normalized field

$$C = -e^{\sqrt{\frac{2}{3}}\phi}, \tag{105}$$

leads to Starobinsky potential and Lagrangian [14, 15]

$$\mathcal{L} = \cdots - \frac{1}{2}(\partial_\mu \phi)^2 - \frac{9}{8}g^2(1 - e^{-\sqrt{\frac{2}{3}}\phi})^2, \tag{106}$$

so that the supersymmetric generalization just reproduces the single-field Starobinsky model with $\alpha \propto g^{-2}$. It is interesting to observe that the particular form of $J(C)$ corresponds to an $SU(1,1)/U(1)$ symmetric Kähler manifold with a parabolic isometry being gauged. For a Kähler potential

$$K = -3\alpha \log Re\, T, \tag{107}$$

the curvature is

$$R(C) = \frac{J'''(C)^2 - J''(C)J^{IV}(C)}{2J''(C)^2} = -\frac{2}{3\alpha} \tag{108}$$

and for $\alpha \to \infty$, the curvature vanishes $R(C) \to 0$. The α-depended potential becomes [14]

$$V(\phi) = \frac{9}{8}g^2(1 - e^{-\sqrt{\frac{2}{3\alpha}}\phi})^2 = \frac{9}{8}g^2 P(C)^2, \tag{109}$$

where $P(C) = J'(C)$. Note that the canonical variable ϕ is related to the C variable by the equation

$$J''(C) = -\left(\frac{d\phi}{dC}\right)^2 = P'(C). \tag{110}$$

It follows then, with $P(\phi) = P(C(\phi))$, $P'(C) = dP/dC$ and $P'(\phi) = dP/d\phi$

$$P'(C) = P'(\phi)\frac{d\phi}{dC} = -\left(\frac{d\phi}{dC}\right)^2, \tag{111}$$

so that

$$P'(\phi) = -\frac{d\phi}{dC} \tag{112}$$

and

$$C(\phi) = \int d\phi \frac{dC}{d\phi} = -\int d\phi \frac{1}{P'(\phi)}, \tag{113}$$

$$J(C) = \int dC J'(C) = \int P(\phi) \frac{dC}{d\phi} d\phi = -\int \frac{P(\phi)}{P'(\phi)} d\phi. \tag{114}$$

In addition, the curvature in the ϕ-variable is

$$R(\phi) = -4 \frac{P'''(\phi)}{P'(\phi)}, \tag{115}$$

whereas, the kinetic term of the Kähler manifold is [30]

$$\frac{1}{2} J''(C) \Big(\partial_\mu a \partial^\mu a + \partial_\mu C \partial^\mu C \Big)^2 = -\frac{1}{2} \Big[(\partial_\mu \phi)^2 + P'(\phi)^2 (\partial_\mu a)^2 \Big]. \tag{116}$$

The previous equations allow us to compute $C(\phi)$ once $P'(\phi) = -d\phi/dC$ is solved.

The one-field supergravity model for inflation can be deformed in two ways:

1. Simply change $J(C)$, i.e., change the Kähler manifold.

2. Do not change the manifold but change its gauged isometry.

For the case of symmetric spaces, this procedure generates five models. Three with constant curvature depending whether a parabolic, elliptic or hyperbolic isometry is gauged and two with vanishing curvature where the parabolic or elliptic isometry is gauged [30].

4.1 Chaotic Inflation

With an "F-term" multi-field potential term, it is hard to obtain (at most in some directions of the field space) a quadratic potential. One way is to impose a shift symmetry on the Kähler potential [17, 23, 24, 31]. In terms of the (T, S) chiral fields this exchange the role of (ImT, ReT) since it is now the ImT which plays the role of the inflaton. It is then natural, in the supergravity dual to cal this scenario "imaginary Starobinsky model" [20] even if a coupling to matter is needed in order to stabilize the ReT component.

In the case of chaotic inflation in the b) single-field supergravity formulation, an exact model is possible since we can take a flat Kähler space where we gauge a parabolic isometry (translations). The alternative gauging of an elliptic isometry would give a quartic potential. For this case,

$$J''(C) = const., \quad J(C) = -\frac{m^2}{2} C^2 + \xi C, \tag{117}$$

but the FI term is irrelevant. Then

$$P(\phi) = \phi, \quad \text{and} \quad V(\phi) = \frac{1}{2} m^2 \phi^2. \tag{118}$$

This model can also be obtained from the constant curvature case by taking the limit [18]

$$\alpha \to \infty, \quad g^2 \to \infty, \quad \text{with} \quad m^2 \propto \frac{g^2}{\alpha} \quad \text{fixed}, \tag{119}$$

so that

$$g^2(1 - e^{-\sqrt{\frac{2}{3\alpha}}\phi})^2 \to \frac{1}{2}m^2\phi^2 . \tag{120}$$

4.2 Integrating out the sgoldistino multiplet: The Volkov-Akulov-Starobinsky Supergravity

We observe that the above potential is a D^2 term so during inflation D is large and the gaugino is the goldstino. The decoupling of the other (chiral) component occurs when $J''(C) \to 0$ and we get an unbroken gauge symmetry in de Sitter space (Freedman model [32]). Supergravity can be formulated in different conformal gauges (different Jordan functions). In the type *a*) formulation of inflation [12], there are three basic fields (chiral superfields), the conformon multiplet S_0, which is not physical, the scalaron multiplet T, which contains the inflaton and the goldstino multiplet S which contains the sgoldstino. The latter is the goldstino partner and it is just $S|_{\theta=0}$ as supersymmetry is linearly realized. The superpotential $W = ST$ has F-terms

$$\frac{\partial W}{\partial S} = T \neq 0 \quad \text{during inflation}, \tag{121}$$

$$\frac{\partial W}{\partial T} = S = 0 \quad \text{during inflation and later}. \tag{122}$$

In fact, this explains why in the *a*) formulation two chiral fields are needed.

A new efective Lagrangian, since supersymmetry is badly broken during inflation, can be obtained replacing the sgoldstino multiplet S by the Volkov-Akulov superfield X which satisfies the constraint

$$X^2 = 0 \tag{123}$$

and allows to express X in terms of the golstino G_α as [33]

$$X = \frac{G_\alpha G^\alpha}{2F_X} + \sqrt{2}\theta^\alpha G_\alpha + \theta^\alpha\theta_\alpha F_X . \tag{124}$$

In the dual supergravity theory, this corresponds to the chiral scalar supercurvature $\mathcal{R}$ to become nilpotent [34]

$$\mathcal{R}^2 = 0 . \tag{125}$$

Let us recall that the Volkov-Akulov Lagrangian is

$$\mathcal{L}_{VA} = f^2 \det V_{\alpha\mu}\,, \tag{126}$$

where f is the SUSY breaking parameter and

$$V_{\alpha\mu} = \delta_{\alpha\mu} + \frac{i}{f^2}\bar{G}\gamma_\alpha\partial_\mu G\,. \tag{127}$$

Supersymmetry is non-linearly realised and eq. (126) is invariant under the transformation

$$\delta G = f\epsilon + \frac{i}{f}\bar{G}\gamma^\mu\epsilon\partial_\mu G\,. \tag{128}$$

In superspace, we can write

$$\mathcal{L} = X\bar{X}\Big|_D + fX\big|_F, \quad (X^2 = 0)\,. \tag{129}$$

When coupled to supergravity, one gets a theory of a massive gravitino coupled to gravity with Kähler potential and superpotential

$$K = -3\log(1 - X\bar{X}) = 3X\bar{X}, \quad W = fX + W_0\,. \tag{130}$$

The vacuum energy and the gravitino mass turn out to be

$$V_0 = \frac{1}{3}|f|^2 - 3|W_0|^2, \quad m_{3/2} = |W_0|\,, \tag{131}$$

whereas the Noether current is

$$J_{\alpha\mu} \sim f\gamma_\mu G + \cdots \tag{132}$$

Now we may couple V-A to supergravity and to the scalaron multiplet. The massive spin 3/2 action is of the form

$$\frac{1}{2}R + \psi\partial\psi + m_{3/2}\psi\sigma\psi + \psi^4 - V(f, m_{3/2})\,. \tag{133}$$

The constrained V-A superfield X is then coupled to the scalaron leading to an almost standard supergravity with kähler potential and superpotential

$$K = -3\log(T + \bar{T} - X\bar{X}), \quad W = MXT + fX + W_0, \quad X^2 = 0\,. \tag{134}$$

The above data give rise to a potential (with no-scale structure $V \geq 0$) for the scalar T

$$V = \frac{|MT + f|^2}{3(T + \bar{T})^2}\,. \tag{135}$$

By defining

$$T = e^{\sqrt{\frac{2}{3}}\phi} + ia\sqrt{\frac{2}{3}}, \tag{136}$$

one gets [34]

$$\begin{aligned}\mathcal{L} &= \frac{1}{2}R - \frac{1}{2}(\partial\phi)^2 - \frac{M^2}{12}(1 - e^{-\sqrt{\frac{2}{3}}\phi})^2 - \frac{1}{2}e^{-2\sqrt{\frac{2}{3}}\phi}(\partial a)^2 - \frac{M^2}{18}e^{-2\sqrt{\frac{2}{3}}\phi}a^2 \\ &= \mathcal{L}_{Starobinsky} + \mathcal{L}_{axion} + (\text{fermionic terms})\,. \end{aligned} \tag{137}$$

The dual supergravity action is

$$\mathcal{L}(S_0, \mathcal{R}) = -[S_0\bar{S}_0 - \frac{\mathcal{R}\bar{\mathcal{R}}}{M^2}]_D + (W_0 + \xi\frac{\mathcal{R}}{S_0})S_0^3 + \sigma\mathcal{R}^2 S_0\,. \tag{138}$$

The bosonic part of this action can be obtained by dualizing the previous action (137) having set

$$e^{\sqrt{\frac{2}{3}}\phi} = 1 + 2\chi\,, \tag{139}$$

and Weyl rescaling

$$g_{\mu\nu} \to (1 + 2\chi)g_{\mu\nu}\,. \tag{140}$$

The result is the Lagrangian

$$\mathcal{L} = \frac{1}{2}(1 + 2\chi)R - \frac{1}{2}\frac{(\partial a)^2}{1 + 2\chi} - \frac{M^2}{3}(\chi^2 + \frac{a^2}{6})\,, \tag{141}$$

which can be written equivalently as

$$\mathcal{L} = \frac{1}{2}(1 + 2\chi)R - \frac{M^2}{18}a^2 + A^\mu\partial_\mu a + \frac{1}{2}(1 + 2\chi)A_\mu^2\,. \tag{142}$$

The dual Lagrangian is then obtained by integrating over a and χ. The result is ($A_\mu \to \sqrt{\frac{2}{3}}A_\mu$)

$$\mathcal{L} = \frac{1}{2}\left(R + \frac{2}{3}A_\mu^2\right) + \frac{3}{4M^2}\left(R + \frac{2}{3}A_\mu^2\right)^2 + \frac{3}{M^2}(\nabla^\mu A_\mu)^2\,. \tag{143}$$

This is the $R + R^2$ Lagrangian with $S = P = 0$. Note that the axion field is much heavier than ϕ during inflation where $\phi = \phi_0$ is positive and large

$$m_\phi^2 = \frac{M^2}{9}e^{-2\sqrt{\frac{2}{3}}\phi_0} \ll m_a^2 = \frac{M^2}{9}\,. \tag{144}$$

5 Conclusions

In these lectures, we have presented the supergravity embedding of higher curvature models of inflation. The prototype of such models is the Starobinksy $R + R^2$ gravity. This theory does not describe only the GR degrees of freedom, i.e. the helicity-2 massless graviton, but in addition it propagates a scalar degree, the scalaron. It also predicts a tiny value for the tensor-to-scalar ratio r due to an addition 1/N suppression with respect to the scalar tilt n_S, which is in perfect agreement with the Planck data. However, this prediction is in conflict with the BICEP2 results. This tension between Planck and BICEP2 is expected to be resolved soon. We should stress that the inflationary predictions of the Starobinsky model is identical to leading order with that of Higgs inflation [4]. As it has been shown in [35], this is due to the fact that Higgs inflation and Starobinsky model are identical during slow-roll, where the kinetic terms of the Higgs and the scalaron are subleading with respect to their potentials. We have presented how the bosonic Starobinsky model can be embedded in $\mathcal{N} = 1$ minimal supergravity. In fact, since it is a higher curvature theory, it is described both in old-minimal (formulation a)) as well as in new-minimal ((b) formulation) $\mathcal{N} = 1$ supergravity. The supergravity formulation of the chaotic model has also been presented.

Concluding, we would like to stress that the Starobinsky model and its descendants as well as their supergravity avatars cannot be excluded as long as BICEP2 results are not independently confirmed.

Acknowledgment

Parts of these lectures are based on collaborations with A. Antoniadis, E. Dudas, P. Fre, R. Kallosh, A. Linde, M. Porrati, A. Riotto, A. Sagnotti, A. Sorin and A. van Proeyen, whom we would like to thank. This research was implemented under the ARISTEIA Action of the Operational Programme Education and Lifelong Learning and is co-funded by the European Social Fund (ESF) and National Resources. It is partially supported by European Unions Seventh Framework Programme (FP7/2007-2013) under REA grant agreement n. 329083. S.F. is supported by ERC Advanced Investigator Grant n. 226455 Supersymmetry, Quantum Gravity and Gauge Fields (Superfields).

References

[1] P. A. R. Ade *et al.* [Planck Collaboration], arXiv:1303.5082 [astro-ph.CO].

[2] P. A. R. Ade *et al.* [BICEP2 Collaboration], Phys. Rev. Lett. **112** (2014) 241101 [arXiv:1403.3985 [astro-ph.CO]].

[3] A. A. Starobinsky, Phys. Lett. B **91**, 99 (1980); V. F. Mukhanov and G. V. Chibisov, JETP Lett. **33**, 532 (1981) [Pisma Zh. Eksp. Teor. Fiz. **33**, 549 (1981)]; A. A. Starobinsky, Sov. Astron. Lett. **9**, 302 (1983).

[4] F. L. Bezrukov and M. Shaposhnikov, Phys. Lett. B **659**, 703 (2008) [arXiv:0710.3755 [hep-th]].

[5] A. Linde, Phys. Lett B **129**, 177 (1983).

[6] D. H. Lyth and A. Riotto, Phys. Rept. **314**, 1 (1999) [hep-ph/9807278].

[7] B. Whitt, Phys. Lett. B **145**, 176 (1984).

[8] S. Cecotti, Phys. Lett. B **190**, 86 (1987).

[9] S. Cecotti, S. Ferrara, M. Porrati and S. Sabharwal, Nucl. Phys. B **306**, 160 (1988).

[10] S.Ferrara, M.T.Grisaru and P.Van Nieuwenhuizen, Nucl. Phys. B 138(1978) 430.

[11] S. Ferrara and B. Zumino, Nucl. Phys. B **134**, 301 (1978).

[12] R. Kallosh and A. Linde, JCAP **1306**, 028 (2013) [arXiv:1306.3214 [hep-th]].

[13] J. Ellis, D. V. Nanopoulos and K. A. Olive, Phys. Rev. Lett. **111**, 111301 (2013) [arXiv:1305.1247 [hep-th]]; J. Ellis, D. V. Nanopoulos and K. A. Olive, JCAP **1310**, 009 (2013) [arXiv:1307.3537];

[14] S. Ferrara, R. Kallosh, A. Linde and M. Porrati, Phys. Rev. D 88 (2013)085038, arXiv:1307.7696 [hep-th]; S. Ferrara, R. Kallosh, A. Linde and M. Porrati, JCAP **1311**, 046 (2013) [arXiv:1309.1085 [hep-th]].

[15] F. Farakos, A. Kehagias and A. Riotto, Nucl. Phys. B **876**, 187 (2013) [arXiv:1307.1137];

[16] S. Ferrara, R. Kallosh and A. Van Proeyen, JHEP **1311**, 134 (2013) [arXiv:1309.4052 [hep-th]].

[17] R. Kallosh, A. Linde and A. Westphal, arXiv:1405.0270 [hep-th].

[18] R. Kallosh, A. Linde and D. Roest, Phys. Rev. Lett. **112**, 011303 (2014) [arXiv:1310.3950 [hep-th]]; JHEP **1311**, 198 (2013) [arXiv:1311.0472 [hep-th]].

[19] T. Li, Z. Li and D. V. Nanopoulos, arXiv:1405.0197 [hep-th].

[20] S. Ferrara, A. Kehagias and A. Riotto, Fortsch. Phys. **62**, 573 (2014) [arXiv:1403.5531 [hep-th]]; arXiv:1405.2353 [hep-th].

[21] R. Kallosh, A. Linde, B. Vercnocke and W. Chemissany, arXiv:1403.7189 [hep-th].

[22] K. Hamaguchi, T. Moroi and T. Terada, arXiv:1403.7521 [hep-ph].

[23] J. Ellis, M. A. G. Garca, D. V. Nanopoulos and K. A. Olive, JCAP **1405**, 037 (2014) [arXiv:1403.7518 [hep-ph]].

[24] J. Ellis, M. A. G. Garcia, D. V. Nanopoulos and K. A. Olive, arXiv:1405.0271 [hep-ph].

[25] S. V. Ketov and A. A. Starobinsky, Phys. Rev. D **83**, 063512 (2011) [arXiv:1011.0240 [hep-th]]; S. V. Ketov and A. A. Starobinsky, JCAP **1208**, 022 (2012) [arXiv:1203.0805 [hep-th]]; S. V. Ketov and S. Tsujikawa, Phys. Rev. D **86**, 023529 (2012) [arXiv:1205.2918 [hep-th]]; S. V. Ketov and T. Terada, JHEP **1312**, 040 (2013) [arXiv:1309.7494 [hep-th]].

[26] S. Ferrara, A. Kehagias and M. Porrati, Phys. Lett. B **727**, 314 (2013) [arXiv:1310.0399 [hep-th]].

[27] S. Cecotti and R. Kallosh, JHEP **1405**, 114 (2014) [arXiv:1403.2932 [hep-th]].

[28] A. Van Proeyen, Nucl. Phys. B **162** (1980) 376.

[29] P. Fayet, Nuovo Cim. A **31**, 626 (1976).

[30] S. Ferrara, P. Fre and A. S. Sorin, Fortsch. Phys. **62**, 277 (2014) [arXiv:1401.1201 [hep-th]]; JHEP 04(2014)095 arXiv:1311.5059 [hep-th].

[31] M. Kawasaki, M. Yamaguchi and T. Yanagida, Phys. Rev. Lett. **85**, 3572 (2000) [hep-ph/0004243].

[32] D. Z. Freedman, Phys. Rev. D **15** (1977) 1173.

[33] M. Rocek, Phys. Rev. Lett. **41** (1978) 451; U. Lindstrom, M. Rocek, Phys. Rev. D **19** (1979) 2300; R. Casalbuoni, S. De Curtis, D. Dominici, F. Feruglio and R. Gatto, Phys. Lett. B **220** (1989) 569; Z. Komargodski and N. Seiberg, JHEP **0909** (2009) 066 [arXiv:0907.2441 [hep-th]].

[34] I. Antoniadis, E. Dudas, S. Ferrara and A. Sagnotti, Phys. Lett. B **733**, 32 (2014) [arXiv:1403.3269 [hep-th]].

[35] A. Kehagias, A. M. Dizgah and A. Riotto, Phys. Rev. D **89**, 043527 (2014)

LECTURER: S. FERRARA

Higher Curvature Supergravity, Supersymmetry Breaking and Inflation

Scientific Secretaries: A. Liashyk, D. Kalinov

DISCUSSION

\- *Ponomarev:*

At the beginning of your lecture you said that inflation is controversial. What did you mean?

\- *Ferrara:*

I did not say that inflation itself is controversial, even though there are some people who are not favorable to it. Indeed, it seems to me that recent data support an inflationary explanation for the evolution of the universe. What is controversial are the different models used to explain inflation, which provide different values for the slow-roll parameters, such as chaotic inflation or the Starobinsky model. It depends on your trust in the data from BICEP2.[1]

\- *Ponomarev:*

Could you explain your statement that de Sitter space cannot be a background for a supersymmetric theory? Is this a general statement about supersymmetry?

\- *Ferrara:*

What I am saying is that in the scenario I have been talking about, with two chiral multiplets parametrizing the potential which gives rise to inflation, de Sitter space appears as a phase where supersymmetry is broken. In fact there is no superalgebra in de Sitter (contrary to anti-de Sitter) space. During the inflationary era in the de Sitter space supersymmetry is spontaneously broken, as can be seen from the potential and from the fact that the F-term (for the sgoldstino) in the superpotential develops a large vacuum expectation value, which is a signal of supersymmetry breaking. We have a supersymmetric model, where in some phase supersymmetry is spontaneously broken, while outside of de Sitter space you reach a time when the vacuum expectation value of the potential is zero and supersymmetry is restored.

\- *'t Hooft:*

At the beginning of the lecture you have argued that canonical gravity with a scalar field Φ is dual to a theory with a $R+R^2$ term. What kind of duality transformation?

\- *Ferrara:*

Actually, this is a work done in the 80's by Whitt [7] (B. Whitt, Phys. Lett. B 145, 176 (1984)). The starting point is to introduce new fields σ and Λ and to write the Lagrangian $\mathcal{L}=-\frac{1}{2}R+\frac{1}{2}\sigma(\Lambda-R)+\alpha\Lambda^2$ (eq. 74). Then depending on which field is integrated out, two different descriptions can be found. Variation with respect to σ yields $\Lambda=R$, and plugging this

[1] Update 2015: With the Planck 2015 results the r parameter has again decreased and is now compatible with the predictions of the Starobinsky model.

back in allows to recover the Lagrangian $\mathcal{L}=-\frac{1}{2}R+\alpha R^2$ (eq. 73). On the other hand, the dual theory is obtained by collecting the two terms multiplying R in the Lagrangian (eq. 74) to give the expression $\mathcal{L}=-\frac{1}{2}(1+\sigma)R+\frac{1}{2}\sigma\Lambda+\alpha\Lambda^2$ (eq. 75). In this way the presence of a Jordan frame function $(1+\sigma)$ is highlighted, thus suggesting to perform a conformal rescaling $g'_{\mu\nu}=g_{\mu\nu}(1+\sigma)^{-1}$, which leads to canonical gravity with a scalar field Φ.

- *'t Hooft:*

I was wondering where the kinetic term for the field Φ comes from.

- *Ferrara:*

Indeed, there is no kinetic term for the scalar field in the Lagrangian (75), but it originates from the presence of an Einstein term in the Jordan frame, which, after the conformal transformation I have just explained, in the Einstein frame yields a kinetic term for the scalar field Φ defined by $1+\sigma=e^{\sqrt{\frac{2}{3}}\Phi}$.

- *Korthals-Altes:*

You introduced the Weyl tensor into Starobinsky's model. This is a conformal invariant tensor, so after a conformal transformation it would still be there. What role would it play?

- *Ferrara:*

I did not consider this term, because it contains a spin 2 field, which is assumed to acquire infinite mass, otherwise it would violate unitarity. But, indeed, it would stay as it is under conformal transformation. In some sense I am assuming that it is decoupling from the other terms, in particular the inflaton, in order to preserve unitarity. Without going into technical details, an interesting aspect I would like to mention is that there is a supersymmetric generalization of this duality transformation, going from a $R+R^2$ theory to a standard supergravity, i.e. a supersymmetric version of Einstein gravity with an extra scalar degree of freedom, which is the inflaton. The nice feature about this $R+R^2$ theory is that the potential comes out exactly to be of the form needed to explain inflation.

- *Minkowski:*

In general the Weyl tensor of g is not the same as the Weyl tensor of g'.

- *Ferrara:*

But this term is conformal invariant, so it would not change under a rescaling. In any case it does not couple with the inflaton. The reasons to introduce it are only pedagogical, in order to count the degrees of freedom. If both terms are taken into account, it can be seen that the off-shell degrees of freedom of the metric become massive states. One is connected to the physical inflaton, and the other to a spin 2 ghost related to the W^2 term.

- *'t Hooft:*

As a follow-up question to my previous question, are you sure about the sign of the term $(\partial\Phi)^2$?

- *S. Ferrara:*

Yes, I'm sure. There is another situation similar to this, where the sign seems to be the wrong one. When you have a conformally coupled scalar, it is equivalent to the Einstein theory. But in order to recover the correct sign of the Einstein term, it is necessary to consider a ghost; in that case there is no problem because it is not a physical degree of freedom, and it is possible to get rid of it by conformal gauge fixing. But this is a completely different situation: a term which consists of R times a Jordan frame function is equivalent through a conformal rescaling to the Einstein term plus a physical kinetic term for the scalar and the sign is the right one. It should not be confused with the case of a conformally coupled scalar.

- *Korthals-Altes:*

Could you comment about Shaposhnikov model of Higgs inflation? The coupling needs to be non-minimal?

- *Ferrara:*

Shaposhnikov's Higgs inflation is a completely different scenario. It is a model for Higgs driven inflation, where a Higgs coupling to the curvature is introduced, modifying the scalar potential. Such a coupling has to be non-minimal. Starting with a standard Higgs potential $V(\Phi) = (\lambda + \Phi^2)^2$, a term is added which couples the Higgs field to the curvature: $H^2 R + V(\Phi)$. Then a transformation to the Einstein frame yields the Higgs potential which drives the inflation. And what is interesting is that even though it is different from the Starobinsky model, it gives the same predictions for the slow-roll parameters. In other words from the point of view of the slow-roll parameters, the Shaposhnikov model of Higgs inflation and the Starobinsky model of scalaron inflation give the same result.

- *Korthals-Altes:*

Which is in contradiction with the result from BICEP2. [2]

- *Ferrara:*

Yes, in contradiction by one order of magnitude. BICEP2 is compatible with a potential which is just a massive scalar field. Actually, you can embed such a potential in a supergravity scenario as well. By supersymmetrizing the mass term with the introduction of a massive vector multiplet, it is possible to recover a supersymmetric extension of the chaotic model, which is discussed in my paper [14] (S. Ferrara, R. Kallosh, A. Linde and M. Porrati, Phys. Rev. D 88 (2013)085038, arXiv:1307.7696 [hep-th]; JCAP 1311, 046 (2013) [arXiv:1309.1085 [hep-th]).

- *Millington:*

In the framework of the Shaposhnikov Higgs inflation scenario, I have a question regarding the possibility of frame independent cosmological observables, i.e. of a quantization frame. Suppose I perform a plane-wave decomposition of the Higgs field in the Jordan frame and build my initial conditions out of some mode decomposition in terms of Fock states, and obtain one set of cosmological perturbations. The claim is that if I do the same calculation in the Einstein frame after making the conformal transformation and doing a field redefinition to ensure that I get a canonical kinetic term, and then repeat the calculation using a Bunch-Davies vacuum and the same mode decomposition, I will get similar predictions. But I am not convinced that this is true; not because of the conformal transformation, but because the field redefinition is not a canonical transformation of the field, which does not preserve the commutation relations of the fields. This means that the vacuum in one frame will be an excited state relative to the other. In other words, while I agree at the classical level, I see a problem with linear perturbation theory. So I'm wondering if you could comment about that, and whether you believe that there exists a frame which is better for model building and for quantization?

---

[2] Update 2015: These predictions have been retreated in 2015.

- *Ferrara:*

I cannot comment about the Shaposhnikov model, which I am not considering here. The Starobinsky model only involves a conformal rescaling of the metric. This is a classical formula, so I don't have an answer to your question. The field redefinition from Φ to σ is something which is ordinarily done and it does not seem problematic to me.

- *Hughes:*

In string theory you have a whole series of high derivatives corrections, and, presumably, for a generic compactification you end up with many more higher derivative terms. How sensitive is Starobinsky model to adding these extra terms?

- *Ferrara:*

The higher derivative corrections you are talking about are very particular, they change the kinetic term. In string theory there are many corrections with higher curvature which do not change this term. It depends on the kind of the correction. I believe that in string theory, for instance, the W^2 term is absent if you have enough supersymmetry. If you have higher supersymmetry some of the quadratic terms are absent in the theory. But in any case, I don't precisely know how to incorporate Starobinsky model in string theory. In string theory there are other scenarios for inflation.

- *Antoniadis:*

In string theory the higher derivative corrections are not considered in same way as they were here. They are considered on-shell and therefore they don't add any degrees of freedom. Here the R^2 term is considered from the beginning as the part of the action, and therefore it contains additional scalar degrees of freedom. So it introduces a particular scalar with a particular kinetic term and a particular potential.

- *Gould:*

In the Starobinsky model it is possible to intuitively understand the slow-roll from the flat part of the potential, but in the chaotic inflation model that is not the case. How do you recover slow-roll inflation in that instance?

- *Ferrara:*

In the chaotic model the behavior of the slow-roll parameters is different from the Starobinsky model and, as I have shown at the beginning, the result is $r = \frac{8}{N} \approx 0.2$ (eq. 5). From the potential you can compute the number N of e-foldings, which is about 50-60, and this gives you the above value $r \approx 0.2$, compatible with that of BICEP2.[3]

[3] Update 2015: The value of r has decreased in 2015 after the last release of Planck data.

Problems with Ultrahigh-energy Neutrino Interactions[1]

Dieter Schildknecht[2]

Fakultät für Physik, Universität Bielefeld,
Universitätsstraße 25, 33615 Bielefeld, Germany
and
Max-Planck-Institute for Physics, Föhringer Ring 6,
80805 Munich, Germany

Abstract

The IceCube collaboration has recently identified events due to ultrahigh-energy neutrino interactions. Predictions of the neutrino-nucleon cross section at ultrahigh energies require a huge extrapolation of the cross sections experimentally measured at laboratory energies. Upon relating neutrino scattering to deep inelastic electron scattering, we show that the empirically verified color dipole picture is well suited for such an extrapolation. The dominant contribution to the total neutrino-nucleon cross section, even at ultrahigh energies, is due to the kinematic range where color transparency is valid for the color dipole interaction. We deviate from various claims in the literature on the presence of screening effects due to non-linear evolution at ultrahigh neutrino energies.

1 The IceCube Experiment

The IceCube experiment, located at the South Pole, covers a cubic kilometer of Antarctic glacial ice. It detects neutrinos by observing Cherenkov light from secondary charged particles produced in neutrino-nucleon interactions. In 2013, the IceCube collaboration announced the detection of 28 neutrino events of ultra-high energy:

- 26 events in the energy range of 50 TeV to 1 PeV $= 10^{15}$ eV,
- 2 events in the energy range of 1 PeV to 2 PeV,
- zero events above 2 PeV.

In fig. 1, I show the flux (e.g. [1]) of cosmic rays. The ultrahigh energies of the observed neutrino events are significantly below the upper end of the energy spectrum of cosmic rays.

---

[1]Presented at the International School of Subnuclear Physics, 52nd Course, Erice-Sicily, 24 June - 3 July 2014

[2]Email: schild@physik.uni-bielefeld.de

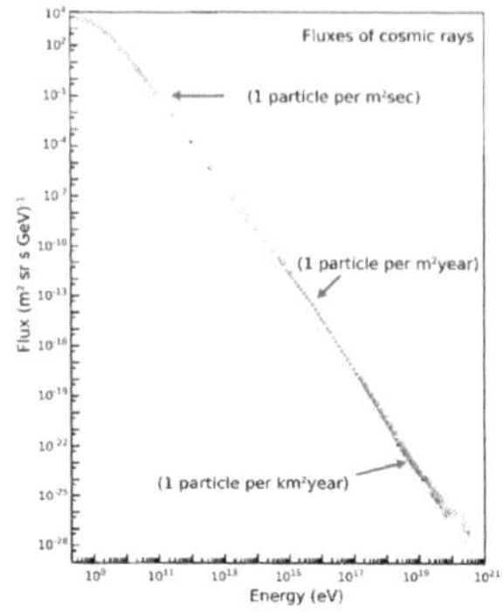

Figure 1: Flux of cosmic rays as a function of their energy.

A recent, eighty-page review paper, entitled "Cosmic Neutrino Pevatrons: A Brand New Pathway to Astronomy, Astrophysics, and Particle Physics" [2], contains a thorough discussion on the interpretation of the IceCube events. It covers potential sources of the neutrino events, galactic and extragalactic ones, as well as probes of fundamental physics. Cosmic probes include e.g. superheavy Dark Matter, lepto-quarks produced via $\nu_\tau + q \to LQ \to \tau + q$, and others.

Let me only mention a single recent speculative remark [3] on the possibility of an exotic neutrino property. The suggestion is as simple, as it is far reaching at the same time. The afore-mentioned energies of the IceCube events (in distinction from the upper end of the cosmic-ray flux in fig. 1) do not extend beyond approximately 2 PeV. In particular, the enhancement of the neutrino signal due to the so-called Glashow resonance [4]

$$\nu_e e^- \to W^- \to \text{anything},$$

corresponding to an energy of about $E \simeq 6.3$ PeV in the rest frame of the interacting electron e^-, has not been observed (up to now). Interpreting this non-observation as the consequence of a strict universal bound on the neutrino energy, E^ν_{Max}, located below 6.3 PeV, from

$$E^\nu_{Max} = \frac{m_\nu}{\sqrt{1-\beta_\nu^2}} = \gamma_\nu m_\nu,$$

one concludes that $\gamma_\nu = E^\nu_{Max}/m_\nu < \infty$ with the dramatic consequence of a violation of Lorentz invariance for neutrinos.

With respect to future searches for, and for the detailed analysis of ultrahigh-energy neutrino events, a knowledge of the neutrino-nucleon cross section is indispensable. Predictions of the cross section for ultrahigh energies of neutrinos require huge extrapolations to energies far beyond the energies where experimental data from laboratory experiments are available. The highest energies in laboratory experiments on lepton scattering were reached for charged-lepton scattering at the electron-proton collider HERA in Hamburg, which was running during the years from 1990 to 2007. By relating neutrino scattering to charged-lepton scattering, the HERA experimental data can be used as a reliable base for the extrapolation to the ultrahigh energies observed for cosmic neutrinos.

2 The Neutrino-Nucleon Cross Section at Ultrahigh Energies

2.1 Connection to Deep Inelastic Electron-Nucleon Scattering

The total charged-current neutrino-nucleon cross section is given by (e.g. [5])

$$\sigma_{\nu N}(E) = \frac{G_F^2}{2\pi} \int_{Q^2_{min.}}^{s-M_p^2} dQ^2 \left(\frac{M_W^2}{Q^2 + M_W^2} \right)^2 \int_{M_p^2}^{s-Q^2} \frac{dW^2}{W^2} \sigma_r(x, Q^2). \tag{2.1}$$

In (2.1), E denotes the neutrino energy in the nucleon rest frame. It is related to the neutrino-nucleon center-of-mass energy by

$$s = 2M_pE + M_p^2 \cong 2M_pE, \tag{2.2}$$

where M_p denotes the proton mass. The integrations in (2.1) run over the four-momentum-transfer squared from the neutrino to the nucleon, Q^2, and the square of the total energy, W^2, of the produced hadronic system. The Fermi coupling is denoted by G_F, and the W-boson mass by M_W. The reduced cross section, $\sigma_r(x, Q^2)$ in (2.1), depends on the three nucleon structure functions $F_{2,L,3}^{\nu N}(x, Q^2)$,

$$\sigma_r(x, Q^2) = \frac{1 + (1-y)^2}{2} F_2^{\nu N}(x, Q^2) - \frac{y^2}{2} F_L^{\nu N}(x, Q^2) + y(1 - \frac{y}{2}) x F_3^{\nu N}(x, Q^2), \tag{2.3}$$

where x is the Bjorken variable,

$$x = \frac{Q^2}{2qP} = \frac{Q^2}{W^2 + Q^2 - M_p^2} \cong \frac{Q^2}{W^2}, \tag{2.4}$$

and y denotes the squared fraction of the total neutrino energy producing hadrons in the final state

$$y = \frac{Q^2}{2M_pEx} \cong \frac{W^2}{s}. \tag{2.5}$$

According to (2.1), at ultra-high energies, $s \gg M_W^2$, the contribution to the total cross section due to $Q^2 \gg M_W^2$ is strongly suppressed. The dominant contribution to the cross section is due to $Q^2 \cong M_W^2$ and due to small values of $x \cong M_W^2/s \ll 0.1$.

At low values of $x \lesssim 0.1$, neutrino-nucleon and electron-nucleon interactions proceed via fluctuations of the virtual W-boson and virtual photon, respectively, into quark-antiquark pairs that propagate and interact via gluon exchange with the nucleon. For a given number of n_f actively contributing quark flavors, the flavor-independent QCD interaction implies the proportionality

$$\frac{1}{n_f} F_{2,L}^{\nu N}(x, Q^2) = \frac{1}{\sum_q Q_q^2} F_{2,L}^{eN}(x, Q^2). \tag{2.6}$$

It allows to predict the neutrino-nucleon structure functions, $F_{2,L}^{\nu N}(x, Q^2)$, from the electron-nucleon structure functions, $F_{2,L}^{eN}(x, Q^2)$. For $n_f = 4$ flavors, the proportionality factor in (2.6) becomes $n_f / \sum_q^{n_f} Q_q^2 = 5/18$.

The dominant term in (2.3) is due to $F_2^{\nu N}(x, Q^2)$. In what follows, we shall frequently use the relation between the proton structure function, $F_2^{ep}(x, Q^2)$, and the (virtual) photoabsorption cross section, $\sigma_{\gamma^* p}(W^2, Q^2)$, that, for $x \ll 0.1$, is given by

$$F_2^{ep}(x, Q^2) = \frac{Q^2}{4\pi^2 \alpha} \sigma_{\gamma^* p}(W^2, Q^2). \tag{2.7}$$

2.2 Electron-Proton Deep Inelastic Scattering, Experimental Results

As noted, neutrino scattering at ultrahigh energies is dominantly due to interactions in the kinematic range of $x \ll 0.1$. This is the region of the kinematic variables Q^2 and W^2, where the HERA experimental results show scaling in the low-x scaling variable [6],

$$\eta(W^2, Q^2) \equiv \frac{Q^2 + m_0^2}{\Lambda_{sat}^2(W^2)}, \qquad \Lambda_{sat}^2(W^2) \sim (W^2)^{C_2} \tag{2.8}$$

i.e.

$$\sigma_{\gamma^* p}(W^2, Q^2) = \sigma_{\gamma^* p}(\eta(W^2, Q^2)) \sim \sigma^{(\infty)} \begin{cases} \frac{1}{\eta(W^2, Q^2)} & , \quad \text{for } \eta(W^2, Q^2) \gg 1, \\ \ln \frac{1}{\eta(W^2, Q^2)} & , \quad \text{for } \eta(W^2, Q^2) \ll 1. \end{cases} \tag{2.9}$$

The photoabsorption cross section, in the approximation where the (hadronic) cross section $\sigma^{(\infty)}(W^2) = \sigma^{(\infty)} \cong const$, depends on the single scaling variable $\eta(W^2, Q^2)$ defined in (2.8). Compare fig. 2.

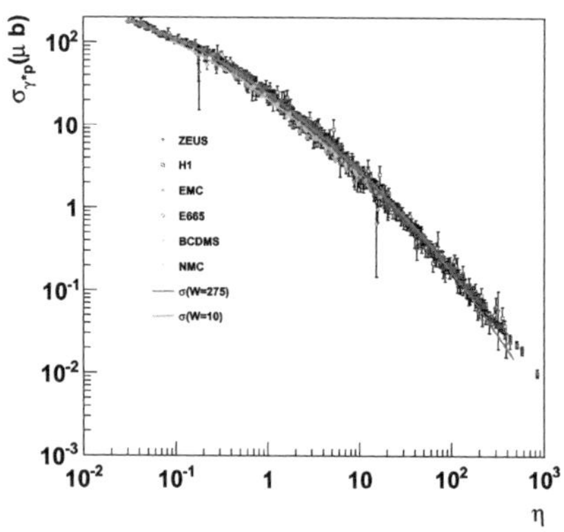

Figure 2: The total photoabsorption cross section as a function of the low-x scaling variable [6] $\eta \equiv \eta(W^2, Q^2)$.

The "saturation scale", $\Lambda_{sat}^2(W^2)$, grows as $\Lambda_{sat}^2(W^2) \sim (W^2)^{C_2}$, where $C_2 \simeq 0.3$, and the constant m_0, as expected from quark-hadron duality [7] for light quarks, is somewhat smaller than the ρ^0-meson mass, m_{ρ^0}. The functional dependence (2.9) of the photoabsorption cross section on $\eta(W^2, Q^2)$ can be read off from fig. 2.

We remark that the theoretical curves in fig. 2 were obtained in the color-dipole picture (CDP) to be discussed in the next section. The slight spread in the predictions for $W = 275$ GeV and $W = 10$ GeV, also visible in the experimental data, is due to the deviation of $\sigma^{(\infty)}(W^2)$ in (2.9) from $\sigma^{(\infty)}(W^2) = \sigma^{(\infty)} = \text{const.}$

The significant differences in the $\eta(W^2, Q^2)$ dependence for $\eta(W^2, Q^2) \gg 1$ and $\eta(W^2, Q^2) \ll 1$ in fig. 2 suggest a subdivision of the (Q^2, W^2) plane into (only and precisely) two regions subdivided by $\eta(W^2, Q^2) = 1$, compare fig. 3.

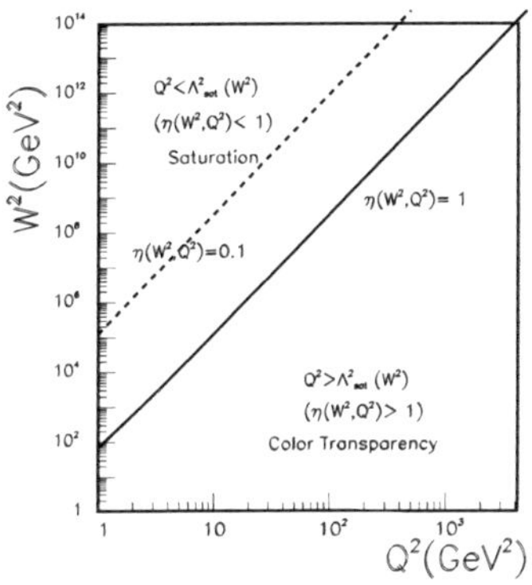

Figure 3: The (Q^2, W^2) plane with the line $\eta(W^2, Q^2) = 1$ separating the large-Q^2 color-transparency region from the saturation domain.

The logarithmic dependence of $\sigma_{\gamma^* p}(\eta(W^2, Q^2))$ for $\eta(W^2, Q^2) < 1$ implies the important limiting behavior [6, 8],

$$\begin{aligned} \lim_{\substack{W^2 \to \infty \\ Q^2 \text{ fixed}}} \frac{\sigma_{\gamma^* p}(\eta(W^2, Q^2))}{\sigma_{\gamma^* p}(\eta(W^2, Q^2 = 0))} &= \lim_{\substack{W^2 \to \infty \\ Q^2 \text{ fixed}}} \frac{\ln \left(\frac{\Lambda^2_{sat}(W^2)}{m_0^2} \frac{m_0^2}{(Q^2 + m_0^2)} \right)}{\ln \frac{\Lambda^2_{sat}(W^2)}{m_0^2}} \\ &= 1 + \lim_{\substack{W^2 \to \infty \\ Q^2 \text{ fixed}}} \frac{\ln \frac{m_0^2}{Q^2 + m_0^2}}{\ln \frac{\Lambda^2_{sat}(W^2)}{m_0^2}} = 1. \end{aligned} \tag{2.10}$$

For $W^2 \to \infty$, and any fixed value of the photon virtuality, Q^2, the ratio of the photoabsorption cross section, $\sigma_{\gamma^* p}(\eta(W^2, Q^2))$, to the $Q^2 = 0$ photoproduction cross section $\sigma_{\gamma p}(W^2) \equiv \sigma_{\gamma^* p}(\eta(W^2, Q^2 = 0))$, becomes equal to unity. The photoabsorption cross section "saturates" to a unique limit that coincides with the hadronlike $Q^2 = 0$ photoproduction cross section.

The scaling behavior of $\sigma_{\gamma^* p}(\eta(W^2, Q^2)) \sim 1/\eta(W^2, Q^2)$ for $\eta(W^2, Q^2) \gg 1$, in terms of the proton structure function $F_2 \equiv F_2^{ep}(x, Q^2)$ at sufficiently large Q^2, corresponds to a single curve in the plot of the experimental data for F_2 against the variable $1/W^2$. The theoretical curve in fig. 4 is based on [9]

$$F_2(W^2) = f_2 \cdot \left(\frac{W^2}{1\text{GeV}^2} \right)^{C_2 = 0.29}, \tag{2.11}$$

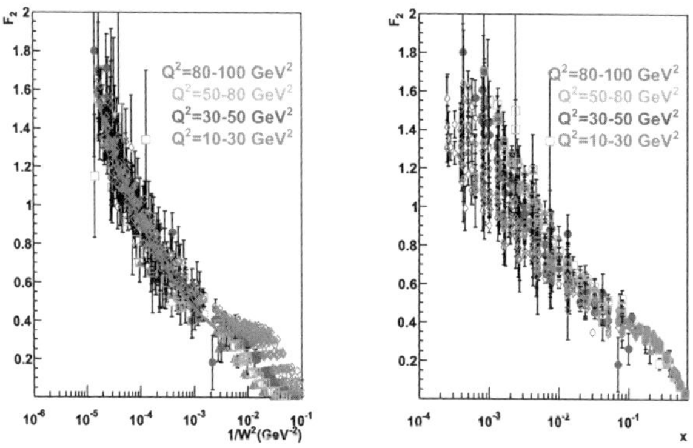

Figure 4: Experimental results for the proton structure function $F_2 \equiv F_2^{ep}(x, Q^2)$ as a function of $1/W^2$ and as a function of $1/x$ [9].

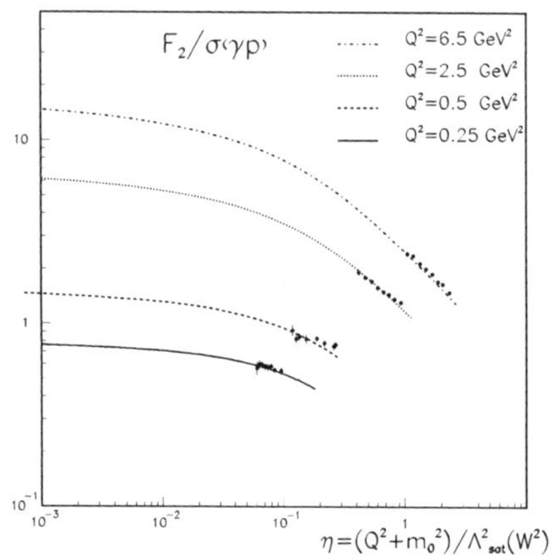

Figure 5: The approach to the saturation limit [9]. With decreasing $\eta(W^2, Q^2)$, the proton structure function $F_2(x, Q^2)$ approaches the saturation limit given by $Q^2\sigma_{\gamma p}(W^2)$, where $\sigma_{\gamma p}(W^2)$ denotes the $(Q^2 = 0)$ photoabsorption cross section.

where $f_2 = 0.063$. For comparison, in fig. 4, we also show F_2 as a function of the Bjorken variable x for fixed values of Q^2.

The transition to the saturation limit for $\eta(W^2, Q^2) \ll 1$, based on the CDP [9] is illustrated in fig. 5. Even for values of Q^2 as small as $Q^2 \cong 1\text{GeV}^2$, a strict and detailed empirical verification of the saturation limit requires "ultrahigh" energies far beyond the energies that were available at HERA.

The afore-mentioned experimental results on low-x DIS find a coherent *qualitative as well as quantitative* explanation in the color dipole picture (CDP) to which we shall turn next.

2.3 The Color Dipole Picture: Theory

In DIS at low $x \lesssim 0.1$, the photon virtually dissociates (or "fluctuates" in modern jargon) into hadronic vector states that subsequently interact with the nucleon (generalized vector dominance (GVD) [10], [11]). In QCD, the hadronic vector states are quark-antiquark $(q\bar{q})$ color-dipole states, which interact with the gluon field in the nucleon. Compare fig. 6 for the imaginary part of the associated forward-Compton-scattering amplitude.

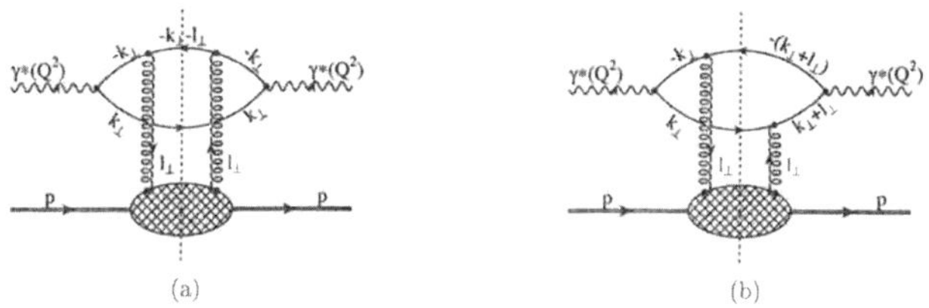

Figure 6: Two of the four diagrams for the $q\bar{q}$ dipole interaction with the gluon field in the nucleon. The diagrams (a) and (b) correspond to channel 1 and channel 2 respectively.

In terms of the transverse size, $\vec{r}_\perp$, of the $q\bar{q}$ pair, and the partition $0 \le z \le 1$ of the longitudinal momenta of quark and antiquark, the photoabsorption cross section, $\sigma_{\gamma^*_{L,T}p}(W^2, Q^2)$, for longitudinally and transversely polarized photons takes the form [12, 13]

$$\begin{aligned}\sigma_{\gamma^*_{L,T}p}(W^2, Q^2) &= \int dz \int d^2\vec{r}_\perp |\psi_{L,T}(\vec{r}_\perp, z(1-z), Q^2)|^2 \\ &\times \; \sigma_{(q\bar{q})p}(\vec{r}_\perp, z(1-z), W^2). \end{aligned} \tag{2.12}$$

The integrand in the photoabsorption cross section (2.12) shows the afore-mentioned factorization into the Q^2-dependent probability for a $\gamma^* \to q\bar{q}$ fluctuation and the W-dependent $(q\bar{q})p$ interaction cross section.

The gauge-invariant two-gluon interaction of the $q\bar{q}$ color-dipole state requires the representation [12, 13]

$$\sigma_{(q\bar{q})p}(\vec{r}_\perp, z(1-z), W^2) = \int^{\vec{l}_\perp^{\,2}{}_{Max}(W^2)} d^2\vec{l}_\perp \tilde{\sigma}\left(\vec{l}_\perp^{\,2}, z(1-z), W^2\right)\left(1 - e^{-i\vec{l}_\perp \vec{r}_\perp}\right). \tag{2.13}$$

The CDP, accordingly, rests on the general form (2.12) of the photoproduction cross section at low x, combined with the constraint on the dipole cross section (2.13)[3]

Employing the explicit (QED) expression for the $\gamma^* \to q\bar{q}$ transition probability, the photoabsorption cross section (2.12) may be written as [9]

$$\sigma_{\gamma^*_{L,T}p}(W^2, Q^2) = \frac{\alpha}{\pi} \sum_q Q_q^2 Q^2 \int dr'^2_\perp K^2_{0,1}(r'_\perp Q) \sigma_{(q\bar{q})^{J=1}_{L,T}p}(r'^2_\perp, W^2). \tag{2.14}$$

[3]We stress the dependence of the dipole cross section (2.13) on the energy W, [10], [13], [6], [11], [14] as implied by the fluctuation of the photon into on-shell $q\bar{q}$ dipole states, forbidding the Q^2 dependence (via Bjorken $x \sim Q^2/W^2$ dependence) frequently assumed in the literature.

In the transition from (2.12) to (2.14), massless quarks were assumed, and, by the appropriate projection, the cross sections $\sigma_{(q\bar{q})^{J=1}_{L,T}p}(r'^2_\perp, W^2)$ for the scattering of $(q\bar{q})^{J=1}_{L,T}$ states were introduced. The sum $\sum_q Q_q^2$ in (2.14) runs over the actively contributing quark flavors, and $K_0(r'_\perp Q)$ and $K_1(r'_\perp Q)$ denote modified Bessel functions. In terms of the $J = 1$ projections, $\sigma_{(q\bar{q})^{J=1}_{L,T}p}(r'_\perp, W^2)$, of the dipole cross section in (2.12), the two-gluon structure of the dipole cross section from (2.13) becomes [9]

$$\sigma_{(q\bar{q})^{J=1}_{L,T}p}(r'_\perp, W^2) = \pi \int d\vec{l}'^2_\perp \bar{\sigma}_{(q\bar{q})^{J=1}_{L,T}p}(\vec{l}'^2_\perp, W^2) \left(1 - \frac{\int d\vec{l}'^2_\perp \bar{\sigma}_{(q\bar{q})^{J=1}_{L,T}p}(\vec{l}'^2_\perp, W^2) J_0(l'_\perp r'_\perp)}{\int d\vec{l}'^2_\perp \bar{\sigma}_{(q\bar{q})^{J=1}_{L,T}p}(\vec{l}'^2_\perp, W^2)} \right), \tag{2.15}$$

where $J_0(l'_\perp r'_\perp)$ is the Bessel function with index 0.

The representation of the photoabsorption cross section (2.14) together with the constraint (2.15) allows one to *derive* [9], [15] the important *qualitative* behavior of the photoabsorption cross section (2.9) seen in the experimental data in fig. 2.

For any fixed value of Q^2, the strong decrease of the functions $K^2_{0,1}(r'_\perp Q)$ restricts the range of $r'_\perp$ relevant for the integration over the dipole size $r'^2_\perp$ in (2.14). The behavior of the dipole cross section in (2.14) for this Q^2-dependent effective dipole size $r'^2_\perp$, limited according to (2.15), is determined by the interaction energy, W, of the $q\bar{q}$ dipole state. Assuming that the effective upper limit of the integration over $d\vec{l}'^2_\perp$ in (2.15), given by $\vec{l}'^2_{\perp Max}(W^2)$, increases with the energy W, two different limiting behaviors of the dipole cross section at a fixed value of $r'_\perp$ can be discriminated.

For the case of relatively low energy, namely for $0 < l'_\perp r'_\perp < l'_{\perp Max}(W^2) r'_\perp \ll 1$, the expansion of the Bessel function in (2.15) implies that the dipole-proton cross section (2.15) vanishes proportional to the dipole size squared ("color transparency" limit),

$$\sigma_{(q\bar{q})^{J=1}_{L,T}p}(r'^2_\perp, W^2) \sim r'^2_\perp. \tag{2.16}$$

The proportionality factor in (2.16) is given by the first moment of $\bar{\sigma}_{(q\bar{q})^{J=1}_{L,T}p}(\vec{l}'^2_\perp, W^2)$, and is identified with the saturation scale $\Lambda^2_{sat}(W^2)$ introduced in (2.8).

For the case of very high energy, such that $l'^2_{\perp Max}(W^2) r' \gg 1$, rapid oscillations of the Bessel function $J_0(l'_\perp r'_\perp)$ in (2.15) imply ("saturation" limit).

$$\lim_{\substack{r'^2 fixed \\ W^2\to\infty}} \sigma_{(q\bar{q})^{J=1}_{L,T}p}(r'^2_\perp, W^2) = \lim_{\substack{r'^2_\perp\to\infty \\ W^2=const}} \sigma_{(q\bar{q})^{J=1}_{L,T}p}(r'^2_\perp, W^2) = \sigma^{(\infty)}_{L,T}(W^2). \tag{2.17}$$

At fixed dipole size, for W^2 sufficiently large the dipole cross section converges to the hadronic cross section that at most depends weakly on W^2, i.e. $\sigma^{(\infty)}_{L,T}(W^2) \simeq \sigma^{(\infty)} = \text{const}$.

Evaluating the photoabsorption cross section (2.12) for the two different limits, (2.16) and (2.17), we find [9, 15]

$$\begin{aligned} \sigma_{\gamma^* p}(W^2, Q^2) &= \sigma_{\gamma^* p}(\eta(W^2, Q^2)) \\ &= \frac{\alpha}{\pi} \sum_q Q_q^2 \begin{cases} \frac{1}{6}(1+2\rho)\sigma^{(\infty)}_L(W^2)\frac{1}{\eta(W^2,Q^2)} & , \quad (\eta(W^2,Q^2) \gg 1) \\ \sigma^{(\infty)}_T(W^2) \ln \frac{1}{\eta(W^2,Q^2)} & , \quad (\eta(W^2,Q^2) \ll 1) \end{cases} \end{aligned} \tag{2.18}$$

The form (2.18) agrees with the empirical result (2.9)[4]. The form of the $q\bar{q}$ color-dipole interaction (2.13) dictated by gauge invariance, *implies scaling* in $\eta(W^2, Q^2)$ of the total photoabsorption cross section. The empirical functional dependence, as $1/\eta(W^2, Q^2)$ and as $\ln(1/\eta(W^2, Q^2))$ in (2.9), is traced back to *color transparency and saturation* of the $(q\bar{q})p$ color dipole interaction.

2.4 A Remark on the Gluon Distribution

The proton structure function in the CDP, according to (2.18) as well as (2.11) and fig. 4, is given by

$$F_2^{ep}(x, Q^2) \sim \begin{cases} \sigma^{(\infty)} \Lambda^2_{sat}(W^2) & , \quad (\eta(W^2, Q^2) \gg 1) \\ Q^2 \sigma^{(\infty)} \ln \frac{\Lambda^2_{sat}(W^2)}{Q^2 + m_0^2} & , \quad (\eta(W^2, Q^2) \ll 1) \end{cases} \tag{2.19}$$

Noting that in the perturbative-QCD-improved parton model the structure function $F_2^{ep}(x, Q^2)$ is proportional to the gluon-distribution function, we have to identify [9]

$$\sigma^{(\infty)}(W^2) \Lambda^2_{sat}(W^2) \sim \alpha_s(Q^2) x g(x, Q^2)|_{x = \frac{Q^2}{W^2}} . \tag{2.20}$$

The W dependence of the structure function, $F_2^{ep}(x, Q^2)$ at large values of $\eta(W^2, Q^2) \gg 1$, compare (2.19) and fig. 4, implies that the gluon structure function depends on the single variable W^2. Since the CDP uniquely fixes the Q^2 dependence of $F_2^{ep}(x, Q^2)$ for $\eta(W^2, Q^2) \ll 1$, a measurement of $\Lambda^2_{sat}(W^2)$ for $\eta(W^2, Q^2) \gg 1$ determines the structure function in the saturation limit of $\eta(W^2, Q^2) \ll 1$, compare the (Q^2, W^2) plane in fig. 3.

With respect to saturation, our approach of the CDP differs significantly from the point of view frequently put forward in the literature: therein, the proportionality of the structure function $F_2(x, Q^2) \sim \alpha_s(Q^2) x g(x, Q^2)$ is *assumed to persist* in the limit of $W^2 \rightarrow \infty$ with $Q^2 = \text{const}$, or $x \rightarrow 0$, leading to an infinite rise of $F_2(x, Q^2)$ for $x \rightarrow 0$ at Q^2 fixed. Non-linear effects in the gluon-distribution function [16], [17] associated with quark splitting and recombination, must then be invoked to obtain agreement with experiment.

Such an approach ignores the long life time, or the long longitudinal distance involved in the interaction of a $\gamma^* \rightarrow q\bar{q}$ fluctuation correctly contained in the CDP. Saturation according to the CDP corresponds to a transition from a proportionality to $\Lambda^2_{sat}(W^2)$ to a logarithmic dependence on this one and the same function, $\Lambda^2_{sat}(W^2)$, when passing from $\eta(W^2, Q^2) \gg 1$ to $\eta(W^2, Q^2) \ll 1$. This transition is traced back to the different interplay, for $\mu(W^2, Q^2) \gg 1$ and $\mu(W^2, Q^2) \ll 1$, between the two interaction paths of fig. 6 corresponding to the color-gauge-invariant dipole interaction (2.13).

2.5 Color Dipole Picture: Quantitative Representation of the Photoabsorption Cross Section

The CDP yields a remarkably simple form for the photoabsorption cross section, essentially an interpolation between the general $1/\eta(W^2, Q^2)$ and $\ln(1/\eta(W^2, Q^2))$ dependences

[4]The quantity ρ yields the longitudinal-to-transverse ratio of the respective photoabsorption cross sections, $R = 1/2\rho$, for $\eta(W^2, Q^2) \gg 1$. The ratio $\rho \gtrsim 4/3$ [9] gives the relative transverse sizes of transversely-versus-longitudinally polarized $q\bar{q}$ dipole states in the color-transparency limit (2.16).

given in (2.18). For details we refer to refs. [6, 9]. The closed explicit expression (for $\rho = 1$) reads

$$\sigma_{\gamma^* p}(W^2, Q^2) = \frac{\alpha R_{e^+e^-}}{3\pi} \sigma^{(\infty)}(W^2) I_0(\eta(W^2, Q^2))$$
$$\times \frac{\frac{\xi}{\eta(W^2,Q^2)}}{1 + \frac{\xi}{\eta(W^2,Q^2)}} + O\left(\frac{m_0^2}{\Lambda^2_{\rm sat}(W^2)}\right) \quad (2.21)$$

where

$$\frac{\frac{\xi}{\eta(W^2,Q^2)}}{1 + \frac{\xi}{\eta(W^2,Q^2)}} \cong \begin{cases} 1, & \text{for } \eta(W^2, Q^2) \ll \xi = 130 \\ \frac{\xi}{\eta(W^2,Q^2)}, & \text{for } \eta(W^2, Q^2) \gg \xi = 130 \end{cases} \quad (2.22)$$

and $I_0(\eta(W^2, Q^2))$ is given by

$$I_0(\eta(W^2, Q^2)) = \frac{1}{\sqrt{1 + 4\eta(W^2, Q^2)}} \ln \frac{\sqrt{1 + 4\eta(W^2, Q^2)} + 1}{\sqrt{1 + 4\eta(W^2, Q^2)} - 1} \cong$$

$$\cong \begin{cases} \ln \frac{1}{\eta(W^2,Q^2)} + O(\eta \ln \eta), & \text{for } \eta(W^2, Q^2) \to \frac{m_0^2}{\Lambda^2_{\rm sat}(W^2)}, \\ \frac{1}{2\eta(W^2,Q^2)} + O\left(\frac{1}{\eta^2}\right), & \text{for } \eta(W^2, Q^2) \to \infty \end{cases}. \quad (2.23)$$

The hadronic dipole-proton cross section, $\sigma^{(\infty)}(W^2)$, in (2.21) is determined by requiring consistency of (2.21) with $Q^2 = 0$ photoproduction.

$$\sigma^{(\infty)}(W^2) = \frac{3\pi}{\alpha R_{e^+e^-}} \frac{1}{\ln \frac{\Lambda^2_{sat}(W^2)}{m_0^2}} \begin{cases} \sigma_{\gamma p}^{Regge}(W^2), \\ \sigma_{\gamma p}^{PDG}(W^2). \end{cases} \quad (2.24)$$

In (2.24), $\sigma_{\gamma p}^{Regge}(W^2)$ asymptotically grows as a small power of W^2 in distinction from $\sigma_{\gamma p}^{PDG}(W^2)$ that grows as $(\ln W^2)^2$. For the explicit expressions for these photoproduction cross sections, we refer to refs. [9, 18].

The parameter ξ, empirically determined as $\xi \cong 130$, restricts the masses of $q\bar{q}$ fluctuations, $M_{q\bar{q}}$, via $M^2_{q\bar{q}} \leq m_1^2(W^2) = \xi \Lambda^2_{sat}(W^2)$. With $m_0^2 = 0.15 GeV^2, \Lambda^2_{sat}(W^2) = C_1(W^2/1GeV^2)^{C_2}$, where $C_1 = 0.34 GeV^2$ and $C_2 \cong 0.27$ to 0.29, one obtains the theoretical results shown in fig. 2.

2.6 The Neutrino-Nucleon Cross Section in the CDP

Finally, I come back to the neutrino-nucleon cross section to be predicted at ultrahigh energies. Inserting the expression for $F_2^{ep}(x, Q^2)$, or rather the photoabsorption cross section (2.21) of the CDP, into the expression for the neutrino-nucleon cross section (2.1), we find [18]

$$\sigma_{\nu N}(E) = \frac{G_F^2 M_W^4}{8\pi^3 \alpha} \frac{n_f}{\sum_q Q_q^2} \int_{Q^2_{Min}}^{s-M_p^2} dQ^2 \frac{Q^2}{(Q^2 + M_W^2)^2}$$
$$\times \int_{M_p^2}^{s-Q^2} \frac{dW^2}{W^2} \frac{1}{2}(1 + (1-y)^2) \sigma_{\gamma^* p}(\eta(W^2, Q^2)). \quad (2.25)$$

Upon numerical evaluation of (2.25), based on the photoabsorption cross section from Section 2.5, we find the results shown in fig. 7 [18].

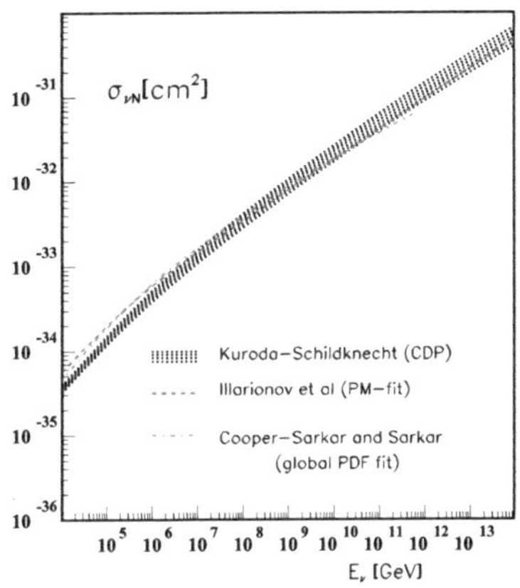

Figure 7: The prediction [18] of the neutrino-nucleon cross section from the CDP. For comparison, we also show the results based on the pQCD improved parton model fit, the "global PDF fit", to the HERA experimental data. The PM-fit is a convenient parameterization of the global PDF fit.

The shaded area in fig. 7 corresponds to a change of the exponent C_2 between $C_2 = 0.27$ and $C_2 = 0.29$. In fig. 7, we see that our prediction, based on a small number of parameters, essentially C_1 and C_2 to determine $\Lambda^2_{sat}(W^2)$ supplemented by the bounds m_0^2 and $m_1^2(W^2) = \xi\Lambda^2_{sat}(W^2)$ on the actively contributing $q\bar{q}$ dipole states, coincide with the very elaborate pQCD fits to the HERA data, the global PDF fit, obtained by the big pQCD collaborations with a huge number of fairly arbitrary fit parameters. The results of the PM fit in fig. 6 are based on a six-arbitrary-parameter fit to the global PDF fit for the parton distribution functions at low x.

2.7 Comparison with the Froissart-inspired Ansatz

As early as in 1953, Heisenberg [19] considered the question of the behavior of cross sections among strongly-interacting particles at asymptotically high energies. Picturing the proton as a Lorentz-contracted sphere with exponentially decreasing edge, and requiring a minimum blackness necessary for particle production to occur, he predicted an increase of strong-interaction cross sections with energy as $(\ln(W^2))^2$. In 1961, it was shown by Froissart [20] that a growth as $(\ln W^2)^2$ is the maximum growth allowed by unitarity and analyticity in quantum field theory.

In a series of papers by Block et al [21] - [24], it was recently shown that a hadron-like "Froissart-inspired" ansatz of the form $F_2^{ep}(x, Q^2) \sim \sum_{n,m=0,1,2} a_{nm}(\ln Q^2)^n(\ln(1/x))^m$, with seven free fit parameters provides an excellent representation of the low-x DIS data

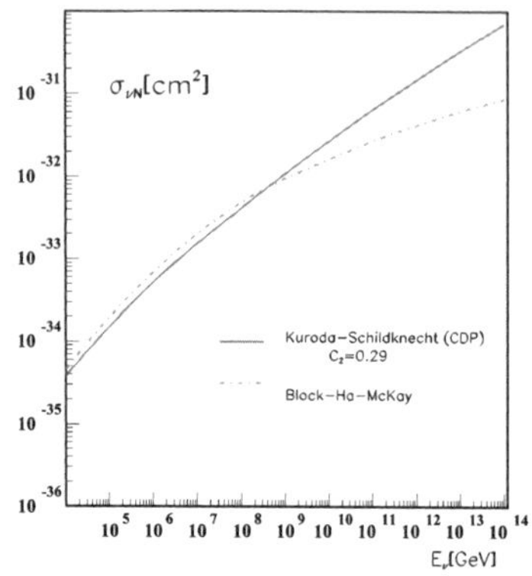

Figure 8: Comparison [18] of the predictions from the CDP with the results from the "Froissart-inspired" ansatz.

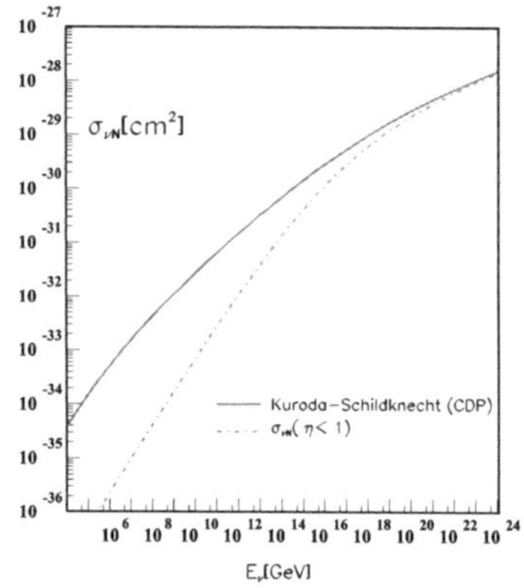

Figure 9: The neutrino-nucleon cross section compared [18] with the part of this cross section that is due to the saturation region characterized by the kinematic constraint of $\eta(W^2, Q^2) < 1$.

from HERA. The prediction for ultrahigh-energy neutrino scattering [23], based on employing in (2.1) with (2.6) the Froissart-inspired ansatz, is shown in fig. 8. There is satisfactory agreement with the predictions from the CDP (and accordingly with the predictions from the pQCD-improved parton model) for energies up to $E \lesssim 10^8\text{GeV} = 100\text{PeV}$, two orders of magnitude above the energy of the recently observed IceCube events.

The Froissart-inspired ansatz, according to fig. 8, for energies $E \gtrsim 10^8\text{GeV}$ implies a weaker growth with energy than the prediction from the CDP. The CDP, according to (2.21) with (2.23) and $\sigma_{\gamma p}^{PDG}(W^2)$ in (2.24), in the saturation region of $\eta(W^2, Q^2) \ll 1$ contains the same Froissart-like $(\ln W^2)^2$ behavior as the Froissart-inspired approach.

The differences in fig. 8 must accordingly be due to the differences in the Q^2 dependences between these two approaches. According to the CDP, as shown in fig. 9, at energies even far beyond $E \cong 10^8$GeV, for energies below $E \cong 10^{12}$GeV, the contribution to the neutrino-nucleon cross section from the saturation region of $\eta(W^2, Q^2) < 1$ is suppressed by at least one order of magnitude. Contributions from the $(\ln W^2)^2$ behavior in the saturation region only start to dominate the neutrino-nucleon cross section at unrealistic ultra-ultrahigh energies beyond the upper end of $E \cong 10^{12}$GeV of the cosmic-ray spectrum in fig. 1.

Our results from the CDP differ significantly from various results in the literature incorporating non-linear evolution effects [16] in the pQCD improved parton-model-approach. The results imply [25, 26, 5, 27] neutrino cross sections that at $E \cong 10^{12}$GeV are approximately a factor of two to three below our predictions in figs. 7 and 8.

3 Conclusions

Inspired by cosmic-neutrino-search experiments, we thoroughly examined the neutrino-nucleon cross section at ultrahigh energies. Our results, based on the simple closed analytic expression for the photoabsorption cross section at low x in the CDP, in the full range of energies $E \lesssim 10^{12}$GeV, are consistent with the results based on the extrapolation of the elaborate multi-parameter global pdf fits to the HERA experimental data.

Below energies of $E \lesssim 10^8$GeV, two orders of magnitude above the energy $E \approx 10^6$GeV of the recently detected IceCube events, our CDP results also agree with the predictions from the Froissart-inspired approach to DIS.

Above energies of $E \gtrsim 10^8$GeV, the predictions from the CDP disagree with both, the predictions from the Froissart-inspired approach, and the predictions incorporating non-linear evolution of parton-distribution functions. These predictions lead to a suppressed cross section, when compared with the CDP, the suppression reaching factors of approximately three at $E \approx 10^{12}$GeV.

The CDP-based analysis shows that the ultrahigh-energy neutrino-nucleon cross section, in the full range of energies, up to $E \approx 10^{12}$GeV, of the cosmic-ray spectrum, is dominated by contributions from the kinematic region, where color transparency determines the interaction. The dominance of color transparency is responsible for the consistency of the CDP neutrino-nucleon cross section with the global pdf fits, while excluding the presence of screening effects due to non-linear evolution.

Acknowledgement

The author thanks Kuroda-san for fruitful collaboration on the color dipole picture.

It is a great pleasure for the author to thank Professor Antonino Zichichi for the invitation to the 52nd Course of the International School of Subnuclear Physics with the lively scientific atmosphere in the magnificent environment of the Ettore Majorana Foundation and Centre for Scientific Culture at Erice, Sicily.

References

[1] W. Bietenholz, arXiv:1305.1346 [physics.hist-ph]

[2] L.A. Anchordoqui, V. Barger et al.
arXiv:1312.6587 [astro-ph.HE], JHEAp 1-2 (2014)1.

[3] L.A. Anchordoqui, V. Barger et al.
arXiv: 1404.0622 [hep-ph]

[4] S.L. Glashow
Phys. Rev. 118 (1960) 316.

[5] V.P. Goncalves and P. Hepp
Phys. Rev. D83 (2011) 014014.

[6] D. Schildknecht, in *Diffraction 2000, Cetraro, Italy, September 2-7, 2000*, Nucl. Phys. B, Proc. Supplement 99 (2001) 121;
D. Schildknecht, B. Surrow, M. Tentyukov, Phys. Lett. B499, 116 (2001);
G. Cvetic, D. Schildknecht, B. Surrow, M. Tentyukov, EPJC 20, 77 (2001)

[7] B. Gorczyca, D. Schildknecht, Phys. Lett. 47B (1973) 71;
J.J. Sakurai, Phys. Lett. B46 (1973) 207;
J.J. Sakurai, in *Laws of Hadronic Matter, 1973 Internat. School of Subnuclear Physics, Erice, Sicily* (ed. A. Zichichi, Academic Press, New York) p. 291;
D. Schildknecht and F. Steiner, Phys. Lett. B56, 36 (1975);
E. Poggio, H.P. Quinn, S. Weinberg, Phys. Rev. D13, 1958 (1976);
M. Genovese, N.N. Nikolaev, B.G. Zakharov, Phys. Lett. B380, 213 (1996);
A.D. Martin, M.R. Ryskin, T. Teubner, Phys. Rev. D55, 4329 (1997).

[8] D. Schildknecht, in *DIS 2001, The 9th International Workshop on Deep Inelastic Scattering, Bologna, Italy, 2001*, G. Brassi et al. (Eds.), World Scientific, Singapore, 2002, p. 798;
D. Schildknecht, B. Surrow and M. Tentyukov, Mod. Phys. Lett. A16 (2001) 1829.

[9] M. Kuroda and D. Schildknecht, Phys. Rev. **D85** (2012) 094001.

[10] J.J. Sakurai and D. Schildknecht, Phys. Lett. **40B** (1972) 121; ibid **41B** (1972) 489; ibid **42B** (1972) 216;
B. Gorczyca, D. Schildknecht, Phys. Lett. **47B** (1973) 71;
H. Fraas, B.J. Read and D. Schildknecht, Nucl. Phys. **B86** (1975) 346;
R. Devenish and D. Schildknecht, Phys. Rev. **D19** (1976) 93.

[11] D. Schildknecht, in *Proc. of Photon 2005, Int. Conf. on the Structure and Interaction of the Photon*, Warsaw 2005; Acta Phys. Polon. B **37** (2006) 595.

[12] N.N. Nikolaev, B.G. Zakharov, Z. Phys. C49, 607 (1991).

[13] G. Cvetic, D. Schildknecht, A. Shoshi, Eur. Phys. J C13, 301 (2000).

[14] C. Ewerz und O. Nachtmann, Ann. Phys. **322** (2007) 1635; **322** (2007) 1670.

[15] D. Schildknecht, in *What we would like LHC to give us, 2012 International School of Subnuclear Physics, Erice-Sicily, June 23-July 2, 2012*, ed. A. Zichichi (World Scientific 2014), pp. 45-65, also published in Modern Physics Letters A Vol. 29, **No. 25** (2014) 1430028

[16] I. Balitzki, Nucl. Phys. **B463** (1996) 99,
Yu. V. Kovchegov, Phys. Rev. **D60** (1999) 034008; ibid **D61** (2000) 074018.

[17] J. Kuokkanen, K. Rummukainen and H. Weigert, arXiv: 1108.1867, Nucl. Phys. **A875** (2012) 29.

[18] M. Kuroda, D. Schildknecht, arXiv:1305.0400
Phys. Rev. **D88** (2013) 053007.

[19] W. Heisenberg, *Vorträge über kosmische Strahlung* (Springer, Berlin, 1953), p. 155, reprinted in W. Heisenberg, *Collected Works, Series B* (Springer, Berlin,1984) p.498; Die Naturwissenschaften **61** (1974), 1, reprinted in *Collected Works, Series B*, p. 912.

[20] M. Froissart, Phys. Rev. **123** (1961) 1053. .

[21] M.M. Block, E. Berger and C.-I. Tan, Phys. Rev. Lett **97** (2006) 252003;
E. Berger, M. Block and C.-I. Tan, Phzs. Rev. Lett **98** (2007) 242001.

[22] E. Berger, M.M. Block, D. McKay and C.-I. Tan, Phys. Rev. **D77** (2008) 053007.

[23] M.M. Block, P. Ha and D. McKay, Phys. Rev. **D82** (2010) 077302.

[24] M.M. Block, L. Durand, P. Ha, D.W. McKay, arXiV:1302.6119v2 [hep-ph], Phys. Rev. **D88** (2013) 014006.
M.M. Block, L. Durand, P. Ha, D.W. McKay, arXiV:1302.6127v1 [hep-ph], Phys. Rev. **D88** (2013) 013003.

[25] K. Kutak, J. Kwiecinski, arXiv: hep-ph/0303209, Eur. Phys. J. **C29** (2003) 521.

[26] M.H. Reno, arXiv: hep-ph/0410109, Nucl. Phys. Proc. Suppl. **143** (2005) 407.

[27] M.V.T. Machado, arXiv: 1112.0555 [hep-ph], Int. J. Mod. Phys. Proc. Suppl. **E20** (2011) 189.

LECTURER: D. SCHILDKNECHT

Problems with Ultrahigh-Energy Neutrino Interactions

Scientific Secretaries: S. Ahmad, S. Mamedov, A: Rebesh

DISCUSSION

\- *Tannenbaum:*

I have a question. Suppose in the USA we spend one and a half billion dollars to build an Electron-Ion Collider. What would you expect to learn? What would you be most interested in?

\- *Schildknecht:*

The Electron-Ion Collider... You see, unfortunately, first of all I am not completely aware of the properties and so on: now, energies and the luminosities and all these things, right? Now if I have an Electron-Ion Collider, how does the cross-section behave as a function of the nuclear number: A^(2/3) in the extreme case? So there are certainly questions which are worth answering in this field. If you do it with ions instead of protons.

\- *Tannenbaum:*

No, we will have protons also. We have protons and ions. I am saying: do you think there is enough new physics? And the energy is $\sqrt{s}$ of about 200.

\- *Schildknecht:*

The energy is comparable to HERA. Is it?

\- *Tannenbaum:*

Lower.

\- *Schildknecht:*

Lower? Yes. I see. That is what I thought.

\- *Tannenbaum:*

Considerably lower.

\- *Schildknecht:*

Considerably lower, I see. What I always regret somewhat is that HERA actually did its measurement on the longitudinal-transverse separation, so to speak at the last minute, in the last two months of running or something like that. This was from my point of view a very big mistake, because the most direct access to the gluon structure function would certainly be to do these separation measurements. Now, of course the separation measurements are much more complicated, that is also well-known. But I think, if one wants to have new instruments, detectors and all that, then right from the very beginning one should very much proceed in such a way that the longitudinal-transverse separation has high priority, right during the construction time. That's probably it. What else can I learn? Well, the shadowing phenomenon, that is the dependence of the cross section on the nuclear number. Deuterons, shadowing on complex nuclei and, in particular, the longitudinal-transverse separation. That is what I, off hand without much thinking about that, would say.

- t'Hooft:

In this connection, would it be helpful to also collide with nuclei which are very deviating from sphericity – these oblong nuclei, ring-shaped nuclei – to collide those instead of the spherical ones like, I presume, lead is. Would that help?

- Schildknecht:

You mean the influence of the shape of certain selected nuclei, changing the shape and then doing electron scattering on them...

- t'Hooft:

Which in practice would mean: just do scattering on different kinds of nuclei

- Schildknecht:

Yes of course, different kinds of nuclei. In some way I would be a bit doubtful whether the selection of specific non-spherical nuclei would give particularly interesting results. There should be a dedicated study of such questions. I mean, one cannot measure everything which is measurable in principle in a finite life time, let me say.

- t'Hooft:

We had surprises many times, so maybe again.

- Schildknecht:

Yes, in principle I agree, of course. A dedicated study may be helpful, but I am not aware of one.

- Korthals-Altes:

You started your lecture by reporting on the ultra-high-energy neutrinos found in the ICECUBE experiment. In the main part of your lecture you elaborated on the photoabsorption cross section of nucleons. How precisely does the photoabsorption cross section give you an answer on neutrino-nucleon scattering at extreme energies?

- Schildknecht:

The connection is very trivial. Actually, I wanted to predict neutrino-nucleon scattering at 10^{15} eV or whatever ultra-high energy. The relevant kinematic range corresponds to very small x in terms of the Bjorken variable x. For small x, there is a well known simple formula that says that the nucleon structure function measured with neutrinos and the nucleon structure function measured with electrons are related by a factor that depends on the number of actively contributing quark flavors. For four flavors, the factor is 5/18. So, if somebody gives you the electron-nucleon scattering cross section at 10^{15} eV, I would not have to do much of a calculation. Now, the highest-energy precise data on deep inelastic scattering are the electron-scattering data from HERA, and that is why you start from that. So, the reason, why I elaborated on the theoretical description of the low-x HERA data is that these data are the only ones that allow us to reliably extrapolate to neutrino-nucleon interactions at ultrahigh energies.

- Zennamo:

You show how the cross-section does not drop off, but continues to rise up. I was wondering how this will affect the search for GZK neutrinos at IceCube?

- Schildknecht:

I think these calculations are important more for the future than just for this moment. At this moment everyone is happy at ICECUBE to find the events and get their energies right from the observed Cherenkov light and so on. However, I certainly believe that knowledge of that cross-section is important, because it's essential, of course, whether the cross-section is twice as large or only half as large because it takes twice less time, ten years instead of twenty, to accumulate a sufficient number of events.

CALT-TH-2014-149

GAUGE THEORIES ON THE COULOMB BRANCH

John H. Schwarz[1]
Walter Burke Institute for Theoretical Physics
California Institute of Technology
Pasadena, CA 91125, USA

Abstract

We construct the world-volume action of a probe D3-brane in $AdS_5 \times S^5$ with N units of flux. It has the field content, symmetries, and dualities of the $U(1)$ factor of $\mathcal{N} = 4$ $U(N+1)$ super Yang–Mills theory, spontaneously broken to $U(N) \times U(1)$ by being on the Coulomb branch, with the massive fields integrated out. This motivates the conjecture that it is the exact effective action, called a *highly effective action* (HEA). We construct an $SL(2, \mathbb{Z})$ multiplet of BPS soliton solutions of the D3-brane theory (the conjectured HEA) and show that they reproduce the electrically charged massive states that have been integrated out as well as magnetic monopoles and dyons. Their charges are uniformly spread on a spherical surface, called *a soliton bubble*, which is interpreted as a phase boundary.

Lectures presented at the Erice 2014 International School of Subnuclear Physics

[1]jhs@theory.caltech.edu

Contents

1 Introduction

Two recent papers [1] [2] explored the construction and properties of certain actions that describe superconformal gauge theories on the Coulomb branch. The examples that were studied are ones with a lot of supersymmetry. The reason for this choice is that the large symmetry gives us more confidence in the plausibility of the conjecture that we will describe.

Even though [1] discussed several examples, this article only describes one of them, namely $\mathcal{N} = 4$ super Yang–Mills (SYM) theory. Our procedure is to begin by deriving the world-volume action of a probe D3-brane in an $AdS_5 \times S^5$ background of type IIB superstring theory. As will be explained, the probe D3-brane action involves various approximations. However, the resulting formula has a number of exact properties, especially symmetries, that make it an attractive candidate for the exact solution to a different problem. Specifically, we conjecture that it is the exact effective action for $\mathcal{N} = 4$ SYM theory on the Coulomb branch, called a *highly effective action* (HEA).

More generally, we consider the world-volume action of a probe p-brane in an $AdS_{p+2} \times M_q$ background. M_q is a q-dimensional compact space with N units of q-form flux, $\int_{M_q} F_q \sim N$. In addition to a D3-brane in $AdS_5 \times S^5$, ref. [1] also discussed

- M2-brane in $AdS_4 \times S^7/\mathbb{Z}_k$
- D2-brane in $AdS_4 \times CP^3$
- M5-brane in $AdS_7 \times S^4$

The first two of these correspond to ABJM theory in three dimensions, whereas the third one corresponds to the $(2,0)$ theory in six dimensions. As our understanding improves, we intend to extend these investigations to theories with less supersymmetry.

Section 2 provides background material in string theory and quantum field theory that is required to follow the rest of the story. People knowledgable in these subjects can skip this section. Section 3 constructs (the bosonic part of) the formula for the probe D3-brane action – the conjectured HEA – and discusses some of its properties. Section 4, based on [2], describes the construction of BPS soliton solutions of this action. We find exactly the spectrum of soliton solutions expected for the HEA. Moreover, they have an interesting structure in which the charge of the soliton resides on a spherical shell. Even though the theory is nongravitational, the solitons turn out to have unexpected analogies with black holes. This analogy suggests that it may be possible to associate an entropy to solitons with a large charge.

2 Background Material

2.1 $\mathcal{N} = 4$ super Yang–Mills theory

When one says that a theory has $\mathcal{N}$ supersymmetries, one means that the conserved supercharges consist of $\mathcal{N}$ irreducible spinors, as appropriate for the spacetime dimension in question. $\mathcal{N} = 4$ supersymmetry is the maximal amount that is possible for an interacting nongravitational quantum field theory in four-dimensional spacetime (4d). The four spinors may be chosen to be two-component complex Weyl spinors. Thus, there are 16 real supercharges. It is believed that the only 4d nongravitational quantum field theories with this amount of supersymmetry are SYM theories. $\mathcal{N} = 4$ SYM exists for any compact gauge group, but we will focus on the case of $U(N)$, so all of the fields are $N \times N$ hermitian matrices.

$\mathcal{N} = 4$ SYM was originally derived by first constructing SYM theory in 10d [3]. It contains a ten-vector gauge field A_M and a Majorana–Weyl spinor Ψ, both in the adjoint representation of the gauge group. The 4d theory was then obtained by dimensional reduction, which simply means dropping the dependence of the fields on six of the (Euclidean) spatial dimensions. This gives a 4d theory containing a four-vector A_μ, six scalars ϕ^I, and four Weyl spinors ψ^A. The 10d SYM theory is not a consistent quantum field theory (by itself), but the 4d SYM theory obtained in this way is a consistent quantum theory. It has an $SU(4) \sim SO(6)$ global symmetry, which corresponds to rotations of the six extra dimensions. Because this symmetry also rotates the four supercharges, Q^A, it is called an R-symmetry.

$\mathcal{N} = 4$ SYM theories are conformally invariant, which implies that they are ultraviolet (UV) finite. This was the first class of UV finite 4d quantum field theories to be discovered, but many more (with less supersymmetry) are now known. Combining all of the spacetime symmetries, which are Lorentz invariance, translation invariance, supersymmetry, scale invariance, conformal invariance, R-symmetry, and conformal supersymmetry, gives the superconformal supergroup called $PSU(2,2|4)$. (A supergroup has both bosonic and fermionic generators.) Its bosonic subgroup is $SU(2,2) \times SU(4)$. The first factor is the 4d conformal group and the second factor is the R-symmetry. Anticommutators of the supersymmetry charges give the momentum operators, which are generators of spacetime translations. Similarly, anticommutators of conformal supersymmetry charges give operators that generate conformal transformations. Mixed anticommutators give the rest of the bosonic generators. Counting both the Poincaré and conformal supersymmetries, there are a total of 32 fermionic generators.

The parameters of $\mathcal{N} = 4$ SYM with gauge group $U(N)$ are a YM coupling constant g, a vacuum angle θ, and the rank of the gauge group N. For large N and fixed 't Hooft parameter

$$\lambda = g^2 N, \tag{1}$$

the theory has a $1/N$ expansion (for large N) with a nice topological interpretation. The leading term in this expansion (the planar approximation) has an additional symmetry called dual conformal invariance. This amount of symmetry is sufficient to make the theory completely integrable. It is not known whether this only applies to the planar approximation or whether it extends to the complete theory. Defining the complex parameter

$$\tau = \frac{\theta}{2\pi} + i\frac{4\pi}{g^2}, \tag{2}$$

the $U(N)$ theory has an $SL(2,\mathbb{Z})$ duality symmetry under which

$$\tau \to \frac{a\tau + b}{c\tau + d}, \tag{3}$$

where a, b, c, d are arbitrary integers satisfying $ad - bc = 1$. The transformation $\tau \to -1/\tau$ is called S duality. When $\theta = 0$, this gives $g \to 4\pi/g$, which allows one to relate strong coupling to weak coupling. S-duality is an exact nonabelian electric-magnetic equivalence.

2.2 Type IIB superstring theory

Type IIB superstring theory is one of the five distinct superstring theories in 10d. The IIB theory has two Majorana–Weyl supersymmetries of the same chirality, for a total of 32 real conserved supercharges. Its massless bosonic fields are

- the 10d metric g_{MN}
- the dilaton σ
- the Neveu-Schwarz (NS-NS) two-form B_{MN}
- Ramond-Ramond (RR) zero-, two-, and four-forms C, C_{MN}, and C_{MNPQ}

The four-form C_4, constructed from C_{MNPQ}, has a self-dual field strength $F_5 = dC_4 + \ldots$.

Type IIB superstring theory has a few solutions that preserve all of the supersymmetry. The most obvious one is 10d Minkowski spacetime, *i.e.*, $g_{MN} = \eta_{MN}$. In this solution σ and C are constants and the other fields vanish. A less obvious maximally supersymmetric solution has a metric describing the geometry $AdS_5 \times S^5$ – we'll give the formula later. In this solution σ and C are again constants. However, now $F_5 \sim N[\mathrm{vol}(AdS_5) + \mathrm{vol}(S^5)]$. N is the number of units of five-form flux threading the five-sphere, $\int_{S^5} F_5 \sim N$, where the coefficients depend on normalization conventions. The isometry of this solution is given by the supergroup $PSU(2,2|4)$, the same supergroup as before! In particular, $SU(2,2)$ is the isometry of AdS_5 and $SU(4)$ is the isometry of S^5.

The value of $\exp(\sigma)$ is the string coupling constant g_s and the value of C is called χ. In terms of these one can form

$$\tau = \chi + i/g_s. \tag{4}$$

Type IIB superstring theory has an exact $SL(2,\mathbb{Z})$ symmetry under which

$$\tau \to \frac{a\tau + b}{c\tau + d}, \tag{5}$$

as before. More generally, the fields σ and C transform in this manner, but we are concerned with the case when they are constants.

2.3 AdS/CFT duality

The preceding sections have been presented so as to make the AdS/CFT conjecture seem obvious. However, when it is was put forward by Maldacena [4], it came as quite a surprise. Specifically, he proposed that $\mathcal{N} = 4$ SYM in 4d with $U(N)$ gauge group is exactly equivalent ("dual") to type IIB superstring theory in an $AdS_5 \times S^5$ background with N units of five-form flux. There are many other analogous AdS/CFT pairs, some of which are relevant to the other examples of probe-brane theories listed in the introduction.

The evidence for AdS/CFT duality that we have presented so far is that both the gauge theory and the string theory have $PSU(2,2|4)$ symmetry and $SL(2,\mathbb{Z})$ duality. However, by now there is much, much more evidence, which we will not describe here. Even though the evidence is overwhelming, a complete proof is not possible. The reason for saying this is that we lack a complete nonperturbative definition of type IIB superstring theory (other than the one given by AdS/CFT duality). The best one can hope to do, in the absence of such a definition, is to show that everything we know about type IIB superstring theory in an $AdS_5 \times S^5$ background with N units of five-form flux agrees with what can be deduced from $\mathcal{N} = 4$ SYM in 4d with $U(N)$ gauge group.

2.4 D3-branes

Superstring theories contain various supersymmetric (and hence stable) extended objects. Ones with p spatial dimensions are called p-branes. They carry a type of conserved current J that couples to a $(p+1)$-form gauge field A with a $(p+2)$-form field strength $F = dA$. Those p-branes on which strings can end are called Dp-branes. (D stands for Dirichlet here, since such strings satisfy Dirichlet boundary conditions in the directions orthogonal to the p-brane. They also satisfy Neumann boundary conditions in the directions parallel to the Dp-brane.)

In type IIB superstring theory supersymmetric Dp-branes exist for $p = 1, 3, 5, 7$. Only the D3-brane is invariant under $SL(2,\mathbb{Z})$ transformations. This theory has an infinite number of different kinds of strings. A (p,q) string arises as a bound state of p fundamental strings and q D-strings, provided p and q are relatively prime. These strings transform irreducibly under $SL(2,\mathbb{Z})$, and any one of them can end on a D3-brane. From the point of view of the world-volume theory of a single D3-brane, which is an abelian gauge theory, the end of the string appears to carry p units of electric charge and q units of magnetic charge.

The various p-branes also act as sources of gravitational and other fields. Maldacena was led to his conjecture by realizing that N coincident D3-branes give a "black brane," which is a higher-dimensional generalization of a black hole, whose near-horizon geometry is $AdS_5 \times S^5$ with N units of 5-form flux on the sphere. In this way the branes are replaced by a 10d geometry with a horizon, and the 10d string theory is represented "holographically" by a 4d quantum field theory.

A stack of coincident flat Dp-branes has fields that are localized on its $(p+1)$-dimensional world volume. They define a "world-volume theory," which is maximally supersymmetric when the background in which they are embedded is maximally supersymmetric. In particular, we will study the world-volume theory of a single D3-brane embedded in an $AdS_5 \times S^5$ background geometry, which may be regarded as having been created by N other coincident D3-branes.

2.5 The Coulomb branch

A stack of N coincident flat D3-branes has a world-volume theory, which is a $U(N)$ gauge theory. Consider starting with N coincident flat D3-branes and pulling them apart along one of the orthogonal axes (let's call it the x direction) into clumps $N_1 + N_2 + \ldots + N_k = N$.

Then, the world-volume gauge symmetry is broken spontaneously,

$$U(N) \to U(N_1) \times U(N_2) \times \ldots \times U(N_k). \tag{6}$$

The world-volume fields that arise as the lowest mode of a fundamental string with one end attached to the ith D3-brane and the other end attached to the jth D3-brane acquire the mass

$$m_{ij} = |x_i - x_j|T, \tag{7}$$

where T is the fundamental string tension. This is similar to the Higgs mechanism. The main difference is that the scalar fields that are eaten by the gauge fields that become massive belong to the adjoint representation of the original gauge group (in contrast to the $SU(2)$ Higgs doublet in the standard model). To emphasize this distinction, this phase of the gauge theory is called a Coulomb branch (rather than a Higgs branch).

Let us consider $\mathcal{N} = 4$, $d = 4$ SYM theory with a $U(N)$ gauge group. For most purposes one can say that a free $U(1)$ multiplet decouples leaving an $SU(N)$ theory. However, this 'decoupled' $U(1)$ is needed to get the full $SL(2, \mathbb{Z})$ duality group rather than a subgroup [5][6]. In the special case of $U(2)$, if we ignore the decoupled $U(1)$, the remaining $SU(2)$ theory on the Coulomb branch has unbroken $U(1)$ gauge symmetry. Let us refer to the massless supermultiplet as the "photon" supermultiplet and the two massive supermultiplets as $W^{\pm}$.

The $U(2)$ $\mathcal{N} = 4$ SYM theory on the Coulomb branch has a famous soliton solution: the 't Hooft–Polyakov monopole. This solution preserves half of the supersymmetry. One says that it is "half BPS." This monopole is part of an infinite $SL(2, \mathbb{Z})$ multiplet of half-BPS (p, q) states, with p units of electric charge and q units of magnetic charges, where p and q are coprime. The masses, determined by supersymmetry, are

$$M_{p,q} = vg|p + q\tau| = vg\sqrt{\left(p + \frac{\theta}{2\pi}q\right)^2 + \left(\frac{4\pi q}{g^2}\right)^2}$$

where $\langle \phi \rangle = v$ is the vev of a massless scalar field. The W mass is $M_{1,0} = gv$ and the monopole mass is $M_{0,1} = 4\pi v/g$ for $\theta = 0$. As discussed above, a (p, q) dyon is introduced when a (p, q) string ends on a D3-brane.

2.6 The probe approximation

Consider a D3-brane embedded in a 10d spacetime. The probe approximation involves neglecting the back reaction of the brane on the geometry and the other background fields. Since the brane is a source for one unit of flux, this requires that the background flux N is large, so that the distinction between N and $N + 1$ becomes negligible. The world-volume action of a single D-brane contains a $U(1)$ field strength, $F_{\alpha\beta} = \partial_\alpha A_\beta - \partial_\beta A_\alpha$. Since, no derivatives of F are included in the formula, F is required to vary sufficiently slowly so that its derivatives can be neglected. A similar restriction applies to the other world-volume fields, as well.

Despite these approximations, the probe D3-brane action has some beautiful exact properties: It precisely realizes the isometry of the $AdS_5 \times S^5$ background geometry as a world-volume superconformal symmetry $PSU(2, 2|4)$. Only the bosonic subgroup is taken into account in the subsequent discussion, since fermi fields are omitted. (A version of the complete formula with fermions is given in [7].) The brane action also has the duality symmetry of the background, which is $SL(2,\mathbb{Z})$ for the D3-brane example. Furthermore, the D3-brane world-volume action is most naturally formulated with local symmetries, which are general coordinate invariance and a fermionic symmetry called kappa symmetry. It is only after implementing a suitable gauge choice that one is left with a conventional nongravitational quantum field theory in Minkowski spacetime.

In principle, one can integrate out the massive fields of the Coulomb branch theory exactly, thereby producing a very complicated formula in terms of the massless photon supermultiplet only. Very schematically,

$$\exp(iS_{\text{HEA}}) = \int DW^+DW^- \exp(iS) \tag{8}$$

If one could do this integral exactly, which is not possible in practice, the resulting action would capture the entire theory on the Coulomb branch, and it would be valid at all energies. This is in contrast with the more common notion of a low-energy effective action, which only includes the leading terms in a derivative expansion. We have proposed to call such an exact Coulomb branch action a *highly effective action* (HEA). Even though we cannot carry out such an exact computation, in some cases we know many of the properties that the HEA should possess. One can hope that they go a long way towards determining it.

3 The Highly Effective Action

3.1 General requirements for an HEA

An HEA should have all of the unbroken and spontaneously broken global symmetries of the original Coulomb branch theory with the massive W fields included. The conformal symmetry is spontaneously broken as a consequence of assigning a vacuum expectation value (vev) to a massless scalar field that has a flat potential. In other words, this vev spontaneously breaks spacetime symmetries (though not Poincaré symmetry) as well as gauge symmetry. An HEA should have the same duality properties as the Coulomb branch theory containing explicit W fields. The global symmetry and duality groups are $PSU(2,2|4)$ and $SL(2,\mathbb{Z})$ in the example considered here.

A further requirement for an HEA is that it should have the same spectrum of supersymmetry protected (or BPS) states as the original Coulomb branch theory. In particular, the $W^\pm$ supermultiplets, which have been integrated out, should reappear in the HEA as solitons. By contrast, this would not be expected for a low-energy effective action. When we examine these soliton solutions explicitly, we will be led to a certain refinement of the interpretation of this requirement. Specifically, a complete understanding of the BPS soliton solutions also requires knowledge of the original conformal branch gauge theory for which the gauge symmetry and conformal symmetry are unbroken.

Ref. [1] conjectured that the probe D3-brane action, in an $AdS_5 \times S^5$ background with one unit of flux ($N = 1$), is precisely the HEA for the $\mathcal{N} = 4$, 4d SYM theory with $U(2)$ gauge symmetry on the Coulomb branch despite the fact that it is only an approximate solution to a different problem. In fact, $N = 1$ is the worst possible case for the probe approximation. However, this shouldn't matter, because we are solving a different problem. Still, it is helpful to keep the 10d description in mind.

I am not certain that this conjecture is correct. On the one hand, the formula that we will obtain seems to be too simple to be the exact answer for such a complicated path integral. On the other hand, we will find that it has all of the expected properties of the HEA. If it not the conjectured HEA, then it would seem to define a new maximally supersymmetric 4d quantum field theory, whose existence is unexpected. So either conclusion is remarkable.

3.2 The AdS Poincaré patch

The AdS_{p+2} geometry with unit radius can be described as a hypersurface embedded in a $(p+3)$-dimensional Lorentzian space of signature $(p+1, 2)$:

$$y \cdot y - u\phi = -1, \tag{9}$$

where $y \cdot y = -(y^0)^2 + \sum_1^p (y^i)^2$. The equation eliminates one of the two time directions leaving a manifold of signature $(p+1, 1)$. The Poincaré-patch metric of radius R is

$$ds^2 = R^2(dy \cdot dy - dud\phi). \tag{10}$$

Note that this preserves the $SO(p+1, 2)$ symmetry of the embedding equation. Defining $x^\mu = y^\mu/\phi$ and eliminating $u = \phi^{-1} + \phi x \cdot x$,

$$ds^2 = R^2(\phi^2 dx \cdot dx + \phi^{-2} dv^2). \tag{11}$$

It is more customary to write the formula in terms of a coordinate $z = \phi^{-1}$, which has dimensions of length. However, the choice ϕ convenient for our purposes, since it will correspond to a scalar field, also called ϕ, with dimensions of inverse length in the D3-brane action.

The Poincaré patch is not geodesically complete. Rather there exists a different coordinate choice that describes a geodesically complete spacetime, called global AdS, for which the Poincaré patch is just a region. The Poincaré patch contains a horizon, which is not present in the global AdS metric. In this respect it is analogous to the horizon of Rindler spacetime, whose geodesic completion is Minkowski spacetime. The Poincaré patch of AdS is sufficient for our purposes.

The ten-dimensional $AdS_5 \times S^5$ metric $ds^2 = g_{MN}(x) dx^M dx^N$ is

$$ds^2 = R^2 \left(\phi^2 dx \cdot dx + \phi^{-2} d\phi^2 + d\Omega_5^2\right) = R^2 \left(\phi^2 dx \cdot dx + \phi^{-2} d\phi \cdot d\phi\right), \tag{12}$$

where $d\Omega_5^2$ is the metric of a round unit-radius five-sphere, and ϕ is now the length of the six-vector ϕ^I. It will be important later that ϕ cannot be negative.

3.3 The D3-brane in $AdS_5 \times S^5$

According to the AdS/CFT dictionary, the value of the radius R, expressed in string units, satisfies

$$R^4 = 4\pi g_s N l_s^4. \tag{13}$$

The positive integer N is the number of units of five-form flux on the five-sphere. Also, $\int_{S^5} F_5 = 2\pi N$ for an appropriate normalization, and N is the rank of the $U(N)$ gauge group of the dual CFT.

The probe D3-brane action is the sum of two terms: $S = S_1 + S_2$. S_1 is a Dirac–Born–Infeld (DBI) functional of the ten embedding functions $x^M(\sigma^\alpha)$ and a world-volume $U(1)$ gauge field $A_\beta(\sigma^\alpha)$ with field strength $F_{\alpha\beta} = \partial_\alpha A_\beta - \partial_\beta A_\alpha$. S_2 is a Chern–Simons-like (CS) term, which is linear in the RR fields. S_1 and S_2 also depend of the fermi fields that we are omitting.

The general formula for the bosonic part of the DBI term is

$$S_1 = -T_{D3} \int \sqrt{-\det\left(G_{\alpha\beta} + 2\pi\alpha' F_{\alpha\beta}\right)}\, d^4\sigma, \tag{14}$$

where $G_{\alpha\beta}$ is the induced 4d world-volume metric

$$G_{\alpha\beta} = g_{MN}(x)\partial_\alpha x^M \partial_\beta x^N. \tag{15}$$

As usual, $\alpha' = l_s^2$ and the D3-brane tension is

$$T_{D3} = \frac{2\pi}{g_s(2\pi l_s)^4}. \tag{16}$$

In fact, only two dimensionless combinations of parameters occur in the brane action:

$$R^4 T_{D3} = \frac{N}{2\pi^2} \quad \text{and} \quad 2\pi\alpha'/R^2 = \sqrt{\pi/g_s N}. \tag{17}$$

Thus, S_1 only depends on the dimensionless string coupling constant g_s and the integer N. We shall see shortly that the g_s dependence also drops out.

The nonvanishing Chern–Simons terms (aside from coefficients, which can be found in [1]) are

$$S_2 \sim \int C_4 + \chi \int F \wedge F. \tag{18}$$

The RR four-form potential C_4 has a self-dual field strength $F_5 = dC_4$. As mentioned previously,

$$F_5 \sim N\left(\text{vol}(S^5) + \text{vol}(AdS_5)\right), \tag{19}$$

and χ is the value of the RR 0-form C_0. It is proportional to the theta angle of the gauge theory.

S_1 contains a potential term $\int \phi^4 d^4x$, which should not appear in S, since there should be no net force acting on the brane when $\phi > 0$. In fact, this term is canceled by the $\int C_4$ term in S_2. To show this we consider the region M of AdS_5 for which the coordinate ϕ is less than the value of ϕ that specifies the position of the D3-brane. Then, using Stokes' theorem,

$$\int_{D3} C_4 = \int_M F_5 \sim \int_0^\phi \int_{D3'} \text{vol}(AdS_5)', \tag{20}$$

where

$$\text{vol}(AdS_5)' \sim (\phi')^3 d\phi' \wedge dx^0 \wedge dx^1 \wedge dx^2 \wedge dx^3 \tag{21}$$

Thus,

$$\int_{D3} C_4 \sim \int \phi^4 d^4x. \tag{22}$$

The coefficients work perfectly.

The general coordinate invariance of the D3-brane action allows one to fix a gauge. A convenient choice for our purposes is the static gauge in which the four world-volume coordinates σ^α are identified with the four Lorentzian coordinates x^μ, which were introduced previously,

$$x^\mu(\sigma) = \delta^\mu_\alpha \sigma^\alpha. \tag{23}$$

In this gauge $\phi^I(\sigma)$ and $A_\mu(\sigma)$ are the only remaining bosonic world-volume fields. Moreover, due to the gauge choice, they become functions of x^μ.

The complete D3-brane action in the static gauge (aside from fermions), expressed in terms of canonically normalized fields $\phi^I(\sigma)$ and $A_\mu(\sigma)$, is

$$S = \frac{1}{\gamma^2} \int \phi^4 \left(1 - \sqrt{-\det M_{\mu\nu}}\right) d^4x + \frac{1}{4} g_s \chi \int F \wedge F, \tag{24}$$

where

$$\gamma = \sqrt{N/2\pi^2} \tag{25}$$

and

$$M_{\mu\nu} = \eta_{\mu\nu} + \gamma \frac{F_{\mu\nu}}{\phi^2} + \gamma^2 \frac{\partial_\mu \phi^I \partial_\nu \phi^I}{\phi^4}, \tag{26}$$

where $\eta_{\mu\nu}$ denotes the Lorentz metric. It is an important fact that the first term in S is independent of g_s and χ and that they only appear (as a product) in the coefficient of the second term.

The scale invariance of the action (24) is manifest, since all terms have dimension four and all parameters are dimensionless. On the Coulomb branch there is a scale, the vev of the scalar field ϕ. However, the full conformal symmetry is realized on the action. Only the choice of vacuum breaks the symmetry (and all choices are equivalent). The D3-brane action contains inverse powers of the scalar field. However, the vev of this field ensures that these are not singular. Individual terms in the action can be arbitrarily complicated and still end up with dimension four by including an appropriate (inverse) power of ϕ.

Regarded as an HEA, (24) should encode all the quantum effects that arise from integrating out the $W^\pm$ supermultiplets. On the other hand, (24) should have its own loop expansion (or path integral) to take account of the quantum effects of the massless supermultiplet. The loop expansion is expected to be free of UV divergences, because of the (spontaneously broken) conformal symmetry. There are IR divergences, but they can be treated by standard methods. Rescaling all fields in (24) by γ brings the action to a form in which all the N dependence appears as an overall factor of N, so that $S(N) = NS(1)$. This shows that the loop expansion of this theory is a $1/N$ expansion.

3.4 S duality

The $SL(2,\mathbb{Z})$ transformation $\tau \to -1/\tau$ duality is accompanied by an electric-magnetic transformation in the S-duality transformation of the gauge theory. This invariance has not been proved in the nonabelian formulation with W fields, but it was proved for the abelian D3-brane action in [1]. Recall that $\tau = \tau_1 + i\tau_2 = \chi + i/g_s$, but that the action only depends on $g_s\chi = \tau_1/\tau_2$. When $\tau \to -1/\tau$, $g_s\chi \to -g_s\chi$. To show invariance under this sign change, we must also examine the transformation that exchanges electric and magnetic fields.

The procedure is standard. It involves replacing the $U(1)$ gauge field A_μ by a new one such that the Bianchi identity

$$\partial_{[\mu} F_{\nu\rho]} = 0 \tag{27}$$

becomes a field equation, and the field equation

$$\partial_\mu \left(\frac{\partial \mathcal{L}}{\partial F_{\mu\nu}} \right) = 0 \tag{28}$$

becomes a Bianchi identity. The nontrivial fact, which was proved in [1], is that the new action, obtained by this field replacement together with the sign change $g_s\chi \to -g_s\chi$, is identical to the original one.

3.5 Summary and discussion

We have conjectured that the world-volume action of a probe D3-brane in an $AdS_5 \times S^5$ background of type IIB superstring theory, with one unit of flux (N=1), can be reinterpreted as the HEA of $U(2)$ $\mathcal{N} = 4$ SYM theory on the Coulomb branch. An explicit formula for the bosonic part of the action has been presented. It is likely that there is a generalization in which the formula with $N > 1$ plays a role in the Coulomb branch decomposition $U(N+1) \to U(N) \times U(1)$. However, when $N > 1$ certain issues still need to be clarified.

The evidence presented so far for the conjecture that the probe D3-brane action is the desired HEA is that the action incorporates all of the required symmetries and dualities: $PSU(2,2|4)$ superconformal symmetry (when fermions are included) and $SL(2,\mathbb{Z})$ duality. The next section will describe BPS soliton solutions of this action. Their properties will give further support to the conjecture. They will also lead us to refine the conjecture somewhat.

4 BPS Soliton Solutions

The previous section described the construction of the bosonic part of a world-volume action for a probe D3-brane in $AdS_5 \times S^5$. We also discussed $\mathcal{N} = 4$ SYM theory, with gauge group $U(N+1)$, on the Coulomb branch. A suitable scalar field vev breaks the gauge symmetry so that

$$U(N+1) \to U(1) \times U(N). \tag{29}$$

The $2N$ supermultiplets corresponding to the broken symmetries acquire mass gv. The probe D3-brane action was conjectured to be the HEA for the $U(1)$ factor.

In this section we will derive classical supersymmetric soliton solutions of the action (24). Even though they are classical, they are supposed to take account of all quantum effects due to the massive fields that have been integrated out. Furthermore, they will turn out be BPS, which implies that they are protected from additional quantum corrections. The soliton solutions will be independent of N (or γ), since N is an overall factor of the action written in terms of rescaled fields. Thus, this parameter drops out of the classical field equations.

For the purpose of constructing the soliton solutions, it is sufficient to consider a D3-brane that is localized at a fixed position on the S^5. In this case, the S^5 coordinates do not contribute to the D3-brane action. The radial position in the AdS_5 space is encoded in the nonnegative scalar field ϕ. The dependence of the action on this field and the $U(1)$ gauge field are all that are required in this section. The fermions are not required, since they vanish for these solutions.

4.1 Solitons

Let us begin by looking for spherically symmetrical static solutions, centered at $r = 0$, for which the action Eq. (24) is stationary. We require that $\vec{E}$ and $\vec{B}$ (the electric and magnetic parts of $F_{\mu\nu}$) only have radial components, denoted E and B. Also, E, B, and ϕ are functions of the radial coordinate r only. It then follows that

$$\begin{aligned} -\det(M_{\mu\nu}) &= -\det\left(\eta_{\mu\nu} + \gamma^2 \frac{\partial_\mu\phi\partial_\nu\phi}{\phi^4} + \gamma\frac{F_{\mu\nu}}{\phi^2}\right) \\ &= \left(1 + \gamma^2\frac{(\phi')^2 - E^2}{\phi^4}\right)\left(1 + \gamma^2\frac{B^2}{\phi^4}\right). \end{aligned} \tag{30}$$

This results in the Lagrangian density

$$\mathcal{L} = \frac{1}{\gamma^2}\phi^4\left(1 - \sqrt{\left(1 + \gamma^2\frac{(\phi')^2 - E^2}{\phi^4}\right)\left(1 + \gamma^2\frac{B^2}{\phi^4}\right)}\right) + g_s\chi BE. \tag{31}$$

The equation of motion for A_0 is $\frac{\partial}{\partial r}(r^2 D) = 0$, where

$$D = \frac{\partial\mathcal{L}}{\partial E} = E\sqrt{\frac{1 + \gamma^2 B^2/\phi^4}{1 + \gamma^2[(\phi')^2 - E^2]/\phi^4}} + g_s\chi B. \tag{32}$$

For a soliton centered at $r = 0$, with p units of electric charge g and q units of magnetic charge g_m, where $g_m = 4\pi/g$, we have

$$D = \frac{pg}{4\pi r^2} \quad \text{and} \quad B = -\frac{qg_m}{4\pi r^2}. \tag{33}$$

The mass of the soliton is given by the Hamiltonian. $H = \int \mathcal{H} d^3x = 4\pi \int \mathcal{H} r^2 dr$. The Hamiltonian density is $\mathcal{H} = DE - \mathcal{L}$. Eliminating E in favor of D gives

$$\mathcal{H} = \frac{1}{\gamma^2}\left(\sqrt{(\phi^4 + \gamma^2(\phi')^2)(\phi^4 + \gamma^2 X^2)} - \phi^4\right), \tag{34}$$

where

$$X = \sqrt{\tilde{D}^2 + B^2} = Q/r^2. \tag{35}$$

and

$$\tilde{D} = D - g_s\chi B. \tag{36}$$

Thus, $\tilde{D}^2 + B^2 = Q^2/r^4$, where

$$Q = \frac{g}{4\pi}|p + q\tau|. \tag{37}$$

As before,

$$\tau = \chi + i/g_s = \frac{\theta}{2\pi} + i\frac{4\pi}{g^2}. \tag{38}$$

We want to find functions $\phi(r)$ that give BPS extrema of H with the boundary condition $\phi \to v$ as $r \to \infty$. The BPS condition turns out to require that the two factors inside the square root in Eq. (34) are equal, which implies that $\mathcal{H} = (\phi')^2$. The proof goes as follows. One first writes the formula for $\mathcal{H}$ in the form

$$(\gamma^2\mathcal{H} + \phi^4)^2 = (\gamma^2 X|\phi'| + \phi^4)^2 + \gamma^2\phi^4(X - |\phi'|)^2. \tag{39}$$

Thus,

$$(\gamma^2\mathcal{H} + \phi^4)^2 \geq (\gamma^2 X|\phi'| + \phi^4)^2, \tag{40}$$

which implies $\mathcal{H} \geq X|\phi'|$. Saturation of the BPS bound is achieved for $|\phi'| = X$ and then

$$\mathcal{H} = X^2 = (\phi')^2 = Q^2/r^4. \tag{41}$$

The equation $(\phi')^2 = Q^2/r^4$, together with the boundary condition $\phi \to v$ as $r \to \infty$, has two BPS solutions

$$\phi_\pm(r) = v \pm Q/r, \tag{42}$$

where $Q = \frac{g}{4\pi}|p + q\tau|$. The ϕ_+ solution is similar to the flat-space case studied by Callan and Maldacena [8]. It describes a funnel-shaped protrusion of the D3-brane extending to the boundary of AdS at $\phi = +\infty$. This solution gives infinite mass (proportional to $\int dr/r^2$), and thus it is not the solution we are seeking.

The ϕ_- solution is different. $\phi = 0$ corresponds to the horizon of the Poincaré patch of AdS_5. Thus, since ϕ is nonnegative, the ϕ_- solution must be cut off at

$$r_0 = \frac{Q}{v}. \tag{43}$$

Then the masses of BPS solitons are given by

$$M = 4\pi \int_{r_0}^{\infty} \mathcal{H} r^2 dr = \frac{4\pi Q^2}{r_0} = 4\pi v^2 r_0 = vg|p + q\tau|, \tag{44}$$

exactly as was expected. We have obtained a complete infinite $SL(2,\mathbb{Z})$ multiplet, which includes the W particles that were integrated out, as well as monopoles and dyons.

The charge of the ϕ_- solution is uniformly spread on the sphere $r = r_0$, which we call a *soliton bubble*. The interior of the bubble should not contribute to the mass of the soliton. So, how should we think about the interior of the bubble in the QFT? The only sensible interpretation is that the gauge theory is in the ground state of the *conformal phase* of $U(N+1)$ inside the sphere. This implies that the bubble is a *phase boundary*. This interpretation has the advantage that the parameter τ is required to describe the $U(N+1)$ theory in the conformal phase. This would explain how the soliton solutions know what the values of g and θ are. The $U(1)$ action does not contain all the required information. One also needs to know the nonabelian theory in the conformal phase.

4.2 Comparison with the BPS 't Hooft–Polyakov monopole

Let us compare the monopole solution of the D3-brane action that we have obtained with the corresponding BPS 't Hooft–Polyakov monopole solution. The single monopole solution of the nonabelian $SU(2)$ gauge theory on the Coulomb branch (for $\theta = 0$) has a triplet of scalar fields ϕ^a whose internal symmetry is aligned with the spatial directions as follows[2]:

$$\phi^a(\vec{x}) = \frac{x^a}{r}\phi(y) \tag{45}$$

with

$$\phi(y) = v(\text{coth}y - 1/y), \tag{46}$$

where $y = M_W r$ and $M_W = gv$. $\phi(y)$ is strictly positive for $y > 0$, and $\phi^a(\vec{x})$ is nonsingular at the origin. Thus, there is no sign of a soliton bubble in the nonabelian description. Both constructions give the correct mass and charge for the monopole, but the D3-brane solution gives a soliton bubble whereas the nonabelian solution does not. So, which formula more accurately describes what is happening?

Equation (46) differs from

$$\phi_-(r) = v(1 - 1/y), \tag{47}$$

the D3-brane theory result for the monopole, by a series of terms of the form $\exp(-2nM_W r)$, where n is a positive integer. In the context of the $\mathcal{N} = 4$ theory, the effect of integrating out the fields of mass M_W should be to cancel these exponential terms for $y > 1$ and to give $\phi = 0$ for $y < 1$. After all, the $U(1)$ HEA is supposed to incorporate all of the contributions due to W loops, and these exponentials are a plausible form for those contributions. Hence, we conclude that the bubble is real and that the D3-brane solution gives a more accurate description of what is happening than the usual classical solution (46) of the nonabelian theory.

4.3 Black hole analogy

We have found a universal formula relating the mass and radius of BPS soliton bubbles: $M = 4\pi v^2 r_0$, which is valid for all (p, q). For comparison, the radius of the horizon of a 4d extremal Reissner–Nordstrom asymptotically Lorentzian black hole in four dimensions is $r_0 = MG$, where G is Newton's constant, for all (p, q). In the latter case the charge Q should be large for the classical analysis to be valid. Thus, the relation between mass and radius is the same in both cases, with $(4\pi v^2)^{-1}$ the analog of Newton's constant. The BPS condition ensures that the analogy extends to the relation between mass and charge.

This analogy is rather surprising, because the D3-brane theory is a nongravitational theory in 4d Minkowski spacetime. If one tries to pursue this analogy, there is a natural question: Does the Bekenstein–Hawking entropy of the black hole, which is proportional to Q^2 (for large Q), have an analog for the solitons of the D3-brane theory? For example, is there an entanglement entropy between the inside and outside of the soliton bubble with

[2]The $\mathcal{N} = 4$ theory has six such triplets, but only one of them is utilized in the construction. This choice corresponds to the choice of a point on the five-sphere in the D3-brane construction.

this value. If the solitons can be shown to have a well-defined entropy of this sort (for large Q), then one may be tempted to take the black-hole analogy seriously.

Even though the D3-brane theory is defined on a 4d Lorentzian spacetime, we know that the field ϕ can be interpreted as a radial coordinate in AdS_5. From this point of view the soliton solutions have a nontrivial geometry induced from their embedding in AdS_5. From the 5d (or 10d) viewpoint, the bubble is on the horizon of the Poincaré patch of AdS_5, where it intersects the boundary of global AdS. This fact may be useful for understanding the origin of the black-hole analogy.

4.4 Multi-soliton solutions

It is easy to derive the generalization of $\phi_-(r)$ to the case of n solitons of equal charge.[3] Since supersymmetry ensures that the forces between them should cancel when they are at rest, their centers can be at arbitrary positions $\vec{x}_k$, $k = 1, 2, \ldots, n$. Since ϕ satisfies Laplace's equation, the solutions can be superposed. The obvious guess, which is easy to verify, is

$$\phi(\vec{x}) = v - Q \sum_{k=1}^{n} \frac{1}{|\vec{x} - \vec{x}_k|}. \tag{48}$$

The surfaces of the bubbles, which are no longer spheres, are given by $\phi(\vec{x}) = 0$. The fields $\vec{D}$ and $\vec{B}$ are then proportional to $\vec{\nabla}\phi$, with coefficients determined by the charges. Their values at $\phi = 0$ determine the charge densities on the bubble surface. This is much simpler than the usual multi-monopole analysis, which involves nonlinear equations! It should also be more accurate.

4.5 Previous related work

Soliton bubbles like those found here have appeared in the literature previously. We will briefly discuss the examples that we are aware of. There may be others.

Using attractor flow equations [9], Denef [10], [11] found similar structures to our soliton bubbles in the context of supergravity solutions in which a D3-brane wraps a cycle of a Calabi–Yau manifold that vanishes at a conifold point, where the central charge modulus is zero.

Gauntlett *et al.* [12] studied soliton solutions of a probe D3-brane in an asymptotically flat black D3-brane supergravity background. This problem is closely related to the one we have considered, since this geometry has $AdS_5 \times S^5$ as its near-horizon limit. They identified half-BPS solutions, like those found here, "in which a point charge is replaced by a perfectly conducting spherical shell." In [13] the authors examined the DBI action of a D3-brane probe in F theory. They constructed a monopole solution containing a soliton bubble that coincides with a 7-brane. The formation of soliton bubbles may also be related to the *enhançon* mechanism in [14]. This mechanism circumvents the appearance of a class of naked singularities, known as *repulsons*.

[3] Pairs of solitons with different charges are mutually nonlocal and therefore difficult to describe.

In [15], Popescu and Shapere studied the low-energy effective action of $\mathcal{N} = 2$ $SU(2)$ gauge theory without additional matter in the Coulomb phase. The Seiberg–Witten *low-energy* effective action [16] should only be valid below the mass of W bosons. However, the theory has a BPS monopole and a BPS dyon that become massless at points in the moduli space of vacua. Therefore, these particles should be obtainable as soliton solutions of the Seiberg–Witten action. Popescu and Shapere constructed these solutions and discovered that they exhibit a spherical shell of charge, just like what we have found.

There is some interesting related evidence for soliton bubbles in a nonsupersymmetric field theory context [17]–[20]. By considering multi-monopole solutions of large magnetic charge in the 4d $SU(2)$ gauge theory with adjoint scalars on the Coulomb branch, Bolognesi deduced the existence of "magnetic bags" with properties that are very close to those of the soliton bubbles obtained here. Bolognesi's magnetic bags do not have sharply defined surfaces, though they become sharp in the limit of large charge. Bolognesi also pointed out the analogy to black holes [19].

5 Conclusion

The action of a probe D3-brane in $AdS_5 \times S^5$ is a candidate for the HEA for a $U(1)$ factor of $\mathcal{N} = 4$ SYM theory on the Coulomb branch. It incorporates all the required symmetries and dualities, and it gives the expected BPS soliton solutions. Even so, it might only be an approximation to the true HEA that is sufficient for computing susy-protected quantities. This would also be noteworthy, since then it would be a candidate for a new interacting UV finite abelian $\mathcal{N} = 4$ quantum field theory in 4d. The existence of such a theory is unexpected. In either case, it is important to settle this question.

There is much more that remains to be explored. We need to understand the extent to which symmetry and other general considerations determine the HEA and whether the world-volume theory of a probe D3-brane should give this HEA. We should construct other analogous p-brane actions and explore their BPS soliton solutions. (BPS soliton solutions of the M5-brane example are discussed briefly in [2].) It should be illuminating to explore tree-approximation scattering amplitudes of the D3-brane theory. It seems likely that they will exhibit beautiful properties, maybe even a (spontaneously broken) Yangian symmetry. Finally, we would like to generalize the analysis to higher-rank gauge theories on the Coulomb branch, which have multiple abelian factors.

Acknowledgment

The author wishes to acknowledges discussions and communications with Hee-Joong Chung, Frank Ferrari, Abhijit Gadde, Jerome Gauntlett, Sergei Gukov, Nicholas Hunter-Jones, Elias Kiritsis, Arthur Lipstein, Nick Manton, Hirosi Ooguri, Jaemo Park, Nati Seiberg, Savdeep Sethi, Yuji Tachikawa, David Tong, and Wenbin Yan.

This work was supported in part by DOE Grant # DE-SC0011632. The author acknowledges the hospitality of the Aspen Center for Physics, where he began working on this project in the summer of 2013, and where he wrote this manuscript in the summer of 2014. The ACP is supported by the National Science Foundation Grant No. PHY-1066293.

References

[1] J. H. Schwarz, "Highly effective actions," JHEP **1401**, 088 (2014) [arXiv:1311.0305 [hep-th]].

[2] J. H. Schwarz, "BPS soliton solutions of a D3-brane action," arXiv:1405.7444 [hep-th].

[3] L. Brink, J. H. Schwarz and J. Scherk, "Supersymmetric Yang-Mills Theories," Nucl. Phys. B **121**, 77 (1977).

[4] J. M. Maldacena, "The large-N limit of superconformal field theories and supergravity," Adv. Theor. Math. Phys. **2**, 231 (1998) [hep-th/9711200].

[5] D. Belov and G. W. Moore, "Conformal blocks for AdS(5) singletons," hep-th/0412167.

[6] O. Aharony, N. Seiberg and Y. Tachikawa, "Reading between the lines of four-dimensional gauge theories," JHEP **1308**, 115 (2013) [arXiv:1305.0318 [hep-th]].

[7] R. R. Metsaev and A. A. Tseytlin, "Supersymmetric D3-brane action in AdS(5) x S**5," Phys. Lett. B **436**, 281 (1998) [hep-th/9806095].

[8] C. G. Callan and J. M. Maldacena, "Brane dynamics from the Born-Infeld action," Nucl. Phys. B **513**, 198 (1998) [hep-th/9708147].

[9] S. Ferrara, R. Kallosh and A. Strominger, "$N = 2$ extremal black holes," Phys. Rev. D **52**, 5412 (1995) [hep-th/9508072].

[10] F. Denef, "Attractors at weak gravity," Nucl. Phys. B **547**, 201 (1999) [hep-th/9812049].

[11] F. Denef, "Supergravity flows and D-brane stability," JHEP **0008**, 050 (2000) [hep-th/0005049].

[12] J. P. Gauntlett, C. Kohl, D. Mateos, P. K. Townsend and M. Zamaklar, "Finite energy Dirac-Born-Infeld monopoles and string junctions," Phys. Rev. D **60**, 045004 (1999) [hep-th/9903156].

[13] R. de Mello Koch, A. Paulin-Campbell and J. P. Rodrigues, "Monopole dynamics in $N = 2$ super Yang-Mills theory from a three-brane probe," Nucl. Phys. B **559**, 143 (1999) [hep-th/9903207].

[14] C. V. Johnson, A. W. Peet and J. Polchinski, "Gauge theory and the excision of repulson singularities," Phys. Rev. D **61**, 086001 (2000) [hep-th/9911161].

[15] I. A. Popescu and A. D. Shapere, "BPS equations, BPS states, and central charge of $N = 2$ supersymmetric gauge theories," JHEP **0210**, 033 (2002) [hep-th/0102169].

[16] N. Seiberg and E. Witten, "Electric-magnetic duality, monopole condensation, and confinement in $N = 2$ supersymmetric Yang-Mills theory," Nucl. Phys. B **426**, 19 (1994) [Erratum-ibid. B **430**, 485 (1994)] [hep-th/9407087].

[17] S. Bolognesi, "Multi-monopoles, magnetic bags, bions and the monopole cosmological problem," Nucl. Phys. B **752**, 93 (2006) [hep-th/0512133].

[18] K. -M. Lee and E. J. Weinberg, "BPS magnetic monopole bags," Phys. Rev. D **79**, 025013 (2009) [arXiv:0810.4962 [hep-th]].

[19] S. Bolognesi, "Magnetic bags and black holes," Nucl. Phys. B **845**, 324 (2011) [arXiv:1005.4642 [hep-th]].

[20] N. S. Manton, "Monopole planets and galaxies," Phys. Rev. D **85**, 045022 (2012) [arXiv:1111.2934 [hep-th]].

LECTURER: J. SCHWARZ

Gauge Theories on the Coulomb Branch

Scientific Secretaries: I. Rudenok, I. Ivanchenko

DISCUSSION

- *Hughes:*

You described how in the theory with W± integrated out we still find them as solitons. How generic is this feature when we "integrate something out"? For example, if we look at type II SUGRA we find string solutions as solitons. Where else may this occur?

- *Schwarz:*

I'm not sure how generic it is. If you start with the non-abelian formulation, so you keep the W fields, the monopole appears as a non-perturbative soliton solution on the one hand. On the other hand, because of the SL(2,Z) symmetry if you have a monopole soliton you would necessarily get a W-soliton by doing an SL(2,Z)-transformation of this solution. Mostly when people talk about an effective action they don't mean the exact effective action, but usually some low-energy approximation, or something like that, and then it is not clear that massive fields that have been integrated out should reappear as solitons.

- *Gould :*

Can we see the Φ-field as a coordinate in the patch outside the soliton? And could one change "coordinates" to describe the inside of the soliton?

- *Schwarz:*

The total energy and mass of the soliton is given by integrating the energy density outside the sphere. The interior of the sphere, where Φ vanishes, doesn't contribute anything to the energy. I have proposed that the inside of the sphere is in the conformal phase. I don't think there is any change of coordinates that will elucidate that choice. The analogy with black holes is surprising, and I'm not sure that it is serious, but I want to explore how far it can be pursued before it breaks down.

- *Gavrylenko:*

Is it possible to add the fermionic fields and rewrite an effective action in the obviously supersymmetric form?

- *Schwarz:*

I haven't done it yet, but I don't see any obstacle in principle. It's technically somewhat challenging to do that, but I think it should work. A formula containing fermionic fields was proposed a long time ago by Matseev and Tseytlin, but it is not in the most convenient form for my purposes.

- *Schrock:*

Today string theory serves as a great toolbox for calculations within Maldacena's AdS/CFT correspondence. But what is the situation of string theory as the fundamental theory of everything?

- *Schwarz:*

There are many possible goals that you can have for string theory and one of them is to understand elementary particle physics. I think there is a good chance that somewhere in the space of string theory solutions there is one that is realistic, but I'm not optimistic that we are going to find it any time soon. What I'm hoping for is the discovery of supersymmetry leading to a new Standard Model that supersymmetric. That will take us some distance further in enclosing this gap of 16 orders of magnitude towards the Planck scale. And then maybe we will have a better target to aim for in top-down string phenomenology. This is a program that is likely to take a large part of the century.

- *'t Hooft :*

Some remarks from me. Well, in the 1980's, string theorists had quite different opinions. I was looked at as a big spoil-sport when I expressed some doubts.

- *Schwarz:*

What happened in 1984 is that a large part of the theoretical physics community made a phase transition from being too pessimistic to becoming too optimistic about string theory without stopping in the middle. I think that now almost all practitioners have a more realistic appraisal of the situation.

- *'t Hooft :*

Well, I also changed in the other direction. I do like all the connections with mathematical physics.

- *Schwarz:*

The field that has been most influenced by string theory so far is mathematics.

- *Korthals-Altes:*

This is not a question but a remark. Could you show this formula that you have of your multisoliton? There are 4-vectors in the formula and you have a square as for multi-instantons. This form is due to 't Hooft.

- *Schwarz:*

Let me answer by showing you the formula for self-dual strings in the (2,0) theory. These 4-vectors are meant to describe string positions in 6-dimensions. There is an analogous black-string solution. In fact, if you take multiple strings to be coincident, you will have formula like that. There may be instanton-like solutions of the 4-dimensional theory, which is what I think you are asking about, but I haven't explored them yet.

ITP-UU-14/26
SPIN-14/20

Canonical Methods in Deterministic Quantum Mechanics

Gerard 't Hooft

Institute for Theoretical Physics
Utrecht University
and

Spinoza Institute
Postbox 80.195
3508 TD Utrecht, the Netherlands
e-mail: g.thooft@uu.nl
internet: http://www.staff.science.uu.nl/~hooft101/

Erice Lectures June 2014

Version October 29, 2014

1 Lecture I. Deterministic Quantum Mechanics

1.1 Motivations for searching deterministic versions of Quantum Mechanics

These lectures are based on work that recently appeared in the ArXive[?], and we plan to publish in book form.

Deterministic quantum mechanics is a conventional theory of quantum mechanics, but it is described in a very special way, such that it appears to be equivalent to an entirely deterministic system. The educated reader will react immediately with some very basic questions: *How is this possible? Does this not contradict fundamental theorems about quantum mechanics?*, and subsequently: *Why should one try to construct such a quantum theory?*

The first question will be addressed shortly. We begin with the last question. This School is about Subnuclear Physics and our attempts to find physically coherent descriptions of the laws of physics that reign in the subnuclear world. The numerous problems we encounter in our attempts to go beyond the TeV scale might ask for new principles. We do not advocate to abandon principles that were found to be useful, but to see if we can add one that restricts the form our quantum mechanical models can take. It remains to be seen whether our proposal will work, but it seems to lead to potentially interesting approaches to physics at the Planck scale. A better understanding of the physical degrees of freedom *there*, and the way they propagate, is dearly needed.

Besides, there is an ongoing discussion concerning our understanding of what a quantum mechanical calculation actually stands for, or, how quantum mechanics relates to 'reality' [?]—[?]. There exist various schemes for the 'interpretation' of quantum mechanics, but none of these is very appealing to all quantum theorists[?]. Our approach leads to an interpretation that may prove to be pragmatic and useful.

Apart from being a useful tool for model building procedures at the Planck scale, one might suspect that deterministic quantum mechanics will lead to very powerful tools in computer simulations. Our present simulation procedures for QCD, for example, require many time steps in a four-dimensional Euclidean space, where fermionic contributions require non-local interactions. This is very expensive in terms of petaflops; the cellular automaton approach would require a 3 dimensional Euclidean space with only local interactions, where we directly would work with a Minkowski time variable.

As yet, however, we are not even close to such practical applications. These motivations might refer to the distant future. One immediate return is the interesting mathematical physics that is involved. A proper treatment of a theory of this sort requires a systematic, 'canonical' procedure, which we wish to formulate mathematically in such a way that it has a wide area of applications.

1.2 The Copenhagen Interpretation

Somewhat anachronistically, we now use the Dirac 'bra-ket' notation to formulate some of the fundamental principles of quantum mechanics as they were established by the Copenhagen school at the beginning of the 20$^{\rm th}$ century:

- We have a Hilbert space of bra-states $\langle\varphi|$ and ket-states $|\psi\rangle$, where an inner product $\langle\varphi|\psi\rangle$ is defined. This means that
- we have superpositions: $|\psi'\rangle = \alpha|\psi\rangle + \beta|\varphi\rangle$, with the usual rules,
- there is a mapping between the bras and the kets, as always indicated by using the same symbol(s) to describe them. It obeys

$$|\chi\rangle = \alpha|\varphi\rangle + \beta|\psi\rangle \;\rightarrow\; \langle\chi| = \alpha^*\langle\varphi| + \beta^*\langle\psi|\ , \tag{1.1}$$

  and
- a norm $\|\psi\|$ is defined by $\|\psi\|^2 = \langle\psi|\psi\rangle$. If $\|\psi\| = 0$ then $|\psi\rangle = 0$.
- At $t = 0$ one may postulate an initial state $|\psi(0)\rangle$, with $\|\psi\| = 1$, and we have a *Schrödinger equation,*

$$\frac{\mathrm{d}}{\mathrm{d}t}|\psi(t)\rangle = -iH|\psi(t)\rangle \tag{1.2}$$

  H is the Hamilton operator (Hamiltonian)
- Consider at $t = T$ a possible final state $|\varphi(T)\rangle$ with $\|\varphi\| = 1$
- The *probability* that the system will end in that mode will be

$$P = \|\,\langle\varphi(T)|\psi(T)\rangle\,\|^2 \tag{1.3}$$

Observables $\mathcal{O}$ are *operators* such as $\hat{\vec{x}}$ and $\hat{\vec{p}}$, but their form depends on the *basis* chosen, for instance:

· the $\vec{x}$ basis, states $\{|\vec{x}\rangle\}$, operators $\hat{\vec{x}} = \vec{x}\ ,\quad \hat{\vec{p}} = -i\frac{\partial}{\partial\vec{x}}$, or

· the $\vec{p}$ basis, states $\{|\vec{p}\rangle\}$, operators $\hat{\vec{x}} = i\frac{\partial}{\partial\vec{p}}\ ,\quad \hat{\vec{p}} = \vec{p}$.

· The *expectation value* of $\mathcal{O}$ is $\langle\mathcal{O}\rangle = \langle\psi|\mathcal{O}|\psi\rangle$

The outcomes of calculations do not depend on the basis chosen, just as in an ordinary vector space, where physical phenomena do not depend on the coordinate frame chosen.

We mention these rules explicitly because they will still be exactly valid in deterministic quantum mechanics (with one exception, to be explained shortly).

1.3 Deterministic quantum mechanics

The evolution operator $U(T)$ is defined to be

$$U(T) = e^{-iHt} \,. \tag{1.4}$$

In deterministic quantum mechanics, it is assumed that a very special orthonormal set of basis elements can be found such that, for a sufficiently fine mesh of time steps T, the evolution operators are pure permutation operators $P(T)$, that is, they have a *representation* in the form of a set of matrices[1], each containing one 1 in each row and each column, and 0's in all other positions. This basis will be called an *ontological basis.* Can such a basis exist in the real world?

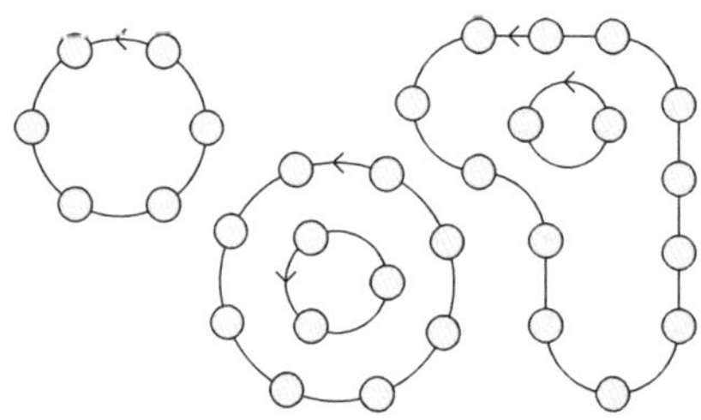

Figure 1: The generic permutator in a finite number of dimensions (here chosen to be 31).

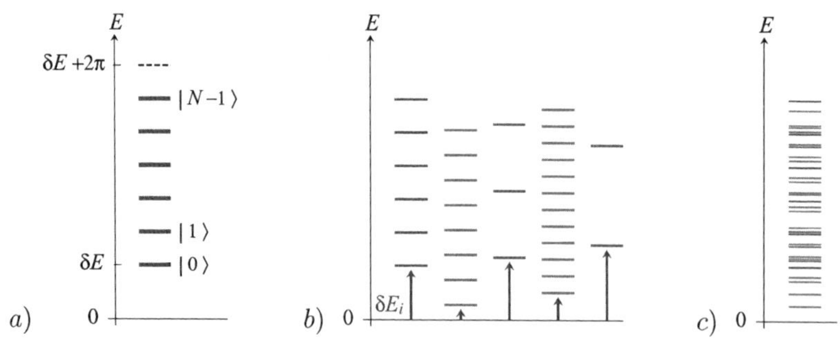

Figure 2: *a*) Energy spectrum of a single 6 component cogwheel; *b*) Energy spectrum of the composite model of Fig ??; *c*) Energy levels combined.

To illustrate the question, consider a general permutation operator in a finite number of dimensions, as illustrated in Figure ??. This example has only 31 states, but we see

[1]The idea of representing a permutation operator in terms of a matrix, may be regarded as the very core of our theory of the interpretation of quantum mechanics; the ultimate law of nature is just the permutation operator itself, while today's quantum mechanics is the science of studying the associated matrices, and the vector space – Hilbert space – on which they act.

that there may be several subunits, invariant classes that transform into themselves. Each class has a simple energy spectrum, see Fig. ??a. If there are N states in a class, then $U(\delta t)^N = 1$, so that the Hamiltonian H in Eq. (??) takes the values $E_n = 2\pi n/(N\,\delta t)$, where n is integer, as can be easily seen. In each of these classes i, we are free to add a constant δE_i to the Hamiltonian, see Fig. ??b. The result may be a fairly complex spectrum, Fig. ??c.

Under a basis transformation, the only true invariant is the spectrum. It is the spectrum that defines all dynamical features of a theory, and from that we conclude that, in principle, the set of models that allow for a basis transformation to an ontological model such as the one displayed in Fig. ?? may be fairly large; it will not always be easy to prove that, given the spectrum, an ontological model is impossible.

Examples of quantum models that do allow for an ontological basis are diverse, see Ref. [?]:

- Atoms with magnetic moment m in a weak homogeneous magnetic field $\vec{B}$ show *Zeeman splitting*; the spectral lines are separated at equal distances:

$$H = \mu\,\vec{B}\cdot\vec{J} = m_j\,\mu\,B\ , \qquad m_j = -j, \cdots, +j\ . \tag{1.5}$$

  This is a spectrum of equally spaced energy levels. By a finite Fourier transform it transforms to a cogwheel with $2j+1$ teeth.

- The harmonic oscillator. It is just the limit of the previous example when $j \to \infty$. This limit has to be taken with some care. It is described in my paper[?].

- Massless non-interacting "neutrinos" (see last year's Erice lecture).

- A massless, non-interacting scalar field in 1+1 dimensions.

Now all these examples are rather trivial; they are integrable systems. Are there examples of non-integrable systems that allow for an ontological basis? Of course, we find examples by starting with non-integrable ontological models.

Our prime example is the *classical, generic, cellular automaton.* A cellular automaton is a D-dimensional array of cells, each containing a few bits or bytes of information, evolving in time in accordance with some local evolution law. We limit ourselves to *time reversible* evolution laws.

Our procedure is: take the evolution laws of such a classical system and re-write it in operator form. Now perform any basis transformation one likes, for instance by going to the energy eigen states. In this basis, local operators take the form of creation and annihilation operators. They form fields, just as in standard quantum field theories. When we do this, however, several difficulties show up:

The theory one finds when starting with 'just any' cellular automaton will not show any rotation invariance or Lorentz invariance. Even, we won't have Galilean invariance, and this means that we can describe static particles, and a few particle types that may be moving with fixed velocities, but not particles that can move with any velocity. A second difficulty is the sign of the Hamiltonian that emerges from this theory: often, one finds it to be either non-local or unbounded from below.

What we really would like to construct is a theory where the Hamiltonian H obeys:

$$H = \int \mathrm{d}^3\vec{x}\,\mathcal{H}(\vec{x}) \ , \qquad \mathcal{H}(\vec{x}) > \mathcal{H}_0 \ , \qquad [\mathcal{H}(\vec{x}), \mathcal{H}(\vec{y})] = 0 \quad \text{if} \quad \vec{x} \neq \vec{y} \ , \tag{1.6}$$

where $\mathcal{H}_0$ is some finite (positive or negative) number, guaranteeing, in a finite volume, the existence of a number H_0 so that $H \geq H_0$.

We found that the equations in (??) are difficult to reconcile with one another, but they do seem to hold *nearly*. How serious are these difficulties? Not only did we not reproduce all symmetries observed in the physical world, the cellular automaton usually also shows no traces of general coordinate invariance. Now the energy density operator $\mathcal{H}(\vec{x})$ is also the generator of local time diffeomorphisms, typical for generally covariant theories. Therefore, one might suspect that our problem with Eqs. (??) will be strongly affected by quantum gravity effects. It may well be that our problems will only be solved *after* we include gravity. This will have to be left as an exercise for the future.[2]

In the mean time, one can decide to ignore what seem to be just technical difficulties, and merely assume that, at the deepest levels, nature is described by some cellular automaton model – or some delicate extension of such a concept.

1.4 The cellular automaton interpretation

As we encountered a number of interesting systems that allow for both a classical and a quantum mechanical description, we were tempted to formulate a conceivable interpretation of quantum mechanics based on the assumption that the real world may be just such a system. Today, a majority of researchers categorically reject such a possibility. However, the mere assumption leads to an extremely clear picture of what might be the real meaning of quantum mechanics. The protests of my colleagues are taken very seriously, but we still claim that it is extremely worth-while to consider this route, while the counter arguments[?][?] can be answered, as we intend to show, see also [?].

Thus, we now *assume* that an ontological basis can be found for the complete Standard Model including the gravitational force. The basis elements are "ontological states", called $|\mathrm{ont}\rangle_i$. Often, we won't know *which* ontological state we are in. We can then describe the state we are in in terms[3] of a *"template state"*,

$$|\varphi\rangle \equiv \sum_i \lambda_i |\mathrm{ont}\rangle_i \ . \tag{1.7}$$

Here, the coefficients λ_i describe probabilities:

$$P_i = |\lambda_i|^2 \ , \qquad \sum_i |\lambda_i|^2 = 1 \ . \tag{1.8}$$

[2]But progress is made on this front, as we plan to report; there it seems that quantum gravity, as yet, may not have to be involved.

[3]Earlier, we called this state a *physical state*, but this was found to be somewhat confusing.

How to treat the final state will be postponed for a moment.

It is now important to note that our ontological states are perfectly standard wave functions that obey the Schrödinger equation. Moreover, the *wave function of the universe* will be assumed to be a *single* ontological state (*never* a superposition of different ontological states).

A *classical state* is typically encountered after a measurement has been made, at which we see the dial of a measuring apparatus, or we see a cat being either dead or alive. Quite generally, one can imagine quantum phenomena at $t = t_1$ having large scale classical consequences at some later time $t = t_2 \gg t_1$.

And what exactly is a classical state? Think of a planet moving in its orbit. One can now imagine that, inside the planet, myriads of our dynamical ontological variables are statistically arranged in a way that may differ very slightly from the statistical distribution they would be expected to have in the pure vacuum. This could then be said of every mm^3 of the insides of the planet, and combining these pieces of information over the entire interior of the planet, will turn the probabilistic information that we have for each mm^3 into a certainty: we have here a planet, not the vacuum. But this observation suffices to argue that the motion of a planet can be covered by studying the ontological state only. This then must hold for all classical observations, and thus we conclude: *classical observables are diagonal in the ontological basis.* A classical state is ontological.

As argued above, it is hard to avoid this conclusion[4], but it is extremely important, and we can introduce it as a new axiom. Classical states are ontological. And now look at the 'template states' we introduced earlier. The λ_i coefficients describe the probability that we have particular ontological states. Therefore, the same λ_i describe the probabilities for the outcomes of experiments. These are now the Born probabilities.

In all respects, we follow the Copenhagen rules for quantum mechanics very strictly. However, the rule that $\langle\psi|\mathcal{O}|\psi\rangle = \langle\mathcal{O}\rangle$ is the expectation value for an observable $\mathcal{O}$, only holds if

- $|\psi\rangle$ is the wave function of the universe (that is, an ontological state), or
- a *template* state, that is, an Ansatz for the state of the universe.
- It does *not* hold if $|\psi\rangle$ is a state defined to be an arbitrary superposition of ontological states, such that $|\lambda_i|^2$ are the probabilities. This would only be so if $\mathcal{O}$ would be an ontological observable (diagonal in the ontological basis). The point is that other operators $\mathcal{O}$ are not truly observable, and the wave function of the universe is never actually a superposition.
- For the same reason, we cannot use arbitrary template states both for the initial state $|\psi\rangle$ and the final state $\langle\psi|$. Only by carefully following the Copenhagen rules, taking the restrictions mentioned in mind, one will unambiguously be lead to the correct expressions.

[4]Indeed, quantum physicists hearing me say this for the first time, tend to frown: it is quite contrary to what they would have expected. I invite them to think again.

We deliberately brought about only one significant modification of the Copenhagen doctrine: *It is not forbidden to ask: what is* really *going on during a typically quantum mechanical process?* And, we claim to know the possible set of wave functions the universe can be in. Superpositions of such wave functions are forbidden, even if we perform superpositions frequently just to enable us to do calculations (just because it is so convenient to diagonalise operators that we encounter frequently).

From the simple assumption that an ontological basis exists, it is easy to deduce that:

- The Bell inequalities are violated (since we still have genuine quantum mechanics);
- The wave function collapses automatically after a measurement[?]. This is because, in the real world, we always are in a single ontological state, and therefore all observed phenomena end in single classical states, not in superpositions.
- The Born rule for the probabilities applies. This happens when we describe the initial state as a 'physical state', as defined earlier. We begin with probabilities $|\lambda_i|^2$ and we end with them (the present argument being more accurate than in Ref.[?]).

The quantum models for the real world that are in use today, describe only the template states; we are unable to identify the ontological states rigorously at present. This may be caused by the fact that the transition from the Planck scale to the Standard Model scale is a renormalization group transformation across more than 20 orders of magnitude, which thoroughly mixes the observable operators.

This is why the real world is an extremely complicated system. In contrast, most models that we have are simple. They obey locality.

We have *superdeterminism*: not only the observed phenomena, but also the measurement processes and the observers themselves are described by deterministic laws. We repeat:

- The universe will refuse ever to go into a superposition of ontological states.
- There is only one world (rather than a 'many world' scenario),
- and there is no need for an ontological 'pilot wave' (as in the DeBroglie-Bohm theory).

This theory is primarily intended to describe the world at the Planck scale, and may be crucial for model building purposes. It is suspected that the gravitational force will play an important role in our understanding of quantum mechanics, but at present we are unsure about this. In any case, the construction of explicit, realistic models is difficult; we often experience problems with locality and positivity of the Hamiltonian, but this is also the case in many other theories of Planck scale physics.

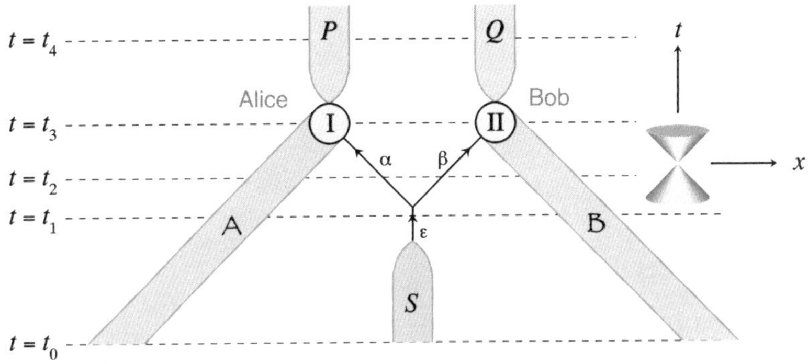

Figure 3: Space-time schematic of a typical Bell-ERP experiment. Coloured bands are classical configurations, single lines are single quantum states. At $t = t_0$, at S, a metastable atom ε is prepared, decaying into entangled objects (typically photons) α and β at time $t = t_1$. Alice at I, and Bob at II do measurements $(t = t_3)$. Their settings are decided upon by looking at distant quasars A and B. The measurements become classically available at P and Q.

1.5 The Bell - CHSH inequality

There is an important reason why theories of the kind described above are often greeted with great scepticism. Quantum mechanics appears to be *fundamentally* different from classical systems. Interference appears to be a really observable phenomenon, which indicates that quantum superpositions *are* possible. Numerous experiments have shown this[?], and it seems that the theory described above denies it.

However, from the very start, our theory is claimed to reproduce quantum mechanics exactly, and so, this should include all interference phenomena, including even all those phenomena that are usually associated with "entanglement", the fact that two or more, possibly quite distant, particles may occur in a quite delicate, combined quantum state. Who is right?

To isolate the problem as sharply as possible, J.S. Bell[?] has rephrased the famous Einstein-Rosen-Podolsky Gedanken experiment, to derive an important theorem:

> *This experiment cannot be mimicked by a model based on only classical observables and classical dynamical laws.*

Although in practice experiments are easiest done with photons (spin 1), the mathematical statement can more conveniently be described as if the entangled particles were electrons with spin $\frac{1}{2}$ Translation from photons to electrons and back is straightforward; in both cases, the two entangled particles each have two states to choose from, to be indicated with an arrow up or down. The atom, ε in Fig. ??, can emit two particles in the total spin $=$ 0 configuration:

$$|\psi\rangle = \tfrac{1}{\sqrt{2}}(|\uparrow\downarrow\rangle - |\downarrow\uparrow\rangle) \tag{1.9}$$

(where photons would have a plus sign). Suppose Alice measures the spin in the direction a, finding $A = 1$ if the spin is parallel to a, and $A = -1$ otherwise. Bob measures the spin of his particle in the direction b, finding $B = 1$ if the spin is parallel to b and $B = -1$ otherwise. A standard quantum calculation gives as a result that, if the state is described by Eq. (??), the correlation between Alice's and Bob's findings will be

$$\langle A\,B\rangle = -\cos(a-b)\ . \tag{1.10}$$

But, according to Bell, this cannot be true if the information were passed on in a classical way. Take the four cases

$$\begin{aligned}(a,b) &= (0^\circ,-45^\circ)\,,\ (90^\circ,-45^\circ)\,,\ (0^\circ,45^\circ)\,,\ (90^\circ,45^\circ)\\ \langle A\,B\rangle &= \quad -\tfrac{1}{2}\sqrt{2}\,,\qquad \tfrac{1}{2}\sqrt{2}\,,\qquad -\tfrac{1}{2}\sqrt{2}\,,\qquad -\tfrac{1}{2}\sqrt{2}\ .\end{aligned} \tag{1.11}$$

Now, suppose that *classical information had travelled together with the particles*, to decide whether A and B are $+1$ or -1, saying $A = A_1$ or A_2 if $a = 0^\circ$ or 90°, while $B = B_1$ or B_2 if $b = \pm 45^\circ$, then, regardless the signs of $A_{1,2}$ and $B_{1,2}$, their averages will all be between -1 and 1. Take the combination $S = \langle A_1B_1\rangle - \langle A_2B_2\rangle + \langle A_1B_2\rangle + \langle A_2B_2\rangle$. Then one easily reads off that

$$-2 \le S \le 2\ , \tag{1.12}$$

while Eq. (??) gives $S = -2\sqrt{2}$. Eq. (??) is the CHSH inequality[?]. Bell concluded that, therefore, a classical theory describing the information flowing from the atom to Alice and Bob, is impossible.

1.5.1 Superdeterminism and correlations

There were two assumptions that went into the tail of the argument. First, it was assumed that both Alice and Bob can decide independently which angle they will set for their polarisation filters a and b, and that they had *free will.* It is obvious, however, that in the completely deterministic models described above, strictly speaking, there is no such free will. Whether Alice will choose a polarisation angle a and Bob will choose b was fixed by laws of nature in their past. The notion that 'free will' is entirely absent, *and that this has to be taken into consideration*, will be referred to as 'superdeterminism'.

Yet, bringing up superdeterminism to overrule Bell's theorem, is often categorically dismissed as a possible resolution of our apparent paradox. Alice and Bob could have based their decisions on incredibly complex algorithms or processes, or they could have used fluctuations of light coming from distant quasars $\mathcal{A}$ and $\mathcal{B}$ (see Fig. ??). How can it be that a decaying atom depends on things taking place outside its light cone? Intuitively, one argues that this should be impossible.

Yet this would be too hasty a conclusion. The point is that there could be, and there are, *correlations* in the physical states that control fluctuating atoms, as well as Alice's and Bob's decisions. Correlations across space-like separated observables *do* occur, not only in quantum systems but also classically. In fact, correlations can be quite strong.

In quantum field theories, space-like correlations in vacuum fluctuations are routinely considered in Euclidean space. These are the N-point functions, and their magnitudes are found to correspond to the scattering matrix elements when analytically continued to Minkowski space. One can calculate the correlations between the polarisation angles a and b, and the angles c of the entangled particles α and β, that would minimally be needed to reproduce the CHSH violating correlation functions (??).

We constructed a simple model, assuming that, if $|c-a|<90^\circ$, then $A=1$ while $A=-1$ otherwise, and $B=-1$ if $|c-b|<90^\circ$ while $B=1$ otherwise. Suppose that c is correlated with a and b as follows:

$$W(c|a,b)=W(x)\ , \quad x=c-\tfrac{1}{2}(a+b)\ , \tag{1.13}$$

We found that the correlation $W(x)$ needed to recuperate the quantum correlations (??) is

$$W(x)=\tfrac{1}{2}|\sin(2x)|\ ; \qquad \int_0^\pi W(x)\mathrm{d}x=1\ . \tag{1.14}$$

At first sight, it looks very strange that such correlations can exist. How could the decaying atom, c, be sensitive to the polarisations a and b chosen by Alice and Bob? Note, that we do have a true 3-point correlation: if we average over all values of a, *or* all values of b, *or* all values of c the distribution W becomes flat.

We know that we are looking at a template state that is very close to the vacuum state, the lowest energy eigen state of the Hamiltonian. In this state, the ontological states must be represented very unevenly. Model calculations show that such uneven distributions can easily emerge. Next, it is crucial to realise that, in deterministic theories, the ontological nature of a state is *preserved in time*: the universe is expected to hop from ontological state to ontological state, through its permutation operators P. If Alice or Bob want to change the settings a and/or b, they have to find an ontological state with these new settings. This new ontological state might have less, or more, probability than the old one. This can be the source of the non-trivial correlations (??).

Most importantly, the question as to what the outcome of a measurement would have been if the setting were changed, assuming that our classical system also propagates this kind of information, is called *counterfactual realism.* Since, in our models, the universe is assumed to exist in only one ontological state, we do not have counterfactual realism.

2 Lecture II, Canonical Methods

Nearly a century of quantum mechanics has given us a set of extremely powerful tools for calculations. We have learned how to relate canonical classical mechanical systems to associated quantum systems, by turning Poisson brackets into commutators. Then we learned to employ various classes of operators, coordinates, momentum operators, angular momentum, harmonic degrees of freedom, creation- and annihilation operators, quantised field operators, and so on.

Without any modifications, much of this can be used for studying deterministic theories that turned into quantum systems. Deterministic theories , however, appear to depend less frequently on real number operators and more heavily on discrete sets. In particular, we would like to see "canonical methods" with completely discrete sets at their basis. This generates new challenges: can one find deterministic versions of quantum mechanics, or new canonical methods to generate quantum field theories such as the Standard Model? More to the point: can we

- include the symmetries of the Standard Model, such as Lorentz invariance, local $SU(3) \times SU(2) \times U(1)$
  (assuming that we will have to restrict ourselves to discretc, lattice-like models)
- generate a Hamiltonian H that is *bounded from below,* $\langle H\rangle \geq E_0$,
- reproduce locality in this Hamiltonian: $H = \int \mathrm{d}^2\vec{x}\,\mathcal{H}(\vec{x})$, $\mathcal{H}(\vec{x}) \geq \mathcal{H}_0$, while $[\mathcal{H}(\vec{x}),\, \mathcal{H}(\vec{x}\,')] = 0$ if $|\vec{x} - \vec{x},'| \neq 0$?

Addressing such questions is like being asked to re-invent the wheel. Our standard quantum mechanical tools are not entirely suited to such purposes.

2.1 PQ theory

A question investigated in great detail in Refs. [?],[?], is how canonical operators $\hat{q}$ and $\hat{p}$, obeying the standard commutation rule

$$[\,\hat{q}_1,\, \hat{p}_j] = i\hbar\,\delta_{ij}\ , \tag{2.1}$$

can be replaced by discrete, lattice-like operators – also to be used as field variables, for instance. We briefly summarise the result here.

It was found that, for each pair $(\hat{q},\, \hat{p})$, we can identify a pair of *commuting* integer-valued operators Q and P, characterising 'wavelet' states $|Q,\, P\rangle$ that are completely orthonormal:

$$\langle Q_1,\, P_1\,|\, Q_2,\, P_2\rangle = \delta_{Q_1\,Q_2}\,\delta_{P_1\,P_2}\ . \tag{2.2}$$

The relation with the original operators $\hat{q}$ and $\hat{p}$ is

$$\begin{aligned}
&\hat{q} = Q + a_Q \quad , \qquad \hat{p} = 2\pi(P + a_P) \quad , \qquad \text{with} \\
&\langle Q_1,\, P_1|a_Q|Q_2, P_2\rangle = \frac{(-1)^{\Delta P+\Delta Q+1}\, i\,\Delta P}{2\pi(\Delta P^2 + \Delta Q^2) + \varepsilon}\ , \\
&\langle Q_1,\, P_1|a_P|Q_2, P_2\rangle = \frac{(-1)^{\Delta P+\Delta Q}\, i\,\Delta Q}{2\pi(\Delta P^2 + \Delta Q^2) + \varepsilon}\ , \quad \text{where} \\
&\Delta Q = Q_2 - Q_1 \quad , \quad \Delta P = P_2 - P_1 \quad , \quad \text{and } \varepsilon > 0 \text{ is infinitesimal.}
\end{aligned} \tag{2.3}$$

A price paid for discreteness is that the commutation relation (??) is not obeyed exactly, but instead we have

$$[\hat{q},\, \hat{p}] = i(\mathbb{I} - |\psi_{\text{edge}}\rangle\langle\psi_{\text{edge}}|) \;, \quad \text{with} \quad \langle Q,\, P\,|\psi_{\text{edge}}\rangle = (-1)^{Q+P} \;. \tag{2.4}$$

$|\psi_{\text{edge}}\rangle$ is an "edge state", a single state for which the commutation rule is anomalous. Demanding physically relevant states to be orthogonal to $|\psi_{\text{edge}}\rangle$ is the easiest way to ensure that this anomaly has no further consequences, but we have to stay aware of this feature.

2.2 An intensive Hamiltonian

Extracting an expression for the Hamiltonian starting from the evolution operator $U(\delta t)$ is not hard. Since U is unitary, its eigen values are on the unit circle, also if we raise U to the nth power: $U^n = \cos nx - i\sin nx$.

Let x be on the segment $-\pi < x < \pi$. Getting x itself from U^n means that we have to find the Fourier coefficients of the function $f(x) = x$ on this segment. These Fourier coefficients are obtained by integrating $\frac{1}{\pi}\int_{-\pi}^{\pi} \mathrm{d}x\, x \sin nx$, and so one finds

$$x = 2\sum_{n=1}^{\infty} \frac{(-1)^{n-1}\sin(nx)}{n} \;, \tag{2.5}$$

or, shifting x by π , so that $0 < x < 2\pi$:

$$x = \pi - 2\sum_{n=1}^{\infty} \frac{\sin nx}{n} \;, \tag{2.6}$$

and from that :

$$H\,\delta t \;=\; \pi - \sum_{n=1}^{\infty} \frac{i}{n}\big(U(n\delta t) - U(-n\delta t)\big) \;\rightarrow\; U(\delta t) \;=\; e^{-iH\,\delta t} \;. \tag{2.7}$$

Note that Eq. (??) fails if there is a state that evolves into itself: $U(\delta t)|\psi\rangle = |\psi\rangle$, implying an eigenvalue $U(dt) = 1$, or $H|\psi\rangle = 0$. Then x is on the boundary $x = 0$ where Eq. (??) does not apply (again, we can call this an *edge state*)[5].

This Hamiltonian is not extensive, since it obeys

$$0 \le H < 2\pi/\delta t \;. \tag{2.8}$$

Fortunately, one can make many other choices, since adding integer multiples of $2\pi/\delta t$ to the eigen values of H does not affect the evolution operator U .

[5]Here, an error in the original slides is corrected.

2.3 The discrete classical Hamiltonian formalism

Again, what we attempt to do in these lectures, is to develop fundamental formalisms with wide, canonical, applications. The standard Hamilton and Lagrange formalisms are beautiful examples, applicable whenever the evolution laws take the form of differential equations in time.

Now, suppose that, for some classical, extensive system (such as a cellular automaton), we have only evolution laws to our disposal that describe finite time steps. How can we generalise the canonical methods for that case? Let us assume that we have integer dynamical variables Q_i and P_i, depending on a discrete (integer) time variable T (as in previous chapters, integer variables are denoted by capital letters, real variables by lower case, and variables taking the form of angles are denoted by Greek letters).

The existence of a positive energy function $H(\vec{Q},\,\vec{P})$ that is strictly conserved by the time evolution law is a very central theme in physics. If we did not have such a function, the evolution law would not be stable, and our world would have looked very different from now, possibly devoid of any systematic structure. In the standard, canonical Hamilton procedure, a very special evolution law is found that guarantees energy conservation. There, one postulates two equal numbers of variables, q_i and p_i, $i=1,\cdots,N$, and the evolution law

$$\frac{\mathrm{d}q_i}{\mathrm{d}t}\equiv\dot{q}_i=\frac{\partial H}{\partial p_i}\,,\quad \dot{p}_i=-\frac{\partial H}{\partial q_i}\quad\rightarrow\quad \dot{H}(\vec{q},\,\vec{p})=\sum_i\left(\dot{q}_i\frac{\partial H}{\partial q_i}+\dot{p}_i\frac{\partial H}{\partial p_i}\right)=0\,. \tag{2.9}$$

How do we do this when Q_i, P_i and time T are all confined to be integers?

The wrong answer is to replace differential equations by finite-difference equations. The Leibniz' rule does not apply, and what would be supposed to be a canonical rule actually comes out being quite messy. In this subsection, we propose a more elegant prescription[?] for updating Q_i and P_i at every step in time T.

There will be one novel element: the values for the index i must by cyclically ordered. One pair of (Q_i,P_i), will be updated for one value of i at the time, in the given cyclic ordering. A prescription must then be searched for doing this in such a way that the total energy is strictly conserved. To describe this procedure, we need to concentrate on one value for the index i only. Therefore, temporarily, we concentrate on a single (Q,P) pair.

Whatever our discrete Hamiltonian function is, it will define closed paths in (Q,P) space on which H is constant. The pairs (Q,P) form a lattice. To avoid that paths of different energies cross one another, we can decide to continue the energy function H across the continuum between the lattice points, by continuous, linear expressions, see Fig. ??. In this figure, the example is shown where $H(Q,P)=T(P)+V(Q)$.

Next, on these contours, we define an arrow of time, simply by demanding that it goes in the same direction as in the corresponding continuum theory (the theory defined in the continuum between the lattice points). Some care must be taken to define exactly which points are where, on which contours, but these details of book keeping can be filled in in a straightforward manner (see caption Figure ??).

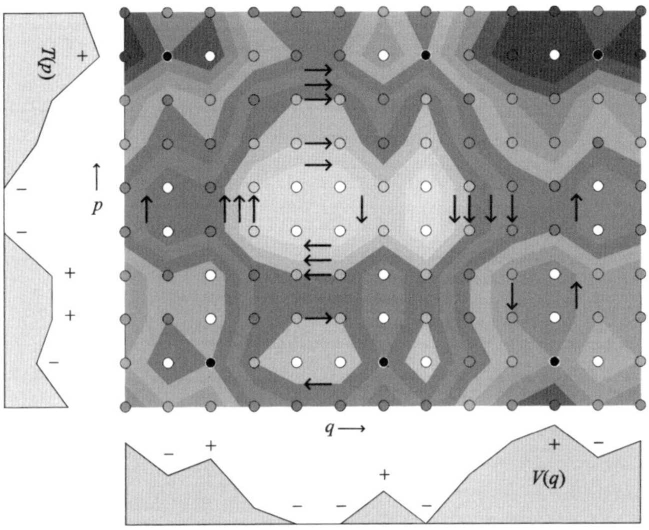

Figure 4: The QP lattice in the 1+1 dimensional case. Constant energy contours are here the boundaries of the differently coloured regions. Points shown in white are local extrema; they are not on a contour and therefore these are stable rest points. Black points are saddle points, where two contours are seen to cross one another. Here, some unique evolution prescriptions must be phrased, such as: "stick to your right", and it must be specified which of the two contours contains the black dot. All these exceptional points are related to local minima ($-$) and maxima ($+$) of the functions T and V.

Finally, our discrete Hamilton evolution law is now straightforward: at every time step, at every value of the index i, the system updates itself by choosing the next available point on its contour. If the contour contains only one point, the system will have to be stationary at that moment.

There is a danger with this prescription: it could happen that all or most of the contours have only one point on them. This can only be avoided if the Hamilton function $H(\vec{Q}, \vec{P})$ varies sufficiently slowly. This then excludes most of the more simple, linear or quadratic functions, since these often go too steeply. One necessarily has to choose the more chaotic functions that contain many straight horizontal stretches in the Hamiltonian.

Only in the latter case, one can ascertain that the conventional, continuum Hamilton prescription Eq. (**??**) is obtained in the continuum limit. This can be derived from the fact that we obtain Liouville flows in the continuum limit, and by checking how fast the system lows along its contours.

Notice that a discrete system controlled by the discrete Hamiltonian now contains two complementary Hamiltonians. The discrete Hamiltonian will be denoted as H_{class} , taking only integer values. Along a given path with energy E_{class}, it defines an evolution law $U(\delta t)$, which we write as

$$U_E(\delta t) = e^{-iH_{\text{quant}}\,\delta t} . \tag{2.10}$$

Since the eigen values of H_{quant} all lie between 0 and $2\pi/\delta t$, while H_{class} is integer, these can be most conveniently combined as

$$H_{\text{tot}} = H_{\text{quant}} + \frac{2\pi}{\delta t} H_{\text{class}} . \tag{2.11}$$

The Hamiltonian (??) can perhaps serve as an extensive Hamiltonian, but this line has not yet been completely pursued.

The discretized version of the Hamilton formalism described here can conveniently be applied to cellular automata. The index i is then replaced by a combination of possible field indices a and space coordinates $\vec{X}$. It works particularly well if the cyclic ordering chosen involves first all even space-lattice sites, and the all odd sites. This avoids the danger that information concerning the states is transmitted across all of 3-space during a single cycle in time T. There is a limiting speed, which could be associated to the speed of light c.

2.4 The AB formalism for the cellular automaton

In this subsection also, a generically applicable tool is displayed. We apply it to the cellular automaton, in particular the type briefly mentioned in the previous subsection.

As was explained, the evolution law is split into two ingredients labelled A and B:

$$U(\delta t) = U_A\, U_B , \tag{2.12}$$

where U_A describes the evolution along all even sites of the space lattice and U_B refers to all odd sites. The rationale of this split is that non-commutativity occurs only between the evolution prescriptions in neighbouring sites, that is, if one site is even and the other a neighbouring odd site. Therefore U_A can be written as a product of actions $U_A(\vec{x})$ where $\vec{x}$ are the even sites, and all $U_A(\vec{x})$ are unitary operations all commuting with one another. The same procedure applies to U_B which only acts at the odd sites.

Consequently, we can write

$$\begin{aligned}
&U_A = e^{-iA} , \qquad A = \sum_{\vec{x}\text{ even}} A(\vec{x}) ; \quad [A(\vec{x}),\, A(\vec{x}\,')] = 0 , \ \forall\, \vec{x}, \vec{x}\,' ; \\
&U_B = e^{-iB} , \qquad B = \sum_{\vec{x}\text{ odd}} B(\vec{x}) ; \quad [B(\vec{x}),\, B(\vec{x}\,')] = 0 , \ \forall\, \vec{x}, \vec{x}\,' ; \\
&\qquad [A(\vec{x}),\, B(\vec{x}\,')] \neq 0 \quad \text{only if} \quad \vec{x} \text{ and } \vec{x}\,' \text{ are neighbours.}
\end{aligned} \tag{2.13}$$

To write U as the exponent of an Hamiltonian, we can use the Baker-Campbell-Hausdorff expansion[?]:

$$\begin{aligned}
&e^P\, e^Q = e^R , \\
&R = P + Q + \tfrac{1}{2}[P,Q] + \tfrac{1}{12}[P,[P,Q]] + \tfrac{1}{12}[[P,Q],Q] + \tfrac{1}{24}[[P,[P,Q]],Q] + \cdots ,
\end{aligned} \tag{2.14}$$

where the ellipses denote a series exclusively built from commutators. Now, we use (??) to observe that the Hamiltonian thus obtained only contains commutator terms built from neighbouring operators. This invites us to write

$$H = \sum_{\vec{x}} \mathcal{H}(\vec{x}) \,, \tag{2.15}$$

and all terms we find for the Hamiltonian density $\mathcal{H}(\vec{x})$ commute with the Hamiltonian density $\mathcal{H}(\vec{x}\,')$ as soon as $\vec{x}$ and $\vec{x}\,'$ are sufficiently far separated from one another. This makes the automaton behave essentially as a genuine quantum field theory.

Now this expansion is still not quite what we want, for two (related) reasons:

1) The BCH series does not always converge. One can easily argue that, if anywhere in the matrices A, B or H eigenvalues separate further than 2π from one another, the expansion passes its radius of convergence, and terminating the series anywhere gives big departures from the desired expression for H.

2) Higher order terms in the expansion involve higher order commutators, such that $\mathcal{H}(\vec{x})$ and $\mathcal{H}(\vec{x}\,')$ already cease to commute when $\vec{x}$ and $\vec{x}\,'$ are farther apart. This gives apparent violations of locality that would make this procedure more questionable.

For this reason, the use of the BCH expansion in this connection should be considered as an illustration of our ideas rather than an explicit prescription for producing the right effective Hamiltonian for a cellular automaton. It so happens that the approach can be approved much more.

2.4.1 The use of conjugacy classes

What we do next, at first sight, may look like merely a cosmetic improvement. We notice that we can still allow ourselves to rotate to different bases for formulating our Hamiltonian. This means that we may take any operator F and replace our Hamiltonian H as follows:

$$\begin{aligned} H \;\rightarrow\; \tilde{H} &= e^F H e^{-F} \;; \\ \tilde{H} &= H + [F,H] + \tfrac{1}{2!}[F,[F,H]] + \tfrac{1}{3!}[F,[F,[F,H]]] + \cdots \,. \end{aligned} \tag{2.16}$$

We choose F also to be a construction using commutators of A and B, to find that the BCH expansion (??) can be replaced by another expansion that converges much faster. Eq. (??) is the expression for the other members of the conjugacy class of H. For instance, the Hamiltonian obtained by switching the order of U_A and U_B is in the same conjugacy class as H itself (just substitute $F = -iB$). This invites us to re-express the series of a member of the conjugacy class in terms of the operators $K = A + B$ and $L = A - B$. The operator K is then the sum of $A(\vec{x})$ at *all* lattice sites, while $L = \sum_{\vec{x}}(A(\vec{x}) - B(\vec{x})) = \sum_{\vec{x}}(-1)^{\vec{x}} A(\vec{x})$, where $(-1)^{\vec{x}}$ denotes a minus sign for all odd lattice sites $\vec{x}$.

To characterise our conjugacy class one now notices that all terms that are even in K and all terms odd in L can be removed by adjusting F.

Let us use short hand notation, writing $[X, [Y, [Z, [U, V]]]]$ as $XYZUV$, etc. Then one finds:

$$\begin{aligned}\tilde{\mathcal{H}}(\vec{x}) &= K(\vec{x}) + \tfrac{1}{96}\textstyle\sum_{\vec{y}_1,\vec{y}_2} L(\vec{x})\,K(\vec{y}_1)\,L(\vec{y}_2) + \\ \tfrac{1}{30720}\textstyle\sum_{\vec{y}_1,\cdots,\vec{y}_4} &L(\vec{x}), \Big(8K(\vec{y}_1)\,K(\vec{y}_2) - L(\vec{y}_1)\,L(\vec{y}_2)\Big)\,K(\vec{y}_3)\,L(\vec{y}_4) + \cdots\end{aligned}$$

Does this series converge better? The next terms were found with some effort. In an even more compact notation the coefficients are

$$\begin{aligned}\tilde{\mathcal{H}} &= K + \tfrac{1}{96}LKL + \tfrac{1}{30720}L(8K^2 - L^2)KL + \\ \tfrac{1}{7741440}\,&L\,(51K^4 + 76LKLK - 33L^2K^2 - 44KL^2K + \tfrac{3}{8}L^4)\,KL \\ &+ \mathcal{O}(K,L)^9 \,. \end{aligned} \tag{2.17}$$

Now one can verify that indeed the convergence in the expansion (**??**) is much faster than that of Eq. (**??**), but this can be attributed to the radius of convergence being 2π rather than π. Since in higher terms more and more neighbours participate, this expansion is definitely not good enough.

Let me add a remark now to this subsection of these lecture notes. All terms in the expansion (**??**) that are just *linear* in the operator K can actually be summed. They correspond to the expansion coefficients of

$$\tilde{\mathcal{H}} = \frac{4\sin\frac{1}{4}L}{L}K \,. \tag{2.18}$$

Not only does this series converge fundamentally better, one can also show that, in these terms alone, *only* nearest neighbours contribute, not the next-to-nearest neighbours[6]. If K and L are given as matrices, and we diagonalize L, then the matrix describing (**??**) can be written in closed form.

Now there may be reasons to suspect that (**??**) is the dominant contribution to the low-energy sector of the Hamiltonian, but a more precise derivation of the effective Hamiltonian, even if we are interested in only the lowest energy sector, is more difficult.

2.5 Conclusion

In these lectures, we hope to have shown that deterministic quantum mechanics is a field that may lead to rich arenas of theoretical particle physics, ranging from pure mathematical physics to the phenomenology of Standard Model Physics. The final sections of the second lecture showed alleys towards constructing quantum field theoretical Hamiltonians

[6] To see this, one writes $L = 2A - K$, and keeps the terms linear in K. We then get $2(\sin(\frac{1}{2}A)/A)K$. Since all $A(\vec{x})$ commute, only their commutators with K matter, and these are non vanishing only for neighbours.

from cellular automata. As yet, our aim has been to set up in a systematic way the mathematical tools needed. Much can simply be copied from standard quantum mechanics, but some new methods are needed as well. These will still require a considerable amount of polishing. We invite people with imagination to participate in these attempts.

References

[1] G. 't Hooft, *The Cellular Automaton Interpretation of Quantum Mechanics, A View on the Quantum Nature of our Universe, Compulsory or Impossible?*, ITP-UU-14/15, SPIN-14/13. http://arxiv.org/abs/1405.1548.

[2] A. Einstein, B. Podolsky and N. Rosen, *Can quantum mechanical description of physical reality be considered complete?*, Phys. Rev. **47** (1935) 777; M. Jammer, "The conceptual Development of Quantum Mechanics", (Mc. Graw-Hill, 1966)

[3] J. S. Bell, *On the Einstein Podolsky Rosen Paradox*, Physics **1** (1964) 195.

[4] J.F. Clauser, M.A. Horne, A. Shimony, and R.A. Holt, *Proposed experiment to test local hidden-variable theories*, Phys. Rev. Lett. **23** (15): 8804, doi:10.1103 / Phys Rev Lett. 23.880.

[5] H.D. Zeh, *On the Interpretation of Measurement in Quantum Theory*, Foundation of Physics, **1** (1970) 69.

[6] M. Schlosshauer, *Decoherence, the measurement problem, and interpretations of quantum mechanics*, Reviews of Mod. Phys., Vol **76**, No. 4 (2004) 1267, arXiv:quant-ph/0312059.

[7] A. Aspect, J. Dalibard and G. Roger, *Experimental Realization of Einstein-Podolsky-Rosen-Bohm Gedankenexperiment: A New Violation of Bell's Inequalities*, Phys. Rev. Letters, **49** (1982) 91; id., *Experimental Test of Bell's Inequalities Using Time-Varying Analyzers*, Phys. Rev. Letters, doi:10.1103 / PhysRevLett. **49**.1804;

[8] J. Conway, in New Scientist, 6 may 2006, p. 8: *Free will - you only think you have it.*

[9] J. Conway and S. Kochen, *The Strong Free Will Theorem*, `arXiv:0807.3286 [quant-ph]`.

[10] G. 't Hooft, *The Free-Will Postulate in Quantum Mechanics*, ITP-UU-07/4, SPIN-07/4, `arXiv: quant-ph/0701097`.

[11] A. Neumaier, *Ensembles and experiments in classical and quantum physics*, arXiv:quant-ph/0303047.

[12] G. 't Hooft, *How a wave function can collapse without violating Schrödinger's equation, and how to understand Born's rule*, ITP-UU-11/43, SPIN-11/34, `arXiv: 1112.1811 [quant-ph]`.

[13] G. 't Hooft, *Relating the quantum mechanics of discrete systems to standard canonical quantum mechanics*, ITP-UU-12/14; SPIN-12/12, arXiv:1204.4926.

[14] G. 't Hooft, *Hamiltonian formalism for integer-valued variables and integer time steps and a possible application in quantum physics*, Report-no: ITP-UU-13/29, SPIN-13/21, http://arxiv.org/abs/1312.1229.

[15] B.C. Hall, *Lie Groups, Lie Algebras, and Representations An Elementary Introduction*, Springer, ISBN 0-387-40122-9 (2003).

LECTURER: G. 't HOOFT

Canonical Methods in Deterministic Quantum Mechanics

Scientific Secretaries: A. Avdoshkin, N. Astrakhantsev

DISCUSSION I

- *Savitskyi:*

Why is the behavior of the superposition principle different at the Planck scale, regarding ontological states?

- *'t Hooft:*

This is not different at the Plank scale, but we have to distinguish the true and only wave function of the universe from the wave functions that we use as "templates". I mentioned the Planck scale because, only if you focus on phenomena at such a tiny scale, you can see exactly the state the universe is in. This is then a single, "ontological" state. The universe cannot interfere or superimpose with itself. We have a classical system there, and classical systems cannot get superimposed. What we do in "ordinary" quantum mechanics is that we hide our ignorance about this single, "true" state, by replacing it with what I call "templates". The template contains superpositions of many different ontological (or would-be-ontological) states. Since these superpositions obey the same Schroedinger equation as the ontological states, one can use the Schroedinger equation to make the best possible guess as to how a system evolves. As I will explain tomorrow, there are lots of correlations. If I change a state here, I can only do it if I change the state somewhere else as well, otherwise I violate the correlation functions. This happens all the time, and if you don't take this into account your act will make what I call a “trial state”, a state that is not a real state of the Universe, because I am not watching what happens at different locations. You have to make a distinction between these ontological states and the states we use as quantum physicists, when we are not working in the ontological basis, where it looks as if the superpositions are all over the place. When we do quantum mechanics, we make superpositions because we don't know the ontological state. If we did, no further superposition would have been necessary. This is a sort of philosophy, but you learn how to use philosophy when you find out how your calculations refer to ontological states.

- *Lukacs:*

When the cellular automaton model is constructed, is a common period T assumed to be stable, in some sense, to perturbations of the system? One gets the expression, that the period of the different cycles might change differently and then for a given T, the automaton structure changes abruptly.

- *'t Hooft:*

Only if you break a system up in small parts, these may have short periods T, perhaps as short as the Planck time. But when these interact, inside a larger system, then that large system will have very long periods, called Poincaré recursion times. The universe itself, starting with a Big Bang, is not periodic at all, we believe. It is just when we break the system up into many small systems, that you get these small periodicities, which disappear when they are coupled together.

- *Yeletskikh:*

Would not be the task of finding the ontological basis too hard, compared to the theory in its conventional formulation?

- *'t Hooft :*

Well, we have to gain experience in finding and describing models that have anontological basis. So that's why I started with the simplest possibilities to see how they can be related to an ontological world. Possibly the standard model must be modified at high energies, at the Plank scale we should add gravity, maybe then we will find an ontological basis. But we could approach the problem from the other end: let us start with ontological models at the Plank scale and see what kind of effective quantum field theories will they give. Then we do have to understand how to construct the Hamiltonian that can be used at very much larger distance scales to describe such things in the language of the standard model. Once we get experience, we will find the rules. I wanted to derive such rules just like what people did in Newton's time when they introduced Lagrange mechanics to streamline their mathematics. Take, for instance, the N-body problem, they handled this problem using Lagrange and Hamilton mechanics. One of my main motivations is to develop a tool kit of instruments just like that.

- *Yeletskikh:*

What makes us believe in the existence of an ontological basis?

- *'t Hooft:*

We don't know for sure, but why should we believe that it does not exist? I claim that such a basis may well exist, and I will say why. If it exists, it will simplify enormously our understanding of the way quantum mechanics works. It if exists, all the mystic descriptions of quantum effects would disappear, everything would become classical with the standard, classical logic. And there are good reasons for thinking why this is the most likely scenario. One is that it will help us to explain the collapse of wave function, which today is only introduced as an afterthought in quantum mechanics. This doesn't sound right to me. In particular, this applies to the Schrödinger cat paradox, where superpositions of states are introduced in which the cat is alive and not. In real life there are no such superpositions, so something has to be done to cure this problem. And I think that my proposal would be the perfect solution, if an ontological basis exists. If we assume this for the moment, all the conceptual difficulties we have with quantum mechanics disappear. These arguments, of course, are up for discussion. If we make this assumption, we will achieve a totally different view of reality; what people said in Copenhagen hardly needs to be corrected, except for one thing: There is a real world behind the quantum one, and we can find it by searching for an ontological basis. If we fail, we fail. If you believe that this basis exists, and you search very well, you will be able to find it. It would add transparency to our view of the world, of the kind that we badly need.

- *Esposito:*

To clarify the last statement in slide 10 – Is the "ordinary uncertainties we had in specifying the initial state" referred to the statistical one or to the intrinsic quantum one?

- *'t Hooft:*

Indeed! Normally, when we talk about this, it's usually claimed that quantum statistics and quantum probabilities are different from classical statistical probabilities, but I now claim they're the same thing. Think of some classical model, a van der Waals gas for example. Usually the origin of the stochastic behavior of that system is that you don't know the laws with infinite precision, or you don't know the initial state. I claim that in quantum mechanics we can use the template description when we set up our initial state. It is a template, which is not the true state of the Universe, because we don't know the true state. The template is a superposition of many ontological states, and I just define that superposition such that all the coefficients squared are defined as being probabilities, I don't know which ontological state I am in. So I construct a template state, a superposition. Let's take a plane wave or an eigenstate of the hydrogen atom as our initial state.

That state is not a single ontological state, it's a superposition. And then those lambdas will survive through our calculation, they won't change, because ontological states go to ontological state. At the very, very end, when you look at the classical final state, it still carries the same lambdas, but now they become Born probabilities. This is how the Born rule reflects the uncertainties of the initial states back in the amplitudes we have that nd up describing the final states.

- *Esposito:*

So, the old superposition of eigenstates and initial states from the quantum point of view just disappeared or is the same thing as the statistics?

- *'t Hooft:*

t disappears, but mathematically it reappears. In the mathematical description I can easily go from the completely statistic classical description to an effective quantum description. If I go from one to the other, the probabilities will be the Born probabilities that we use in quantum mechanics. Before that they were just statistical. Now I am talking about the same things. And that is what people often do in physics: make transitions from one model to another, that is equivalent.

- *Trofymenko:*

Why is it “not good” that momentum operator in quantum mechanics is defined as a derivative d/dx?

- *'t Hooft:*

Oh, I didn't want to say that. In the x-basis, the momentum operator looks like that, but you could use the p-basis, where the x operator is the (i d/dp) derivative. Or you could use a basis of some special functions, Hermite polynomials, for example; you may choose anything you like, it all is equivalent. In field theory we do this, we go from the field representation to Fock space, where we take a basis of particles with certain momenta. And when we use that as the basis to formulate a theory, the Copenhagen interpretation says that it doesn't matter what basis we chose. So I chose the basis I like best, the ontological basis. For doing calculations, all the others are equally good.

- *Schröck:*

Could you please outline the main advantage and disadvantages of your interpretation of quantum mechanics over Bohmian mechanics?

- *'t Hooft:*

The advantage of the CAI would be that the description would be a lot simpler; it would demystify QM. The Many World interpretation, and also Bohm pilot's waves in Bohm's interpretation, both have some mystic elements in their formulations. I want nature to be completely logical and understandable without any magic. Now the fact that I want it, is not a good enough reason to believe that it is true. But the fact that it seems to take away many of the mysteries that quantum mechanics seems to be shrouded in, is a very good sign for me. Secondly, of course, I need to have better leads to build models. Now, one can build models starting from a cellular automaton. However, in addition to that, one needs to impose symmetries such as Lorentz invariance and rotation symmetry and all the other symmetries of the SM, such as local, gauge symmetries. Now that, unfortunately, is quite difficult and hasn't been achieved. If we would succeed in that, we would have important new constraints on model building. Today, we have too much freedom in making a Standard Model, you have the freedom to adjust some 27 free adjustable constants in the SM, we can adjust them in any way we like. That is not good enough if we want to understand where those numbers come from, so one needs extra constraints. That is what I'm proposing. Our theory could lead to such extra constraints. Besides, if it is really successful, we might arrive at classical models for many quantum theories, which would be an enormous benefit if you want to do a computer simulation for instance; this is also at the back of my mind. Maybe this would lead to a new class of computer simulations that are much easier and faster than what we have presently. Now those would be just distant hunches for the future, at best. What was achieved so-far is a picture that removes the mysteries of QM, while you can just go on doing your calculations the way you always did. You have your hydrogen or helium atom or your rare events experiment about the SM, and now you understand much better what they imply, it could mean something as simple as I am proposing. You can eventually use it to explain QM in an easier way to graduates or undergraduate students. Now there are plenty of disadvantages, which I could try to hide, but I can tell you, of course, what they are. One of them is that we haven't found ontological models for any interesting nontrivial quantum system. And the problem has always been that even though my ontological theory is local, even though I believe the quantum theory that comes out of it is local as well, the Hamiltonian I can construct for my cellular automaton nearly never is completely local, and positive-definite. I'll explain more about that tomorrow. I need to have a ground state, a vacuum state, which is absolutely necessary to understand QM properly. To get a vacuum state is pretty difficult in this approach, but not impossible. So I just go on and I postulate the existence of a vacuum state; however, the vacuum that you see in our world seems to be a very special state. For instance, all the existing1-particle states in the vacuum seem to obey Lorenz invariance. Lorentz invariance is extremely difficult to provide in this model. So I may have a hard time ahead of me and maybe I never succeed. Maybe I'm a 1000 years too early with this idea. It hasn't really done any of the tricks which I hope to get out of it in the future. I can say so far I have not really been successful in removing all the difficulties. But I think it does also bring about quite nice mathematical physics. This is a big advantage of string theory. But it is not really the advantage you want. You want it to be a better description of Nature. This hasn't happened yet. So if you want to criticize what I'm saying for that reason you are quite welcome to.

- *Boren:*

If the ontological basis would remove the necessity of superposition, what about mixed states(states defined by a density matrix)?

- *'t Hooft:*

This is a very interesting question. I said that you can't have both the initial and the final states be superpositions of ontological states. If you do that you don't get quite the right QM expression. It is either-or: either you take the initial state or the final state as a physical thing. Then you can interpret that the way it works here. So, the theory tells you what you can have to do. But it doesn't say much about mixed states. So indeed the question is how it comes that the mixed states are different from superimposed quantum states. Because, as you know, mixed states are also a way to deal with statistics. I think that probably what we call quantum states describe the maximum amount of information that you can get in macroscopic systems. You can reduce that and then you are mixing different states. So, as you may know, if you take a single qubit, it occupies a 2-sphere, this gives the possible states a qubit can be in. And if you are on the surface of the 2-sphere, which is a 2-dimensional space, you have a pure quantum state, but can also go inside the sphere and then you have a mixed state. So mixed states contain less than your optimal amount of information. I think that this is the answer to what the mixed states are.

- *Prochazka:*

What is about entanglement? Is the product of the ontological states ontological?

- *'t Hooft:*

You are not allowed to superimpose ontological states, that is, the result won't be an ontological state. But I can use templates, which are superpositions of ontological states of any form. The coefficients then represent probabilities. If I then transform to the basis of my choice, I get ordinary quantum mechanics. I can certainly get entangled states that way. In fact, in terms of the ontological states, this transformation can easily become so contrived that the original ontological states themselves look like entangled states now. Experimentalists can produce such entangled states, while, in terms of the cellular automaton theory, experimental physicists can only produce ontological states, and so, yes, one must conclude that many ontological states will look like entangled states to us. So the answer is yes, or it should be yes. Amazingly, the entangled superpositions that you would expect to be very difficult to obtain must be very common appearances. Having said all this, however, it is still hard to see what happens 'in reality' when you do experiments with entangled photons. This still has to be phrased better.

- *Prochazka:*

The physics should be the same at the end of the day. So, if I talk about the things like entropy or arrow of time it should not depend on the basis I choose, right?

- *'t Hooft:*

Absolutely! Indeed the physics does not depend on the basis chosen. Even when I have entangled photons, I should be able to transform that back to the ontological basis where you don't notice anything being entangled. It is there then, that I will be using a classical description. That classical description is then not in terms of photons, electrons or other particles or even fields, but in terms of what happens at the Planck scale. At the Planck scale very, very complicated things happen. We don't even have the proper words for that today. You would have to describe the degrees of freedom inside Nature's cellular automaton, while the words we have today are not appropriate to describe any experiment in terms of what happens microscopically. Why not? Because the dynamical variables that would be relevant to describe an experiment with entangled states occupy so many billions times billions of volume elements at the Planck scale that if you what to specify what happens in all these volume elements, you won't have the appropriate means to follow that kind of book keeping. The best thing you can do today is indeed to use the templates that we have been using all along, use the basis you've been using all along and then allow those basis elements also to generate what you thought were entangled states. These would also be the original ontological states in disguise. So, the answer is yes, it is inevitable to transform to any of the basis choices that we are familiar with. But still we're not allowed to take the superposition of two ontological states and say it is ontological.

- *Prochazka:*

So can I take a superposition of two non-ontological states and obtain ontological one?

- *'t Hooft:*

Yes, you can take superpositions of the basis we are used to in order to reproduce ontological state. So the outcome of an experiment is a classical state which often is a superposition of the template states we have. And that is an ontological state. So, yes, the ontological state in turn is just the transformation backwards, the transformation from the ontological state to the state we normally use, called template state. All superpositions that we think we see around just must eventually be ontological states. Just take, for example, the Fourier transformation. A localised particle contains lots of complicated superpositions of Fourier waves. And when you want to transform it back you need again many states. So, yes, ontological states are superpositions of that you normally use in a non-ontological basis.

- *Salakh:*

Does this way of looking at QM have a slight chance to produce any predictions, except of being only an interpretation?

- *'t Hooft:*

I don't think you can do an experiment today that distinguishes this theory from any other theory of QM. In the many world interpretation we have the same thing: you cannot do an experiment that shows that it is the correct interpretation. The problem we have is building theories, building models that should be better than the SM. The final word that says: yes, the procedure we use is correct, and useful, if we can use it to make a better model than the Standard Model today. If that could be accomplished then this would be a very strong indication that what I'm doing makes sense. As long as this is not the case, all we have is that the theory generates QM exactly, or even better than what is obtained in the Copenhagen interpretation, where people talk about the collapse of the wave function. Several researchers expect to detect deviations from the Schrödinger equation, because they need deviations from the Schrödinger equation to understand the collapse of the wave function. In our theory no such thing is needed, I don't need to modify the Schrödinger equation at all. Now as yet all experiments only confirm Schrödinger's equation at all stages of any calculation. This agrees with my theory. As yet I'm not going be proven or disproven anyway further by experiments. Note added in proof: there is a negative prediction, which is that if quantum computers become reality, one will find that there are limits: a quantum computer will never outperform a classical computer when the latter is scaled to Planckian dimensions.

- *Salakh:*

I also have a question about the lower bound of the Hamiltonian. How severe are the restrictions? If you follow the cellular automaton theory there is no lower bound.

- *'t Hooft:*

Once you have a Hamiltonian, you can write e^{-iHt} and just check if there is a certain basis where this expression permutes elements. The converse is in principle easier: if you have a permutation operator, you can always find a Hamiltonian, of which the permutation operator is the exponent. That is really easy to do. I'll show that tomorrow, if I have the time. And you will also see that it is not unambiguous: there are actually quite many different Hamiltonians that generate the same theory. This is because I only look at integer time steps. So, what happens at in-between times doesn't matter. One can make Hamiltonians that can make anything that you like happen at in-between times. So there is a lot of freedom in the Hamiltonian. I have to pick one I like most. It should have the correct locality properties, there should exist a vacuum state, which is the lowest energy state, and there are other extra physical requirements.

- *Ponomarev:*

In usual QM, quantum uncertainty is of order of ħ, which is in turn a degree of non-commutativity of x and p. Can you reproduce that with you ontological approach?

- *'t Hooft:*

ħ is by definition 1, or you can call it 1/2pi, depending on how you work with the units. Thus, ħ is just a dimensionless mathematical quantity rather than physical. It essentially looses its significance, or it serves as a simple conversion coefficient that converts energy to frequency and momentum to wave number, so also in this theory, ħ functions as a definition of momentum, and a definition of energy. ħ doesn't appear anymore. I do not equalise it to zero. The uncertainty relation and what happens instead is the part of the talk that I have to skip tomorrow because it takes too much time. But what I will do is replace q's and p's of the theory by integers, and so Δq and Δp become the distances between integers and their product is h. And since I'm working with integers, there is nothing in between then, and these things are constrained. This replaces the usual uncertainties.

LECTURER: G. 't HOOFT

Canonical Methods in Deterministic Quantum Mechanics

Scientific Secretaries: M. Savitskyi, N. Wardle

DISCUSSION II

- *Csanad: (Question about diagram on slide 25)*

You had a plot of the Hamiltonian values, how do the white and black points work?

- *'t Hooft:*

The colored pattern generated from the Hamiltonian plotted on a grid of points P and Q gives the contours. The rule is to follow the contour either anti-clockwise or clockwise depending on whether you are at a minimum or a maximum just as in the continuum theory. The subtlety is what happens at the black points or the white points. On the white points, the rule is simple: you stand still. On the black points two contours seem to cross. There, you have to decide on which of the contours you are. It cannot be both since once you are at a point you have to know which direction to go. As long as the rule is unambiguous there are no problems. This rule must be such that it does not give an extra degree of freedom, for a given P and Q there must be exactly one rule as to where to go next.

- *Millington: (Question about diagram on slide 25)*

The way in which you construct the generators of discrete time translations is remarkably similar to Liouville's theorem. This discretization of the time is reminiscent of hopping between microstates in a statistical sense. Does this approach have something to say about a deeper connection with the statistical interpretation of quantum states, as has been discussed by people like Terrence Rudolf, Mathew Pusey and collaborators in quantum information?

- *'t Hooft:*

Well, this P-Q lattice is indeed typical for quantization. In the case of quantization of continuous p-q space, you can find that Heisenberg's uncertainty relation also tells you that every little square corresponds to one quantum state, which sometimes is called *wavelet*. This quantum state occupies a certain region in P-Q phase space, since the P and the Q are determined as well as possible. And that was a deeper reason why this has such a naturally continuous limit. Once the lattice will be very dense, and a continuous limit will be reached, one finds the continuum Hamilton equations in the limit.

- *Jejelava:*

My question is about the concept of ontological basis. If I understand correctly, you have to rewrite everything in the ontological state, but you can do it only for whole systems, which means for the whole universe, otherwise in case of any finite systems you will end up in ordinary quantum mechanics. Does it mean that this approach is kind of very idealistic and you can do everything altogether or it is just a matter of interpretation in quantum mechanics?

- *'t Hooft:*

I think that you understand it perfectly well. It is a package deal. You have to accept that we have determinism all the way. That's why I have to enter such a notion as *superdeterminism*, which means that also the observer is controlled by deterministic laws, whether he or she likes it or not. That is an essential part of the game. I had some questions here to answer about individual systems like a single atom or a single harmonic oscillator, but theory is not intended for that. Those are just simplistic examples, that you can use as the starting point to build more complex theories. There are no systems only by themselves, so you can't split the universe into smaller parts and say that these are all by themselves ontological systems. If that would be possible, these smaller parts should not communicate, so you could think of two universes sitting next to each other and which don't communicate. You could do that, and you could try to introduce interactions within the systems, which originally were non-interacting, but this is a way to build models. You can't split things completely apart easily without getting into trouble, when one part of the universe can't observe another part of the universe. As soon as you talk about observation, if you talk about the whole thing, you must also have interactions.

- *Jejelava:*

As soon as you started to interact with this independent universe, it is still some kind of superposition, which is not an ontological basis anymore. Am I correct?

- *'t Hooft:*

Yes. You can't superimpose ontological states for the entire universe anymore. Superposition can be done when you have what I called *templates*. A template is a simple-minded approximation of what you think the universe is like. For instance, the hydrogen atom, which we usually use in example calculations. According to this approach, the hydrogen atom, as all the other atoms, are templates, they are simple-minded constructions, where we do use superpositions to get the complete calculation right. On the other hand, the entire universe is completely full of hydrogen atoms and all sorts of superpositions occur in the real world, but altogether they form one ontological state, which by itself does not allow for superpositions. This is the reason why classical states no longer allow for superpositions. I think you understood it pretty well.

- *Haidt:*

How is entropy defined in your formalism?

- *'t Hooft:*

I'm not addressing thermodynamics directly. The question on what you do with entropy can be asked in any theory of quantum mechanics. You could ask in cosmology or the standard model what entropy means, so I will give a similar answer. To do thermodynamics in the theory you have to go from pure states to mixed states and entropy can best be described in terms of mixed quantum states. Then " $\Sigma_i P_i \cdot \log(P_i)$" in the thermodynamic limit is going to represent the entropy of the system, where " P_i" is the probability of possible states *i*, that doesn't change in this picture.

- *Gavrylenko:*

Generally, the discretization leads to the violation of the phase volume conservation law. Is there such problem in your approach?

- *'t Hooft:*

That is indeed an important question. Phase volume has important information of the theory. The case of reducing the phase volume means that quantum states are squeezing out. It means, that quantum states are no longer conserved. In ordinary quantum mechanics such a thing is not allowed to happen. So, in ordinary quantum mechanics regions of volume of phase space are indeed conserved. They can't increase, they can't decrease, unless you add the uncertainty of your calculations or something like that. Normally in quantum mechanics phase space volume is conserved. So, the first deterministic theories have a built-in conservation of phase space, which means that there is no information loss. Let us compare the cellular automaton with what happens in your laptop. The laptop also does computations and evolves states into states, but your laptop has a "delete" button and it can delete states. On the other hand, normally the cellular automaton is constructed in the way that it doesn't have any "delete" button, hence no information can be deleted. However, we can also think of a cellular automaton, that does delete states. In fact, computer experts or people who run cellular automaton models routinely on a computer also note that a cellular automaton often doesn't preserve all the information, well, some information just gets lost on the way. There is the famous example "The Game of Life", which was a popular computer game in the '70s; it is a cellular automaton program, that removes information out of the system. If I would try to use such an automaton to produce quantum mechanical models, at first sight, I could get into big difficulties, because the unitary operation would have to be a permutator. By using the permutator operator you can always permute backwards, so the *permutation group* is the full group, in which every element has an inverse. I can go backwards in time as easily as forwards in time. However, when you think of physics of black holes and subjects like that, you might suspect that information can be drained, for instance into black holes. Then the operator, which we are dealing with, does not have an inverse, because different systems can all go into the same system after we have deleted information. The question appears: is that allowed or not? At first sight, it seems to be not allowed, but then you can say: "Wait a minute! If you think again more carefully, maybe it is allowed?" Let us take all states that lead to just one single state in the not too distant future. All these states are fundamentally equivalent, because no physicist, no observer ever will be able to figure out which state we started off from, if there was a "delete" button. Well, if you will look at your laptop and the information is truly deleted, you can't find it back. So, all states your laptop was in should be equivalent, if you suppose that you don't touch the buttons, but you programmed to delete things so that all states lead to the final configuration. That means that I introduce *equivalence classes* in the theory. Two states are equivalent, if in some reasonably small amount of time they evolve into the same state in the future. I now revise the notion of the *quantum state* or a basis element of the quantum state: it is not longer a single state of the automaton, but is an equivalence class of states. So, each equivalence class forms a separate basis element of my basis. And then I can carry out my program as before. That's much more difficult, but if you think of the physics of black holes it might be something to consider. I am not sure about this, but this is a possibility. It is a possibility that I keep open. Perhaps it is something we need to go into, if you want to really understand quantum gravity.

- *Laszlo:*

What exactly prohibits the restriction of your deterministic cellular automata model to a localized (compact) "domain of dependence" portion of the spacetime?

- *'t Hooft:*

There is a problem with that. If you divide the universe in two domains, you have the choice of whether the two domains interact or not. If they don't interact then the universe falls apart literally; there is no way of going from one region to the other region. If they do interact you have to consider the entire interacting system as your model. In physics we always have to start with something small and start building from there. For instance in the standard model we start with non-interacting particles, each of them form their own universe or domains and we introduce interactions afterwards, so we might have to go through similar steps to get to more sophisticated models. We start with domains in the theory, say leptons and quarks without having them talk to each other. When we introduce interactions, they become one overall system in the standard model. Step-wise one may construct a theory, where during the early stages one may have such domains, but in the final theory we talk about the entire universe, where every part of the universe talks to every other part. There is a delay, since signals need time to travel, but still information can go from one region to the other. There is also the problem of Bell's inequalities. I had to postulate that all parts of the universe are strongly correlated with non-vanishing correlation functions, without which I would not be able to understand how a classical system can violate the Bell's inequalities, so I'm also forced to have the whole universe as one single ontological theory.

- *Semenyakin:*

You have built the theory over integer numbers. Classical quantum mechanics is built over real numbers. Is there some deformational parameter "h" in your approach, which guides deformation between different scales: in case of h approaching infinity we obtain your theory at the Planck scale based on integers, while values of h close to "0" lead us towards ordinary quantum mechanics and real numbers?

- *'t Hooft:*

If I understand well, your question is about integer numbers as opposed to real numbers. Just before we had a talk about number systems in physics. Real numbers are everywhere, as well as integers. The point is:I tend to believe more firmly , that the ultimate model of nature is based on information. And I'm strengthened in that belief by the *holographic principle*: the fact that black hole states have finite amounts of information, residing only on its horizon. It seems to be impossible to squeeze more than one bit of information on every unit of surface area at the Planck scale. So, that suggests that information is discretized. When you have only bits and bytes, you can't very well represent those in terms of real numbers, but every single particle in our universe requires real numbers to describe it, and imagine how many particles there are according to the Standard Model. Every single particle represents real numbers to describe its position. At every point in real number space there could be a field value, which give us us a real number. Real numbers are everywhere. Every real number represents an infinite number of decimal places. A single real number contains an infinite amount of information already. And now I'm postulating, that at the Planck scale the amount of information on the surface is strictly finite. So, I can't realise that with real numbers for that reason, but at some intermediate stage of the model it may be very well possible to work with real numbers. An example can be the neutrino field, which is based on real numbers. In that case you can also think of determinism just like a planetary system, you can think of planet described by real numbers and which are moving around in orbits. Then you don't need the permutator operator, but just an ordinary differential equation with respect to time, that describes everything. So, you can take that continuum limit. Yes, in my theory you can take continuum limits, if you want, but I don't want that, in general. Indeed, it is something that might be considered, but it has to be a part of our toolkit to see how the continuum limit of the certain theory looks. As I said, ultimately, at the Planck scale, where this model is particularly important to think about, everything is discrete. This is our starting point.

STATUS OF EXPERIMENTAL POWER

Dark matter searches

Alessandro Bettini
University of Padua. Department of Physics; INFN. Padua division (Italy)
Laboratorio Subterráneo de Canfranc (Spain)

Abstract

These lectures begin wit a brief survey of the astrophysical and cosmological evidence for dark matter. We then consider the three principal theoretically motivated types of dark matter, sterile neutrinos, axions and SUSY WIMPs.

In chapter 4 we discuss the motivations for the so-called neutrino minimal standard model, nuMSM, an extension of the SM with three sterile neutrinos with masses similar to the charged fermions.

In chapter 5 we briefly recall the strong CP problem of the SM and the solution proposed by Peccei and Quinn leading to the prediction of axions and of their characteristics. We then discuss the experimental status and perspectives.

In chapter 6 we assume that the reader to be acquainted with the theoretical motivations for SUSY and move directly to the direct search for dark matter and the description of the principal detector techniques: scintillators, noble fluids and bolometers.

We conclude with an outlook on the future perspectives.

1. Evidence for dark matter

In 1933[1], Fritz Zwicky, from the measurement of the radial velocities of eight galaxies in the Coma cluster, found an unexpectedly large velocity dispersion of about 1019±360 km/s. The implication was that the total matter density in the cluster should be much larger than that of luminous matter. He concluded that: "If this [overdensity] is confirmed we would arrive at the astonishing conclusion that dark matter {*dunkle Materie*} is present [in Coma] with a much greater density than luminous matter." This was the discovery of dark matter. For a review of the early history of dark matter see[2].

Since then, evidence for the presence of dark matter overwhelming the visible one have been found in the Universe at all the scales and all the epochs of its evolution. Referring to Lars Bergström[3] for a complete review, we recall here two important pieces of evidence.

The "rotation curve" of a galaxy is obtained by measuring the velocity relative to its centre of the gases orbiting around it as a function of the distance from the axis. As found for the first time in 1970 by V. Rubin[4], such curves show, for all spiral galaxies, that velocity remains constant, or even increases, at radii well above the luminous radius.

In the bullet cluster merger 1E 0657-558 two clusters that have just (on astronomical scale) crossed each other are observed[5]. The observation of luminous matter is obtained in X rays and of total matter via gravitational lensing. The resulting picture shows the effects of viscosity on the former, but not on the latter, which extends at farther distances.

More generally, gravitational lensing provides quantitative information on dark matter and on its distribution in the Universe. Indeed, in any case, dark matter in the Universe is detected via gravitational effects.

The accurate measurement of the Cosmic Microwave Background (CMB) temperature fluctuations is a fundamental element of the present precision cosmology. The Planck mission[6], in particular, has measured the cosmological parameters with per-cent accuracy.

We recall here the value of the Hubble constant, $H_0 = (67.3\pm1.2)$ km s^{-1} Mpc^{-1}, of the total matter density (as a fraction of the critical density), $\Omega_m = 0.315\pm0.017$, of the baryon density $\Omega_b h^2 = 0.02205\pm0.00028$ (where $h = H_0$ km s^{-1} Mpc^{-1}/100 is the reduced Hubble constant, and h^2=0.45) and of the cold dark matter density $\Omega_c h^2 = 0.1199 \pm 0.0027$.

2. CMB polarisation

Measurements of the CMB polarisation provide further fundamental information on cosmology. They are very difficult due to the smallness of the signal and to the presence non cosmological contributions in the foreground, which must be understood and subtracted. The vector polarisation field is divided in two components: the irrotational component, called E-mode, and the non-irrotational one, called B-mode.

Quantization of the gravitational field joined to the exponential inflationary expansion leads to a primordial background of stochastic gravitational waves with a characteristic spectral shape. These inflationary gravitational waves (IGW) produce local quadrupole anisotropies in CMB, which appear in the B-mode at degree angular scales. Its amplitude depends upon the tensor-to-scalar ratio, r, which is a function of the energy scale of inflation.

However, gravitational lensing between CMB and us induces a non-irrotational component in the originally irrotational E-mode. The IGW B-mode, however, is predicted to peak at multipole l=80, making it possible to distinguish from lensing effects.

The BICEP2 telescope, located at the South pole, was especially designed to search for such effects. The 26 cm aperture telescope, in a liquid He bath, observed in a unique band at 150 GHz a 380 square degrees of the sky. The refracting optical system equipped with a focal plane of 512 antenna coupled transition edge sensors (TES) at 270 mK. The sub-Kelvin focal plane assembly sits within a superconducting Nb magnetic shield.

Past spring BICEP2 reported[7] the observation of an excess of B-mode power spectrum over the base lensed-LCDM expectation in the range $30 < l < 150$, inconsistent with the null hypothesis at a significance larger than 5σ. The spectrum is well fit by adding a tensor component with tensor to scalar ratio of

$$r = 0.20^{+0.07}_{-0.05}\,. \tag{1}$$

If confirmed the result would for the first time give us the Hubble constant at the time of inflation as

$$H_I = (6.00 \pm 0.91)\times 10^{14}\ \text{GeV}\,. \tag{2}$$

(For comparison, its value now, in GeV units is $H_0 \approx 1.5\times10^{42}$ GeV) and the energy density at the time of inflation

$$V = 2.2\times10^{16}\left(\frac{r}{0.2}\right)^{1/4}\ \text{GeV}\,. \tag{3}$$

of $V \approx \left(10^{16}\ \text{GeV}\right)^4$, a value close to the GUT scale!

However, the result cannot considered established for the following reasons. Spurious B-mode polarization can be induced by powders in the foreground. At the time of the BICEP2 publication, the map of those powders had been measured by PLANCK in the larger fraction of the sky, but not at the high latitudes at which BICEP2 observed. The subtraction of the effect is consequently uncertain.

[Note added. After these lectures were delivered, PLANCK published data[8] showing a contribution of powders to B-mode polarization of the same magnitude as reported by BICEP2. The two collaboration have also decided to perform a joint analysis of their data. The projected sensitivity of PLANK when all the available data will be analysed will reach

r=0.05. In the following discussion we shall assume a value of the energy density at inflation not very different from (3). Notice that it depends only on the fourth root of r].

3. Dark matter candidates

Dark matter particles are inferred due to their gravitational interactions. They do not have either strong or electromagnetic effects. If they have weak interactions we can try to detect them. We shall consider here the direct searches only.
Dark matter particles cannot be any of those of the Standard Model. A large number of theoretical possibilities has been put forward. However, experiments are difficult and demanding and some guidance is necessary. This can be provided by the fact that the SM itself appears to need extensions to solve issues like the hierarchy, leading to SUSY and WIMPs particles, and the strong CP problem, leading to axions. In the following we shall discuss these two possibilities and, in addition, the possible existence of further, sterile, neutrinos.

4. Neutrinos and dark matter

The SM assumes that neutrino flavours are three. This assumption has been experimentally proven with high precision by measuring the line shape of the Z at LEP. Indeed, the width of the resonance increases and its peak height decreases with increasing number of neutrinos into which the Z can decay. The LEP experiments have established $N_\nu = 2.984 \pm 0.008$. Additional information comes from cosmology, which gives two pieces of evidence. The first one is on the "effective neutrino density", namely the density of non-interacting radiation at decoupling (corresponding to masses smaller than about 1 MeV), neutrinos or else. The corresponding effective neutrino number expected in the SM is $N_{\nu\mathrm{eff}}$=3.0395[9]. A fit by the PLANCK mission including their own data on CMB and data from other experiments (In particular on the Baryon Acoustic Oscillation peak) gives $N_{\nu\mathrm{eff}}=3.30\pm0.27$. The second piece of information are upper limits on the sum of all neutrino masses, including hypothetical ones, called sterile, not coupled to the Z. The limits depend on several assumptions but may safely be taken between 230 and 600 meV.

Claims for sterile neutrinos have been advanced both in ν_e or $\bar{\nu}_e$ disappearance and in ν_μ to ν_e oscillation data, without however being supported by sound evidence, as we shall now summarise.
For electron neutrino disappearance, the so-called "Gallium anomaly" is the following. Both Gallium-based solar neutrino radiochemical experiments, GALLEX/GNO and SAGE, performed two runs each with radioactive neutrino sources for checking the overall detection performance. The average of the four ratios between the measured interaction rate and the expected ones is $R=0.86\pm0.05$. However, the expected values depend on the cross section of the detection process, namely $\nu_e+{}^{71}\mathrm{Ga}\rightarrow{}^{71}\mathrm{Ge}+e^-$. This includes non-measured contributions that have been calculated, with partially unknown theoretical uncertainty. The latter is not included in the above-mentioned error bar. This is why neither experiment used R for calibration and why no anomaly can be reliably stated.
For electron antineutrino disappearance, the so-called "reactor anomaly" is the following. In 2011 Th. A. Mueller et al.[10] re-evaluated the neutrino flux from power reactors, finding it higher than the previously accepted value by about 3%. Soon afterwards, G. Mention et al.[11], considered the available measurements of the neutrino flux from several experiments at different distances, shorter than 100 m, from a number of reactors, finding that in average they account for only 0.943±0.023 of the expectations with the new flux calculation. The effect is only 2.5 σ, but was claimed as an anomaly. Notice however, that the fission beta

spectra involve some 6000 transitions, one third of which are forbidden. The latter introduce large uncertainties in the predicted antineutrino flux, which remain also in the Mueller at al. re-evaluation and that is larger than the claimed uncertainty and effect. Moreover, in 2013, when all mixing angles were known including θ_{13}, C. Zhang et al.[12] could fit the known distance dependence of the antineutrino flux to the measured values. The extrapolation to zero distance gives 0.959 ± 0.009 (exp) ± 0.027(syst) of the Mueller re-evaluation. This is a 1.4 σ effect, which is much smaller than the systematic uncertainty of the latter. In conclusion, there is no sound reactor anomaly.

Anomalies have been reported also in appearance experiments, which, notice, look at physical processes different from the disappearance ones. In 1995 the LSND experiment observed 3.8 σ excess in $\bar{\nu}_\mu$ to $\bar{\nu}_e$ appearance. If interpreted as an oscillation it would correspond to Δm^2>0.2 eV^2. However, the effect was based only on the extrapolation of a background model, which is a very unsafe procedure indeed. In 2001, the KARMEN experiment excluded a large fraction of the LSND claimed parameter space. MiniBooNE at FermiLab started as ν_μ to ν_e appearance experiment, which should be equal to in $\bar{\nu}_\mu$ to $\bar{\nu}_e$ for *CPT*. The experiment worked at higher energy but in the same range of *L/E*, which is the relevant quantity, as LSND. In 2009 the experiment published a result, in which no signal was present in the LSND claimed signal region. There was however excess at lower energies, where the background model was declared not to be reliable, and, in any case, not compatible with oscillations. In 2010 MiniBooNE presented results on in $\bar{\nu}_\mu$ to $\bar{\nu}_e$ claiming a signal in oscillation region having observed an excess over calculated background of 20.9±14.0 events. Notice that the statistical significance is only 1.5 σ and that, if real, when combined with the neutrino result of the same experiment would imply a large *CPT* violation. Finally, at the Neutrino conference at Kyoto in 2012, MiniBooNE reported that higher statistics anti-neutrino data were "now much more consistent with neutrino data. It is not yet known whether the excesses [those at low energy] are due to oscillations, some unrecognized neutral currents γ background, or something else"[13]. The lesson is that background models are always uncertain and consequently claims of discoveries based only or mainly on an observed difference between data and expectations from such models are not reliable.

Additional information comes from the searches of anomalies in the ν_μ to ν_e appearance by ICARUS[14] and OPERA[15] at the CERN to Gran Sasso beam (CNGS). These lead to the exclusion of a large fraction of the parameter region claimed by MiniBooNE[9]. The surviving region would imply a "new" neutrino mass between 0.6 and 1 eV. However, such large values of the mass, as we have mentioned above, are excluded by cosmology. To be precise, the particles for which the limits are valid, the "neutrinos", are any weakly interacting particles behaving as radiation at decoupling, when temperature is of the order of the MeV. If heavier neutrinos exist they should be out of thermal equilibrium. Such an assumption is arbitrary and makes completely arbitrary any theory built on it.

In the SM neutrinos are mass-less, but in Nature they are not. We do not have precise values of their masses as yet, but we have the above mentioned upper limits from cosmology and lower limits from oscillations, the frequencies of which are proportional to the absolute values of the relevant square mass differences. The situation is shown in Fig. 1a). Clearly neutrino masses look substantially different from those of the other particles.

The so called see-saw mechanism tries to understand the issue. Assuming neutrinos to be Majorana particles, the "Majorana" mass term $M\nu_R^T\nu_R$ is allowed by the SU2 ⊗ U1 symmetry and clearly induces both lepton number *L* and *L–B* violations at scale *M*, which consequently should be very large. In the model, heavy sterile right neutrinos exist. A "Dirac" mass term $m_D\bar{\nu}_L\nu_R$ should also be included, with the Dirac mass $m_D << M$. The mass matrix is written as $\begin{pmatrix} 0 & m_D \\ m_D & M \end{pmatrix}$ and has two eigenvalues, a light $m_{\text{light}} = m_D^2 / M$ and a heavy one $m_{\text{heavy}} = M$. The Dirac mass should be of the order of the electro-weak scale, namely m_D ; $VEV = \upsilon$; 200 GeV. m_{light} should be the scale of neutrino mass, namely $m_{\text{light}} \approx 100$ meV or smaller and consequently $M \approx 10^{14} - 10^{15}$ GeV. This value is close to the GUT scale and, if BICEP2 is right, or almost so, to the scale of the inflation. The mass spectrum is shown in Fig. 1b), where we assume three heavy neutrinos N_1, N_2 and N_3, of mass M.

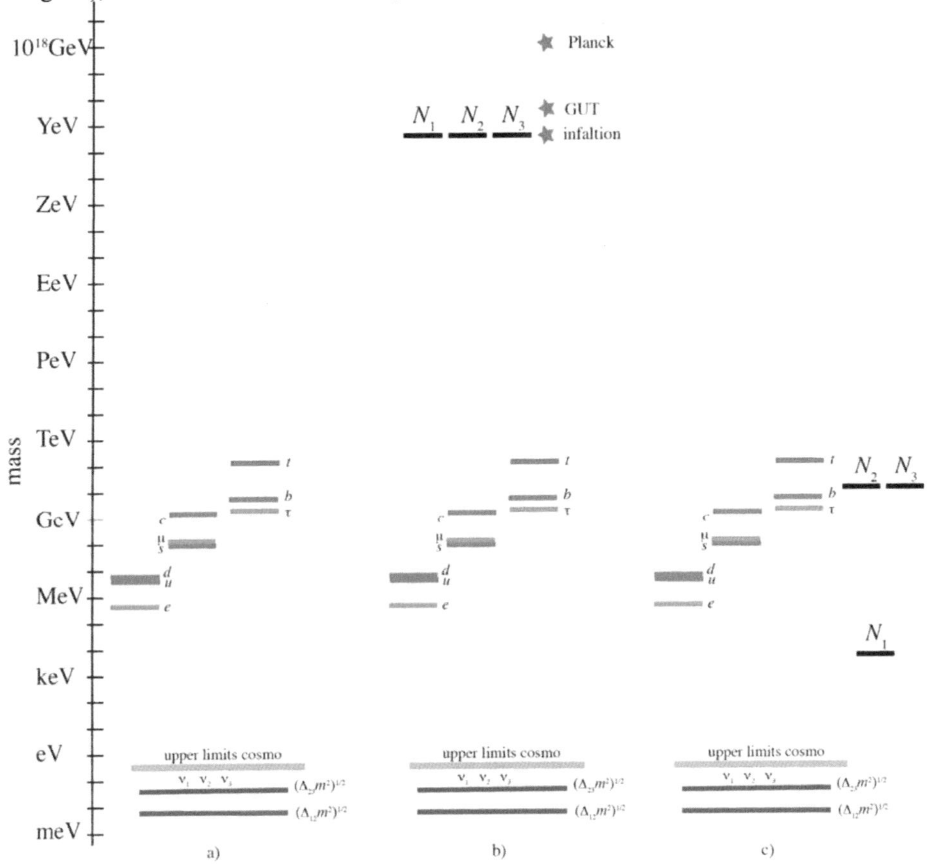

Fig. 1. Mass spectra for a) known fermions, b) and sterile neutrinos in the see-saw, c) and sterile neutrinos in the nuMSM

The "Neutrino Minimal Standard Model" nuMSM[16] extends the SM including three right, sterile neutrinos, as in the see-saw mechanism, but with much smaller masses, a degenerate doublet at about 100 MeV and a singlet, N_1 at the keV scale. No reason is given for the difference. The raison d'être of the sea-saw is lost. In order to have small neutrino masses the model must assume a very small Yukawa coupling ($|y_{\text{light}}^2| < 10^{-13}$), giving $m_{\text{light}} = y_{\text{light}}^2 \upsilon^2 / M$, while the mass of the heavy doublet is $m_{\text{heavy}} = M$. The situation is shown in Fig. 1c).

In the nuMSM the sterile neutrino singlet N_1 is the dark matter. Two additional very strong ad hoc assumption must be made to make this compatible with the available evidence. The LEP limit is evaded assuming an extremely small mixing of N_1 with known neutrinos ($\theta^2<10^{-8}$); the cosmological limits are evaded decoupling N_1 from other particles in order to allow them not to be in equilibrium.

Non the less N_1 is not completely stable and decays, in particular, to normal neutrinos as $N_1 \rightarrow \nu+\gamma$. We can look for this decay in galaxy clusters searching for a monochromatic signal, a line, in the spectrum. We do not know its position, but we know it should be at the keV scale, namely X-rays. Existing data already exclude the largest part of the parameter space, namely the mixing angle θ^2 versus the mass M_1, but an allowed region remains.

"Evidence" for such a line has been reported[17] in a stacked spectrum (i.e. assembling spectra measured by different experiments) of 73 galaxy clusters with red-shifts $0.01 < z < 0.35$, at 3.57±0.02 keV. This would correspond to $M_1 \approx 7.1$ keV; $\theta^2 = 2\times10^{-11}$, which is in the allowed region. Notice however, that the local statistical significant is only about 3 σ, which is marginal, considering that it is not a priori known were to look. In addition, the "signal" appears after subtracting from the spectrum, which is linear in the region, a background, which is evaluated to be concave with a minimum right at the "signal" position (flat–valley=peak). Moreover, the authors report discrepancies between the "line" position and intensity in the different subsamples of the analyzed data. Clearly enough, no reliable claim of an observation can be done under the present circumstances and, as the authors themselves admit, we need to wait for data from the next generation X-rays observatories.

5. Axions

a. The strong CP problem

The QCD **symmetry** properties allow the presence in the Lagrangian of the term (G is the gluon field strength)

$$L_\theta = \frac{\alpha_s}{8\pi}\theta\times\left(\frac{1}{2}\varepsilon^{\mu\nu}_{\alpha\beta}G^{\alpha\beta}\right)G_{\mu\nu} \equiv \frac{\alpha_s}{8\pi}\theta\times\tilde{G}^{\mu\nu}G_{\mu\nu}\,. \tag{4}$$

where θ is an unknown coupling. The term is pseudoscalar, hence, added to the scalar terms, violates P, and changes sign under time reversal, hence violates CP. Notice that the corresponding term in electrodynamics can be eliminated, being the field Abelian

The phases in the mass term of the Lagrangian

$$L_m = \frac{\alpha_s}{8\pi}\mathrm{Arg}(\mathrm{Det}M)\tilde{G}_{\mu\nu}G^{\mu\nu}\,. \tag{5}$$

would violate CP too, but can be eliminated by a phase rotation. The physically meaningful quantity is

$$\bar{\theta} = \theta + \arg\det M\,. \tag{6}$$

This gives a non zero neutron dipole moment of $|d_n|/e$: $10^{-16}\,|\bar{\theta}|$ cm to be compared with the experimental limit $|d_n|/e < 2.9\times10^{-26}$ cm, corresponding to the upper limit

$$|\bar{\theta}| < 10^{-10}\,. \tag{7}$$

The question arises: why is it so small? There should be a reason for that. This is the strong CP problem. Back in 1977 R. Peccei and H. Quinn[18] proposed the following elegant solution. Make the θ parameter dynamical introducing a new pseudoscalar field ϕ_a. Then the term of the Lagrangian can be written

$$L_\theta = \frac{\alpha_s}{8\pi}\left[\tilde{G}^{\mu\nu}G_{\mu\nu}\right]\left(\bar{\theta} - \frac{\phi_a}{f_a}\right). \tag{8}$$

Assume that the Lagrangian originally possesses an U(1) invariance involving the quarks Yukawa couplings, and that the U(1) symmetry spontaneously breaks down (Fig. 2a) at the energy scale f_a and

$$\frac{\phi_a(\mathbf{x},t)}{f_a} \equiv \theta(\mathbf{x},t). \tag{8}$$

Notice that the valley is extremely narrow, with a huge curvature. Very high frequency field oscillations perpendicular to the valley correspond to very large masses and decay immediately. Oscillations along the canyon have zero frequency, corresponding to massless Goldston boson, the axion.

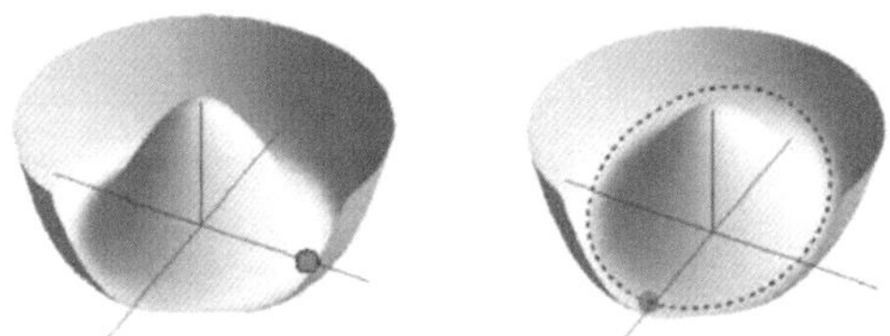

a) b)

Fig. 2. a) PQ symmetry breaking, b) symmetry breaking at the QCD scale

Gerard 't Hooft proved that the QCD axial anomaly (the ABJ anomaly) in non Abelian theories, one vacuum state is chosen among an infinity of possible vacua. The potential corresponds to a particular choice of the phase of the scalar vacuum expectation value. This phase appears in the fermion mass terms.

Peccei and Quinn (PQ) showed that when all fermion masses are made real by rotations of the fermion fields, the resulting θ is zero. The θ-term in the QCD Lagrangian can then be eliminated by absorbing it into the axion field

$$\phi_a = \bar{\phi}_a - f_a\bar{\theta}\,. \tag{9}$$

The topological charge density induced by topological fluctuations of the gluon fields such as QCD instantons, provides a potential for the axion field which is minimized at zero expectation value, as in Fig. 2b)

$$\langle \bar{\phi}_a \rangle = 0\,. \tag{10}$$

This second phase transition happens when the temperature is about Λ_{QCD}

In the original PQ paper f_a was tied to the EW breaking scale. This was subsequently ruled out by the experiments. Models were later developed in which f_a >> VEV. The axion mass, m_a that is inversely proportional to f_a is consequently very small. They were consequently dimmed "Invisible axions".

There are basically two types of extensions of the SM to include PQ axions: the KSVZ model[19], with additional heavy quarks with new charges, called PQ-charges and the DSZ[20] model with two Higgs doublets, and with normal quarks and leptons having PQ charges.

b. Axions. Properties and limits

Axions, the Goldstone bosons of the broken PQ U1 symmetry, have predictable properties, which depend mainly on f_a. As already stated, their mass is inversely proportional to f_a

$$m_a = \frac{\sqrt{m_u/m_d}}{1+m_u/m_d}\frac{f_\pi m_\pi}{f_a} \simeq 6\ \text{meV}\ \frac{10^9\ \text{GeV}}{f_a}. \tag{11}$$

If f_a is very large, the axion is a very weakly interacting sub-eV mass particle. It can contribute to dark matter.

Axions couple to the photons, through the term

$$\frac{G_{a\gamma\gamma}}{4} F_{\mu\nu}\tilde{F}^{\mu\nu}\phi_a = -G_{a\gamma\gamma}\mathbf{E}\cdot\mathbf{B}\phi_a . \tag{12}$$

where, with n a number of the order of unit

$$G_{a\gamma\gamma} \simeq n\frac{\alpha}{2\pi f_a}\frac{m_a}{m_\pi} . \tag{13}$$

Axions decay in to photons via the diagram of Fig. 3

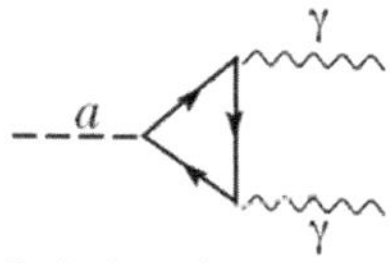

Fig. 3. Axion photon coupling

Calculation of the lifetime gives

$$\tau \sim \frac{1}{G_{a\gamma\gamma}^2 m_a^3} \propto \frac{1}{m_a^5} . \tag{14}$$

With τ of the order of the life of the Universe, $m_a \approx 20$ eV, showing that only very light axions can be dark matter.
Axions couple to fermions through

$$\frac{C_f}{2f_a}\bar{\psi}_f\gamma^\mu\gamma_5\psi_f\phi_a . \tag{15}$$

where C_f is a model dependent coefficient.
Being coupled to photons, axions contribute to the radiation by astrophysical bodies. Too large an energy loss contradicts observed stellar evolution. The non observation of such phenomenon allows fixing upper limits on the coupling and lower limits on the mass. Data from globular clusters, white dwarfs, etc. we have about $f_a < 10^8$ GeV and $m_a > 20$ meV.
From the theoretical point of view, two are the possible scenarios. If PQ symmetry breaks down before inflation, there is only a lower limit for $f_a > 10^{38}$ GeV and m_a can have practically any value smaller than 100 meV. On the contrary, if the break-down is after inflation, as would be the case if BICEP2 is correct, the predictions are rather precise, namely

$$\begin{aligned} f_a &= (8.7\pm 0.2)\times 10^{10}\times(1+\alpha_{dec})^{-6/7}\ \text{GeV} \\ m_a &= (7.1\pm 0.2)(1+\alpha_{dec})^{6/7}\ \mu\text{eV} \end{aligned} . \tag{16}$$

where α_{dec}=0.2-200 takes into account the contributions of cosmic defects that may decay into axions in early epochs. If the Hubble parameter at inflation would have the value (2) from the BICEP2 result, axion mass would lay in a rather narrow range 70 μeV$< m_a <$ 1200 μeV.
Fig. 4[21] shows in yellow (gray) the regions of the parameter space excluded by the above mentioned astronomic data, without considering BICEP2. The two white regions correspond to the two above mentioned scenarios. If BICEP2 is correct, PQ symmetry break down happens after inflation, and the white region on the left is excluded. Even more precisely it predicts the narrow band above $H_I = 10^{14}$ GeV

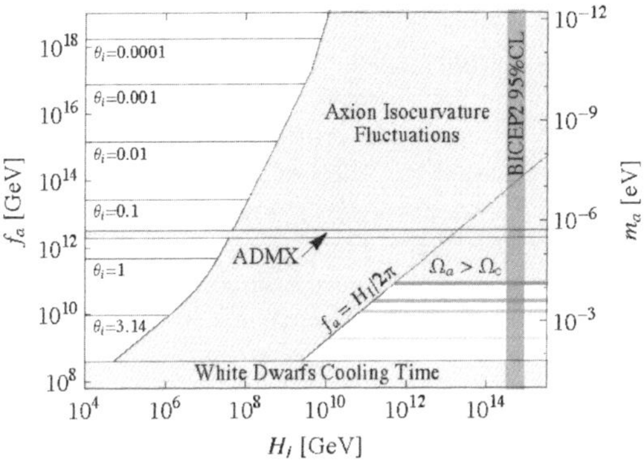

Fig. 4. Cold dark Matter Axion parameter space[22]

c. Axions from the Sun?

In the Sun axions can be produced via Primakoff effect via the diagram of Fig. 5a). The expected flux at Earth is

$$\Phi_a = \left(G_{a\gamma\gamma} \times 10^{10} \text{ GeV}\right)^2 3.75 \times 10^{15} \text{ m}^{-2}\text{s}^{-1}. \quad (17)$$

To detect them we can be converted back to photons (the frequency is of the order of 10^{18} Hz, namely X-rays) through the same mechanism, Fig. 5b) in a strong magnetic (or electric) field. Notice that neither axions or photons are propagation eigenstates. Axion-photon conversion should be viewed as an oscillation.

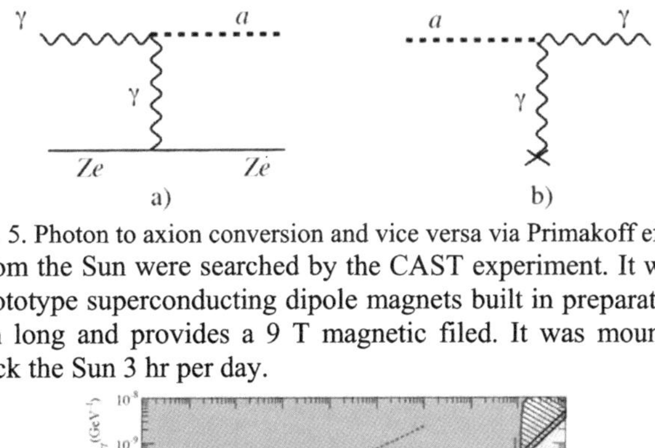

a) b)

Fig. 5. Photon to axion conversion and vice versa via Primakoff effect

Axions coming from the Sun were searched by the CAST experiment. It was using at CERN one of the first prototype superconducting dipole magnets built in preparation of LHC by the INFN. It is 9.6 m long and provides a 9 T magnetic filed. It was mounted on a movable support able to track the Sun 3 hr per day.

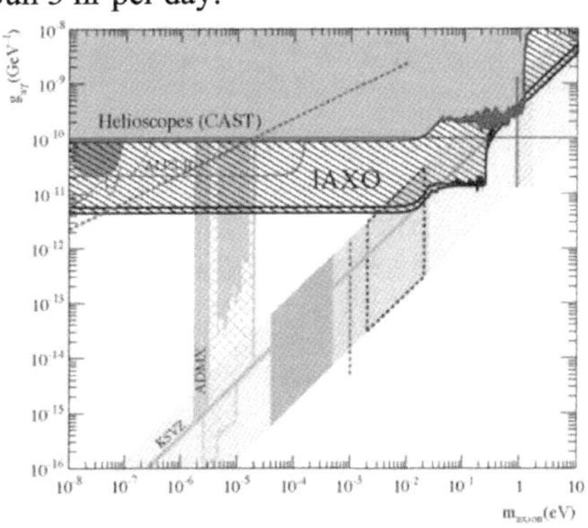

Fig. 6. Regions excluded by CAST and projected sensitivity of IAXO

As we have already discussed, there is a rather well defined correlation between photon axion coupling $g_{a\gamma}$ and axion mass. Axions must be searched in a rather narrow band in these parameters space. This is shown in Fig. 6 together with the region excluded by CAST. This is a small fraction of the allowed band, and, in addition, at fairly large values of the mass.
The IAXO project has been submitted to CERN as a Letter of Intent (CERN SOSC-I-242). It is based on the same principles as for CAST, with several improvements. Unfortunately however, the projected sensitivity is still at pretty high masses, around 0.1 eV. In addition, the explored region in the axion band would be very limited.

d. Axions in the galactic halo

Dark matter axions should be present around us. They can be converted to photons in a strong magnetic field. Considering that the expected signal is extremely small, the sensitivity must be enhanced using a cavity and tuning its resonant frequency to correspond to the photon's one[22]. Being the latter unknown, one needs to scan through the widest possible frequency range. The cavity should be as big as possible with a Q-value as large as possible. The magnetic field should be also large. On the other hand, both thermal and amplifier noise must be minimized.

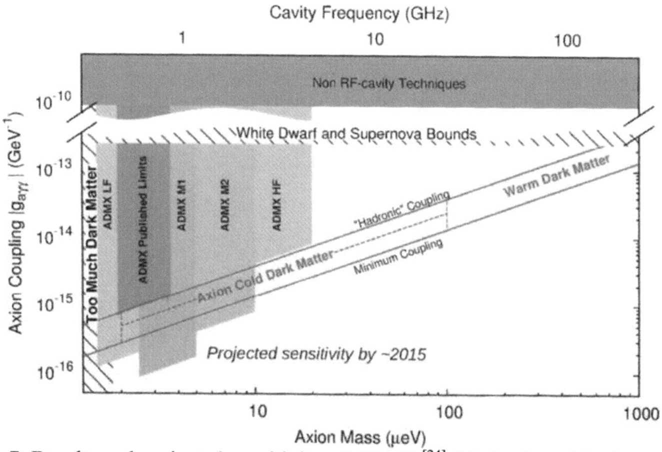

Fig. 7. Results and projected sensitivity of ADMX[24]. Notice logarithmic scales.

This is the principle on which the ADMX experiment works. The pure Cu cavity is tuned by changing the distance between two Cu tuning bars inside it. In the years 1996-2009[23] the achieved tuning range has been between 450 and 850 MHz, corresponding to axion masses between 2 and 4 μeV, but the achieved sensitivity was just touching the axion band, as shown in Fig. 7.
ADMX has already introduced a number of improvements, reducing both physical and amplifier noise. This will allow it to increase the scan velocity by two orders of magnitude, with the projections shown also in Fig. 7[24]. In addition, an R&D programme is ongoing aiming to reach much higher frequencies, 10-100 GHz, which are needed to search up to masses of several hundred μeV.
However, the exploration of all the relevant parameter space, especially in the post-inflationary scenario, namely at several 100 GHz, will require techniques different from the resonant cavity scanning. Their development is still in a preliminary stage.
A possibility under study is the use of magnetic focusing mirrors. Calculations show that dark matter axions around us have an enormous phase space density. Consequently their

propagation can be treated as a classical wave. In addition neither photons nor axions are propagation eigenstates. In a magnetic field the axion wave mixes with the component of the vector potential parallel to **B**. It can also be seen that a sharp boundary surface between a region with and one without magnetic field partially reflects and partially transmits an axion-photon wave.
One can then build a spherical dish in a strong magnetic field to focus the wave in a small region in which to locate a detector[25]. With a 10 m^2 mirror in a 10 T field and a bandwidth $\Delta\omega/\omega$=10^{-6} and a detector thermal noise at the quantum limit a S/N = 14 is expected for one year exposure and an axion mass m_a=10 μeV. This is clearly an extremely difficult situation.
More experiments are clearly needed and much more input of experimental ingenuity. Indeed, this is a sector of fundamental physics in which the experimental art, rather then big experiments, is still needed. Efforts are going on, in particular, in the INFN.
A complementary possibility is the search of monochromatic signals (lines) in the microwave radiation from astronomical sources. Dwarf galaxies are particularly interesting because they are pretty dark and very reach in dark matter. A pioneering search with the Haystack radio observatory, on three dwarf galaxies[26] has provided limits, which however are far from the expectations. Sensitivity needs to be improved by 5-6 orders of magnitudes. This can be achieved by using larger telescopes and with progress in cryogenic detectors techniques.

6. WIMPS. Direct search

The direct searches for WIMPs look at their elastic scattering on nuclei, detecting the energy deposited by the recoiling nucleus.
If dark matter particles are neutralinos, their interaction cross sections can be calculated, in principle, on the basis of SUSY models. However, the unknown parameters of the theory are so many that the results span over many orders of magnitudes in uncertainty. Two basic possible couplings can be distinguished: a) "spin-dependent" (SD), an axial vector coupling to the nucleons spin in which only unpaired nucleons couple; b) "spin-independent" (SI), a scalar interaction with the mass of the nucleons, proportional to A^2. This makes the search in the SI sector more sensitive and we shall concentrate on that. The WIMPs flux is uncertain too, depending on the assumed model of the Galactic halo.
The results of the calculations and of the experiments are usually represented in the plane of the cross-section, reduced to a single nucleon, σ_p vs. the WIMP mass m_W. This corresponds to projecting on a plane the multidimensional parameter space (local W density, SI and SD cross section ratio, ratio of SD coupling to neutrons and to protons, etc.).
There is no firm theoretical upper limit on m_W, but we can reasonably consider a few thousand GeV; the lower limit is about 6 GeV from cosmology. The theoretical expectations for the cross sections have even more uncertainties, ranging from 10^{-5} to 10^{-13} pb. The corresponding rates depend on target nucleus and on energy threshold. Typically, they may range from 1 to 10^{-7} counts/(kg keV yr). The "ultimate" detector should have a sensitive mass of the order of 10 t with a background rate smaller than a few counts per year in a 10 keV window at, say, 30 keV. Today experiments (not looking for modulation) have exposures of the order of 10 – 100 kg d, thresholds at several keV and few events background.
Experimentally, the target coincides with the detector, in which one tries to detect the recoiling nucleus. The energy is deposited in three forms: as ionisation charge, scintillation light and phonons.
The experimental challenges are the extremely low signal rate, small energy deposits (several keV) and a signal spectrum decreasing with increasing energy, similarly to backgrounds. The detectors must be operated in an underground laboratory, but this is not enough. Specific

R&D is needed for achieving the highest radio-purity and for the systematic screening of all the set-up components, including the shields and the detector itself.

We can distinguish three basic backgrounds: 1. electromagnetic, due to γ's hitting electrons, which is the dominant one; 2. neutrons, which give nuclear recoils indistinguishable from WIMPs; 3. radioactive contaminants of the surfaces and their surroundings, which release only part of the energy in the sensitive volume, simulating WIMP signal; surfaces should be avoided as much as possible. The physical processes through which the three types of background deposit energy have differences that can be exploited to identify them. This can be done measuring two quantities amongst ionisation charge, scintillation light and phonons pulse. In addition, the time evolution of each of these signals can be exploited (pulse shape analysis).

Backgrounds can be modelled, but the accuracy and the reliability of the models is limited. Hence, evidence for dark matter cannot be claimed solely by observing a discrepancy with the background model. Consequently, the experiments develop, in a blind analysis, selection criteria to define a "background free" region in the experimental parameter space. Then within the assumed model, for each mass m_χ, the maximum possible signal rate allowed by the data observed when "opening the box" is calculated. The result is reported as an "exclusion plot" in the σ_p vs. m_W plane.

A second method, used by the DAMA collaboration, as we shall see, searches for a specific positive signature, namely the annual modulation of the signal rate expected for the following reason. The solar system moves in the Galaxy with a speed of about 230 km/s. The velocity of our detector in the halo is modulated, with one year period, by the speed of the Earth around the Sun of about 30 km/s. Correspondingly, the signal rate is expected to be maximum in June, when the two velocities are in the same direction, minimum in December when they are opposite. This is a clear signature for the signal. However, several environmental parameters are also annually modulated. The cosmic muon flux underground is maximum in summer and minimum in winter as a consequence of the seasonal variations of the density of the atmosphere. The air temperature is controlled in the experimental hall keeping it constant in time. This implies a seasonal modulation of the humidity, which is usually less accurately controlled, with maximum in summer and minimum in winter. Similar, correlated, seasonal variation of the Rn activity in the air have been measured. Experiments using this approach need to accurately monitor all these parameters.

a. Kinematics

Fig. 8 shows the kinematics of the elastic WIMP nucleus collision defining the variables.

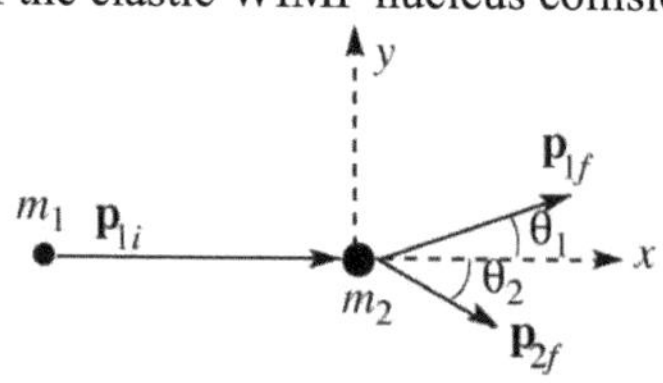

Fig. 8. Kinematic variable of the elastic WIMP nucleus collision

The kinetic energy of the recoiling nucleus is given by

$$T_R = T_W \frac{4m_W m_N}{(m_W + m_N)^2} \cos^2\theta_R = \frac{2m_W^2 m_N}{(m_W + m_N)^2} v_W^2 \cos^2\theta_R \,. \tag{17}$$

The WIMP recoil kinetic energy T_W is largest for m_N=m_W. In this case $T_R = T_W \cos^2\theta_R$. Considering the values of nuclear masses, the sensitivity is usually a maximum for m_W=50-

100 GeV, decreasing rapidly at smaller masses, when the signal goes below threshold, and smoothly at large masses, due too kinematics.

Fig. 9 shows the recoil spectra on Ge nuclei for different WIMP masses. Notice its faster decrease for smaller m_W.

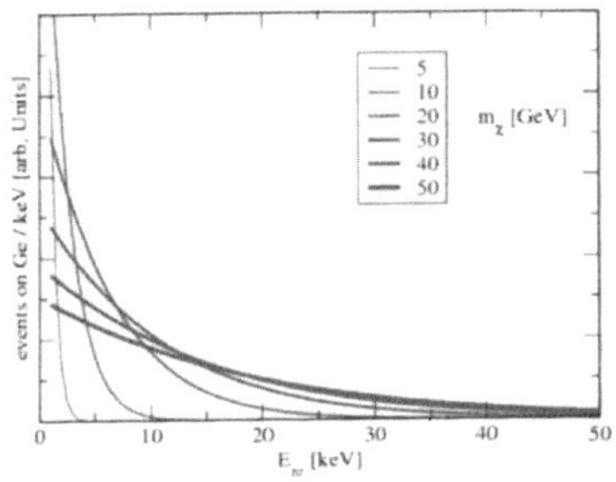

Fig. 9. Ge nuclei recoil spectra for different WIMP masses

The WIMP interaction rate is given by

$$R(E_R)=\frac{\rho\sigma_p A^2 F^2(q)}{2m_W\mu_p^2}\eta(E_R,t). \tag{18}$$

where

$$\eta(E_R,t)=\int d\Omega_v \int_{v_{\min(E_R)}}^{v_{max}} f_{Earth}(\mathbf{v},t)vdv. \tag{19}$$

We see that the WIMP interaction rate is directly proportional to: the local WIMP density ρ [ρ=0.3 GeV/cm^3], the cross section on protons σ_p, the mass of the nucleus squared A^2 for Spin Independent coupling, the nuclear form factor $F(q)$, the time dependent "average" WIMP velocity η and inversely proportional to the WIMP mass m_W and the reduced mass $\mu = M_W m_p / (M_W + m_p)$ squared.

An important quantity, which differs from nucleus to nucleus, is the form factor, i. e. the Fourier transform of the spatial density in the nucleus. Examples, in the Wood Saxon model are shown in Fig. 10. The importance of an accurate knowledge of the form factor especially when comparing different experiments is evident.

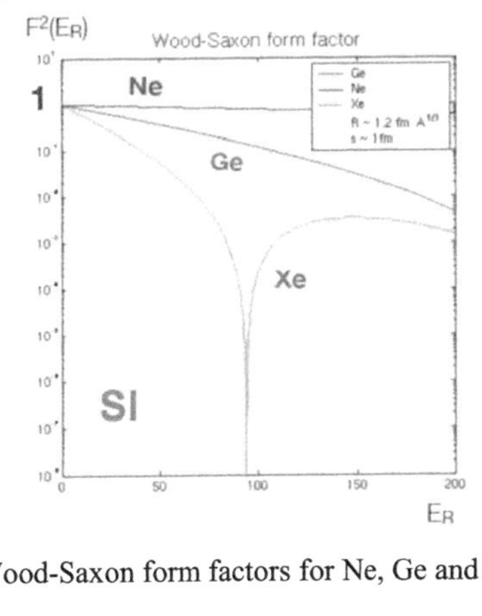

Fig. 10 Wood-Saxon form factors for Ne, Ge and Xe nuclei

The momentum transfer is given by

$$q^2 = 2m_N T_R = T_W \frac{4m_W m_N^2}{(m_W + m_N)^2} \cos^2\theta_R. \tag{19}$$

The A^2 dependence of the cross section is valid for coherent scattering, but not so if the momentum transfer is large enough to resolve the structure of the nucleus, in order of magnitude for $q>1/R$.

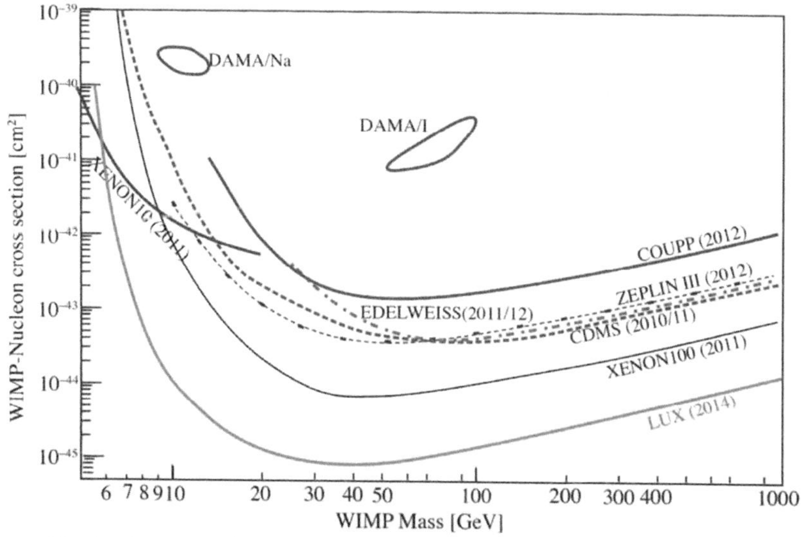

Fig. 11. Summary of the best limits and reported evidence in WIMP searches

Fig. 11 summarises the status of the WIMP search for spin independent coupling. We shall discuss later some examples. Notice that the comparison of different experiments depends somewhat on the assumed halo model and on other assumptions. The open curves are the upper limits of the most sensitive experiments, not searching for modulation, that have found no signal. The two closed curves represent the positive evidence reported by DAMA, the only experiment looking for modulation and not antiselecting the electromagnetic background. The experiment uses NaI(Tl) detectors; the two regions correspond to the assumptions of interaction on one or the other nucleus. Further positive evidence at small WIMP masses, but in regions not compatible with each other either with DAMA, had been reported by two other experiments, CRESST-II[28] and CoGeNT[29]. These results are not shown because they have been later excluded, in a model independent way, respectively by a further run with improved detector by CRESST-II itself[30] and by CDMS[31] using, as CoGeNT, Ge target nuclei.

b. Experiments searching for modulation

DAMA/LIBRA and its predecessor DAMA is the only experiment that has reported a not disproven positive evidence with very high statistical significance. The experiment uses 250 kg of NaI(Tl) detectors produced by St. Gobain on specifications given by DAMA. They are arranged in a 5 x 5 array of rectangular NaI bars, each 10.2 x 10.2 x 25.4 cm^3. Each bar is encapsulated in a radio-pure OFHC copper housing and viewed through special quartz light guides at both end-faces by low-background PMTs. The sensitivity is in the range of 5.5 – 7.5 pe/keV.

The array is enclosed in a graded shielding of radiopure Cu - Pb - Cd - Polyetilene and paraffin. N2 is constantly fluxed in the shielding. All the relevant physical parameters are constantly monitored.

Once the external backgrounds are reduced at the necessary level the internal backgrounds dominate. DAMA developed sophisticated techniques over the years for radio pure NaI (Tl) powders and protocols for crystal growing. Activity of the major radioactive nuclides traces have been measured in the crystals (including cosmogenic ^{125}I, ^{129}I, ^{210}Pb, ^{22}Na, ^{24}Na). In particular: ^{228}Th level is 2 - 30 μBq/kg, ^{238}U several μBq/kg, and ^{nat}K 20 ppb (g/g). The latter is very important due to the presence of ^{40}K, with an abundance $a_{40}=1.17\times10^{-4}$. This nuclide beta decays with a 89% branching to ^{40}Ca (maximum energy Q=1.31 MeV) and 11% EC to ^{40}Ar (E=1.461 MeV), followed by X or Auger de-excitation E_X=3.2 keV, which is just in the WIMP search region. In the majority of the cases the 1.461 MeV gamma is delectated and the event is discarded. But when the gamma ray escapes, the X ray contributes to background. This is in fact the dominating one[32].

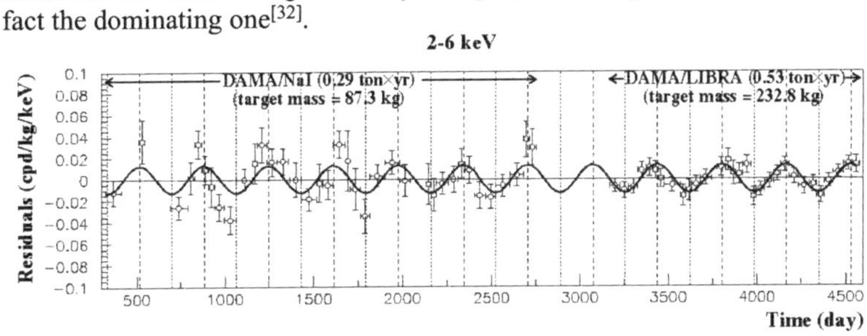

Fig. 12. The DAMA and DAMA/LIBRA modulation signal between 2 and 6 keV recoil energy

The result of an exposure of 0.82 t yr is shown in Fig. 12[33]. The sinusoidal modulation is detected between 2 and 6 keV with 8.2 σ evidence. The fitted amplitude is 0.0131±0.0016, the period is 0.998±0.003 y and the phase of the maximum at 144±8 d (expected 152.5). The result is very clear: the annual modulation with the expected characteristics is there. The question is: is that due to WIMPs or some instrumental effect? On purpose, DAMA has monitored the relevant parameters: temperature, Rn activity, energy threshold, noise level and efficiency. None of them can explain the findings, as cannot do the modulation of the muon flux.

However, some features of the published data (raw data were never made available) which are difficult to understand can be noticed.

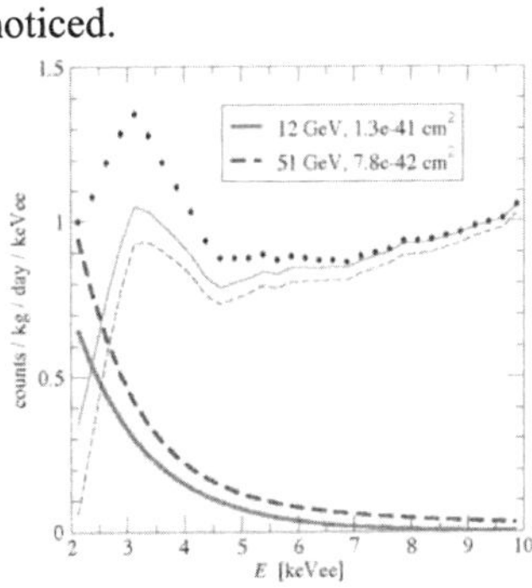

Fig. 13. Energy distribution of the time averaged rate observed in DAMA/LIBRA extracted from Fig.1 of [33] (points), together with the prediction for two examples of WIMP masses and scattering cross sections (thick curves) as well as the corresponding un-identified background (thin curves)[34]

Fig. 13 shows the spectrum, as analysed by Fairbairn and Schwetz[34], were one can see the dominating ^{40}K peak at 3.1 keV, the calculated contribution of the signal, for the two WIMP mass values that best fit the data, and the total background after signal subtraction. The latter appears to give a contribution close to zero at 2 keV to increase to 100% at 6 keV.
Fig. 14. shows the modulated spectrum, which is expected to decrease monotonically, but is instead peaked at 3 keV.

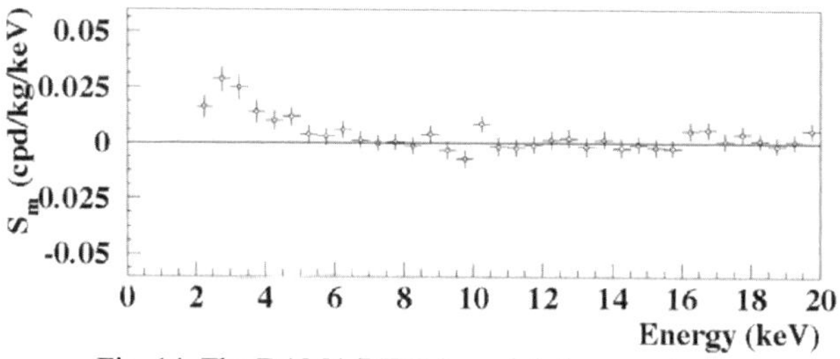

Fig. 14. The DAMA/LIBRA modulation spectrum

Finally, Fig. 15 shows the modulation amplitude measured year by year. While the overall χ^2 is not too large, one notices that almost all the data of DAMA are above the average, vice versa those of LIBRA are lower.

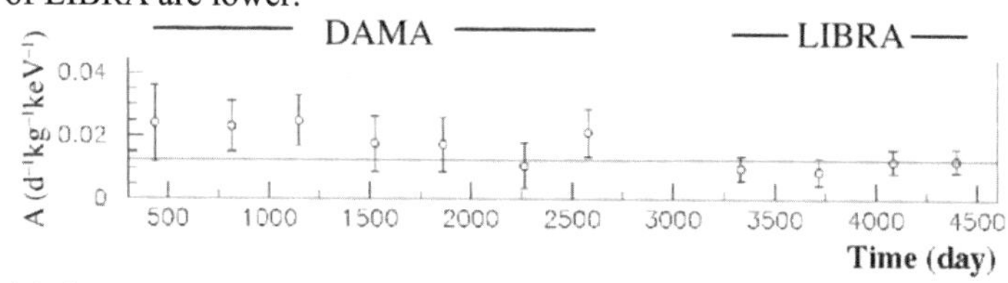

Fig. 15. Modulation amplitudes of each single year of DAMA and DAMA/LIBRA experiments in the (2 – 6) keV energy interval

Looking at Fig.11 one might think that the DAMA result is excluded by several other experiments, but this conclusion is not correct, due to the model dependence of the results as shown. The importance to check the result by independent experiments using the same NaI(Tl) detectors is pretty obvious. Two projects are in an advanced stage ANAIS at LSC in Spain and KIMS in Korea. In both cases prototype crystals are under test in order to reach the needed radiopurity.

c. Noble fluids

Noble fluids, liquids and gases, Xe, Ar, and Ne are employed in the most sensitive detectors. These fluids can be easily assembled in large masses (good scalability), can be cleaned from radioactive traces at extremely high levels. They allow building self-shielding structures with the central part shielded by a large contiguous mass (in the same container) of the same liquid reaching the "no" free surface condition. The shield can be instrumented to act as an active veto form external background sources.
Dark matter detectors must be massive, due to the extremely small scattering cross sections. Consequently the detectors generally employ liquids rather than gases (the exception are direction sensitive detectors, which we shall not discuss here) .
There two basic categories of detectors.

1. Two-phase Time Projection Chambers. Both ionization charges and scintillation light are detected, following the 1993 proposal of P. Picchi and collaborators[35]. The main volume is

liquid inside a uniform electric field used to drift the ionization electrons to the sense wires. Being the electrons very few, charge amplification is needed. This cannot, practically, be implemented in a liquid. Consequently electrons are extracted from the free surface of the liquid in the gas phase above it and here amplified at a grid of sense wires. Scintillation light is also detected by arrays of photomultipliers.

2. Single liquid phase detectors, in which only the scintillation light is detected. This can be done with higher efficiencies than in the TPCs. Electron-nuclear recoil discrimination is made via pulse shape analysis.

Table 1 shows the noble fluids characteristics relevant for ionisation detectors. Notice that Xe has the largest ionisation yield of all liquid rare gases.

Element	Ar	Kr	Xe
Gas			
Ion. Pot, (eV)	15.8	14.0	12.1
W (eV)	26.4	24.2	22.0
Liquid			
Gap energy	14.3	11.7	9.3
W (eV)	23.6	18.4	15.6

Table 1. Noble fluids characteristics

A charged particle moving in a medium loses energy by producing electron ion pairs and by exciting the molecular, or atomic, levels. Not all the produced ion pair remain free however, but a fraction of them recombine. This process is a function of the applied electric field **E**. The recombination can happen between the two original partners, called geminate recombination and originally described, at low energies, by Onsager[36], and between different partners, called columnar recombination, and originally described by Jaffe[37].

Alpha particles ionize heavily and nuclear recoils even more, producing a cylindrical track in which the largest fraction of energy is deposited (delta rays are outside it) with a high recombination rate. Typically 10% of the charge is collected at fields E<20 kV/cm. For comparison, the charge collections for electrons from a typical calibration source of 570 keV gammas is >90% already at E=5 kV/cm.

The effect of electric field on charge recombination and, consequently, on the charge and light yield on electrons measured with a gamma source of 662 keV by E. Aprile et al.[38] is shown in Fig. 16.

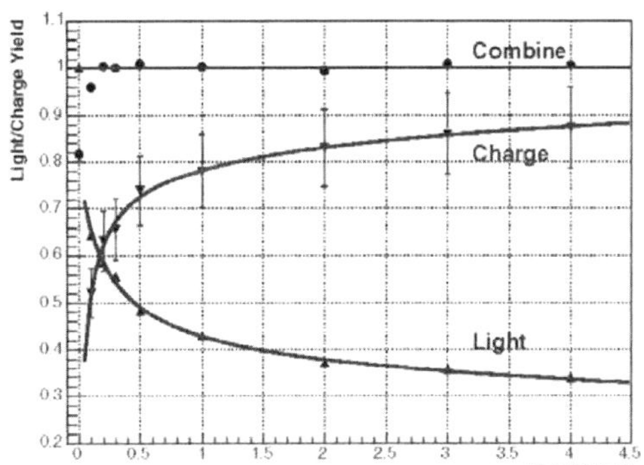

Fig. 16. Light to charge yield in liquid Xe for Compton electron produced by 662 keV γ-rays as a function of the electric field[38]

In the majority of the detectors, the energy deposited in its medium is evaluated from the detected light, which is the measured quantity. However, the efficiency of the process, namely the ratio between scintillation light and energy, is different for electrons and nuclei of the same kinetic energy. This ratio, called light efficiency (or quenching factor), L_{eff}, which is a function both of energy and of the electric field, must be accurately known down to the energy threshold.

To measure L_{eff} one uses a monochromatic neutron source, employing typically the $d+d \rightarrow {}^3He+n$ fusion reaction producing neutrons of 2.45 MeV kinetic energy. The neutron beam illuminates the target under study (liquid Ar, liquid Xe, NaI, etc.). The elastic scattering cross section is measured, with a neutron detector, as a function of the scattering angle θ. For elastic scattering, the measurement of the scattering angle gives the energy of the recoil

$$E_r = 2E_n \frac{m_n m_N}{(m_n + m_N)^2}(1-\cos\theta). \tag{20}$$

which is compared to the separately measured light yield.

Fig. 17 shows an example of L_{eff} measurement in liquid Xe[39].

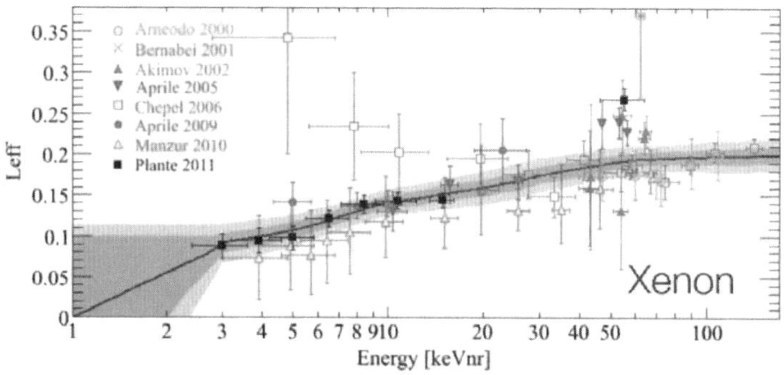

Fig. 17. Light yield for nuclear recoils in liquid Xe[39]

Notice that the unit on the horizontal axis, keV_{nr}, is the detected recoil energy. The units for the total recoil energy, namely as it were for electrons, is usually called keV_{ee}. Notice that at low recoil energies both the scattering angle, which becomes small, and the recoil energy itself become difficult to measure. This is the reason not having data below 3 keV_{nr}. The blue band is a generous estimate of the uncertainty in the lowest energy region.

A further experimental function that is measured with similar calibration programmes is the light yield for electron recoils.

When ion pairs are produced in liquid Ar and liquid Xe, excited dimers (called excimers) and molecular ions are formed by double collisions with metastable atoms as $Xe^*+Xe+Xe \rightarrow Xe_2^*+Xe$ followed by the decay $Xe_2^* \rightarrow 2Xe+\gamma$. At **E**=0 scintillation light in LAr and LXe about 70% due to excited molecules produced by free electrons recombination about 30% due to self trapped excitons.

In liquid Ar the UV luminescence is due to the transitions to the ground level (two separate atoms) of the lowest molecular levels ${}^1\Sigma_u$ (Λ=0, total spin=0) and ${}^3\Sigma_u$ (Λ=0, total spin=1), within picoseconds. The singlet decay is strongly allowed (few ns lifetime). The triplet decay need to flip one electron spin (S=1$\rightarrow$S=0) and cannot go emitting a photon. It needs to wait for collisions. In Xe similar situation. In any case the emitted radiation is not re-adsorbed because too low energy to excite atomic levels.

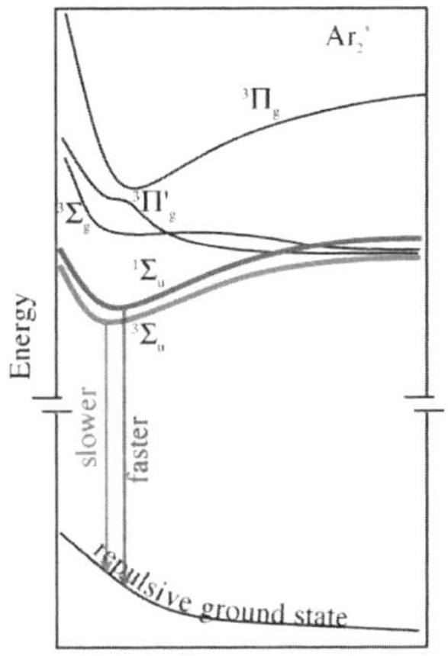

Fig. 18. Ar excimers triplet and singlet energy levels vs. interatomic distance

The dependence of the Ar excimer relevant energy levels is shown, approximately and simplified, in Fig. 18.

	T_1(ns)	T_3(ns)
Ne	5	15400
Ar	7	1600
Xe	4	22

Table 2. Singlet and triplet excimers lifetimes

The de-excitation curves have two components: one fast from the singlet $^1\Sigma$ states, which are populated mainly by the heavy recoils, and one slow from the triplet $^3\Sigma$ states, populated mainly by the light recoils. This behaviour has been discovered and studied several years ago by T. Doke and collaborators[40]. Table 2 shows the singlet (T_1) and triplet lifetime for Ne, Ar and Xe. Clearly the pulse shape analysis is particularly powerful in discriminating heavy from light recoils for Ar and Ne (which is however too light as a WIMP target). A disadvantage of Ar is the presence of cosmogenic ^{39}Ar with beta decays at specific activity of 1 Bq/kg. This must be identified via PSA and suppressed by 8 orders of magnitude or more to reach the present sensitivity of the Xe based detectors.

Fig. 19 shows the two detector concepts: single phase with luminescence read out only and double phase with charge and luminescence read-out.

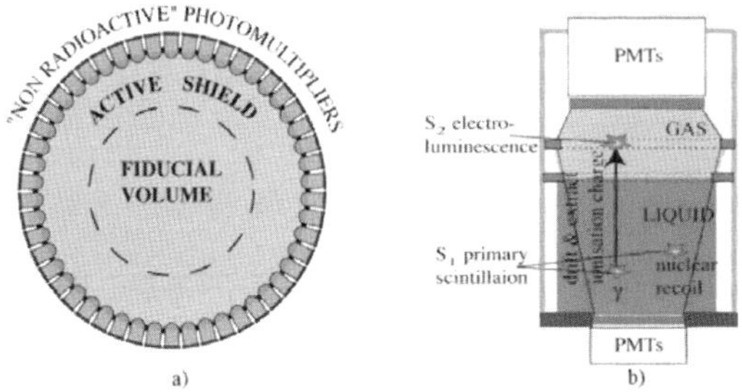

Fig. 19. a) Principle of a self-screening noble liquid detector, b) two phase TPC.

A second tool to discriminate nuclear from electron recoil is the ratio of ionisation to scintillation that is much smaller for the former ones. The ionisation charges can be detected by the proportional scintillation produced in the very strong electric field near to a thin anode wire. As we proposed in 1993[35], with the same photo-sensors one can then measure two light pulses, the first, S_1 due to the primary scintillation, the second, S_2 due to the proportional scintillation. The ratio S_2/S_1 is large for minimum ionising particles (γ background) and small for nuclear recoils, or α particles. The evolution of the method lead to the dual-phase detectors[41],[42]. As shown schematically in Fig. 19b, a uniform electric field is applied to drift the electrons, produced in the liquid, vertically up to the surface and extract them into the gas phase (with almost 100% efficiency); here they are amplified using appropriate grid structures and the proportional scintillation light is detected. A few considerations are in order. The ionisation electrons are very few and must be collected with high efficiency; in particular, the electric field must be accurately designed, in order, for example, not to "lose" electron in the non-uniformities close to the race-track. Photons must be efficiently collected too, because they are so few (e.g. ≈200 at 16 keV in Xe). Another important aspect is the following. Clearly, the liquid in the innermost regions of the detector are shielded against incoming background radiation (typically gammas) by the outer parts of the same liquid. One can then define a central "fiducial volume" that may be background free, or almost so. It is then important to be able to measure the position of the events with enough resolution for an accurate "fiducialisation". This can be achieved basically in two ways: building the detector as an image chamber or using arrays of photomultipliers.

Three ton-scale liquid Ar or Xe single phase detectors are running or have been completed and are starting operation. The detectors are lodged in ultra pure water tanks of about 10 m diameters to reduce by several orders of magnitude the environmental radiation of gamma and neutrons.

XMASS-RFB at the Kamioka Observatory in Japan has 835 kg of liquid Xe, of which the central 100 kg fiducial, the peripheral 735 are used as an active shield.

CLEAN at SNOLab in Canada has 500 kg of liquid Ar of which 150 kg fiducial.

DEAP, also at SNOLab, has 3600 kg (1 t fiducial)

Five double phase TPC detectors are also running or close to do that. Again, the external shields are ultrapure water tanks.

XENON100 at LNGS in Italy has 161 kg of liquid Xe (about 50 kg fiducial). The experiment already published several results, in particular the second best limit (2011) for SI coupling shown in Fig.11. It is still running for more science.

LUX at SURF in the USA has 350 of liquid Xe (100 kg fiducial). Already published the best limit on SI coupling, shown in Fig. 11 and is still running.

PANDAX at CJPL in China has 125 kg of liquid Xe (25 kg fiducial)

ArDM at LSC in Spain has 850 kg of liquid Ar (100 kg fiducial). Will start data taking in 2015

DarkSide at LNGS has 50 kg (33 kg fiducial) has 50 kg of liquid Ar. The very important feature is that the Ar have been obtained by underground extraction and is depleted in ^{39}Ar.

The liquid Ar detectors are expected to produce results comparable to the Xe ones in a few years. Indeed, the two techniques are complementary in several aspects and, joined together, may pave the road to multi-ton detectors. Considering that the dark matter search needs to work in "background free" conditions an increase by, for example, of an order of magnitude in the sensitive mass needs a decrease of the background index by an order of magnitude. DARWIN[43] is a European design study with the goal to complete the necessary research for the construction of the "ultimate" dark matter detector, using several tons of liquid xenon and/or liquid argon for the direct detection of particle dark matter.

d. Bolometers

Bolometers are small devices with 0.1 kg order masses operating at cryogenic temperatures. They are assembled in arrays to reach the hundred kg scale.

Advantages of bolometers are the low energy threshold (keV, sub-keV), the small work function (the energy to produce an electron hole pair), which is W=3 eV/pair for Ge and W=3.8 eV/pair for Si.

A bolometer is a small calorimeter, providing very good energy resolution. The operation principle is shown in Fig. 20 a). The crystal is in contact with a thermal bath at milliKelvin temperature, where the heat capacity is very small, being given by

$$C(T) \propto \frac{m}{M}\left(\frac{T}{\Theta_D}\right)^3. \tag{20}$$

Where m is the absorber mass, M the absorber molar mass and Θ_D the Debay temperature.

The energy deposited by a recoil generates a temperature rise that is measured by a W thermometer. The energy resolution is very good. Being all the energy measured they work with L_{eff}=1.

Good electron recoil vs nuclear recoil discrimination is obtained by measuring light and phonons (CRESST) or charge and phonons (CDMS and EDELWEISS). Given the small size of such detectors, surface backgrounds are very important. Surface vs bulk discrimination is obtained via pulse shape analysis on athermal phonons.

Disadvantages are the need of using milliKelvin technology, the interference between shielding and cryostat and the presence of surfaces facing detectors in vacuum. In addition, they are difficult to scale.

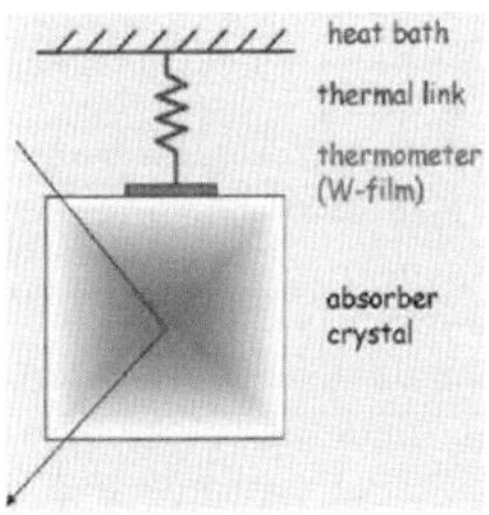

a)

b)

Fig. 20 a) Principle of the bolmeter, b) One ZIP detector of CDMS

Three projects have been developed in different laboratories, each in a series of experiments of increasing mass and complicity and decreasing background in the region of interest.

CRESST-II at LNGS in Italy operating with arrays 300g $CaWO_4$ crystals

CDMS in the past and presently superCDMS, operating 5 towers of advanced iZIP detectors (~9 kg Ge) at the Soudan Underground Laboratory in the USA. Plans for the future foresee a 200 kg array to be operated at SNOLab in Canada. A ZIP detector (of the previous generation) is shown in Fig. 20 b).

EDELWEISS, now at its III stage, operating with 40 detectors, called FID, in a 32 kg array in the LSM in France. Plan for the future is EURECA to be operated in a planned LSM extension.

The three experimental collaborations have developed different techniques to discriminate the nuclear recoil expected signal from the two main backgrounds: electron recoils and surface events. Being the ratio surface to volume much larger than in the noble liquid detectors, the latter background is very important.

For reasons of space we shall limit the discussion of the athermal phonons to CDMS[44].

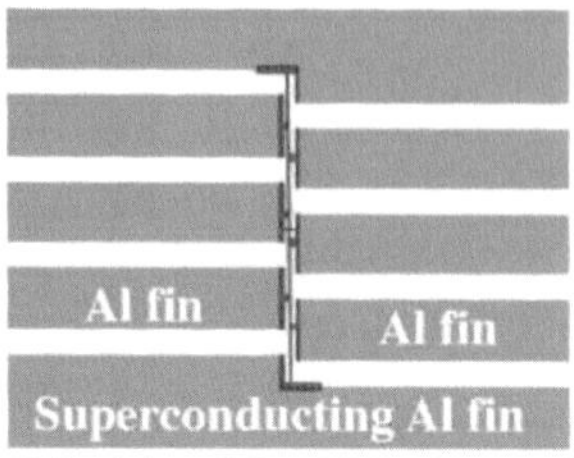

Fig. 21. Principle of the QET phonon detector

Fig. 21 shows schematically the "Quasiparticle-assisted Electrothermal-feedabck Transition-edge sensor", QET[45]. Most of the surface of the crystal is covered with Al films (fins). They are well below the critical temperature $Tc \sim 1$ K. The energy deposited by a recoil produce athermal phonons, which strongly interact with the Al, breaking Cooper pairs and producing quasiparticle excitations. These excitations diffuse from the Al to the Transition Edge Sensors (TES), where they deposit their energy heating the W electron system. The TES are 1 μm wide W strips, which the small ones at the centre of the figure. The quasiparticles are trapped into the tungsten and cause its electrical resistance to increase. The resulting decrease in current through the sensor is measured with a DC SQUID array.
Ballistc phonons (< 1 THZ) travel at speeds of the order of the sound velocity υ_{sound}. They do not scatter frequently. Electron recoils create a higher initial population of ballistic phonons than nuclear recoils. The consequence is a faster phonon leading edge for electron recoils, allowing discrimination. If energy is deposited near the surface there is an excess of ballistic phonons. The resulting phonon signal leading edge rise-time even faster than for electron recoils.
Fig. 11 shows the limits obtained by CDMS2 exposure[46]

e. Future perspectives

A community study of the perspectives of the direct search for WIMPs, from the USA point of view, was performed at Snowmass in 2013[47]. Fig. 22 shows their compilation of WIMP-nucleon spin independent cross section limits (solid curves), hints for WIMP signals (shaded closed contours – including those that were later confuted) and projections (dot and dot-dashed curves) for experiments with important US participation that are expected to operate over the next decade. European (DARWIN), Japanese and Chinese projects are not included. The figure includes an approximate estimate of the limits where coherent scattering of solar, atmospheric and diffuse supernova neutrinos with nuclei will begin to limit the sensitivity of direct detection experiments to WIMPs. Some theoretical model predictions are also indicated.

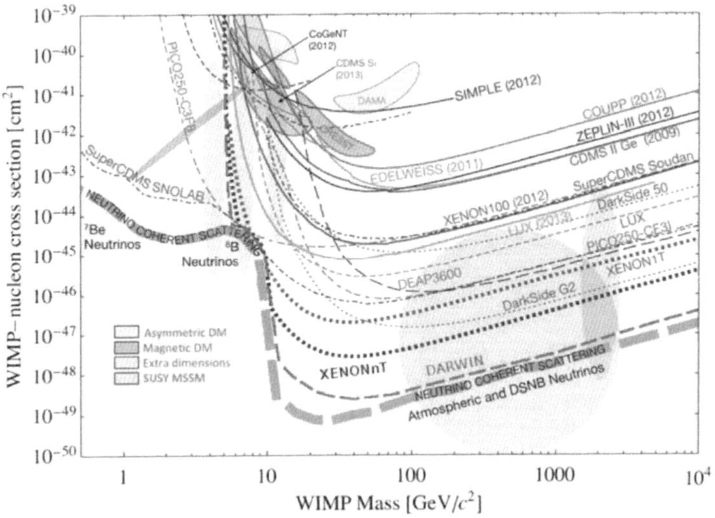

Fig. 22. Limits of the existing experiments and projected sensitivities of their evolution and of new US projects

7. Conclusions

Dark matter search is a major challenge for particle physics. Dark matter may consist of more than one component, but we have no strong guidance on where to search. We should not look at a single candidate. However, the searches require considerable investments in economic and human resources. Their use should be optimised. For that, the following elements should be considered.

To be a substantial part of dark matter, sterile neutrinos require several ad hoc hypothesis.

Axions are well theoretically motivated and we have clear indication on were to look for them. However, only recently efforts are being directed there. Experiments searching for axions outside their parameter space claim to search for ALPs, Axion Like Particles, which do not have any solid theoretical motivation (and consequently can be anywhere).

WIMPs are motivated by SUSY with R-parity conservation. Search is presently a large world wide effort, which will continue.

A number of experimental clever techniques to suppress backgrounds have been developed: single an double phase noble fluids detectors, scintillating crystals, bolometers, superheated detectors and bubble chambers (not treated here, but see COUPP results in Fig.11). Indeed, it is necessary to have several technologies in different locations.

Several projects are already being designed towards the limit sensitivity for the WIMP search. In the coming years, LHC may be so kind to provide evidence of dark matter candidate particles.

8. References

1. F. Zwicky; Helv. Phys. Acta. **6** (1933) 110
2. S. van den Bergh; astro-ph/9904 251
3. L. Bergström. Rep. Prog. Phys. **63** (2000) 793; hep-ph/0002126
4. V. C. Rubin and W. K. Ford; Ap. J. **159** (1970) 379
5. D. Clowe et al.; Astr. Journ, 648:L109-L113, 2006
6. PLANCK collaboration; arxiv:1303.5076v3
7. BICEP2 Collaboration. P. A. R. Ade et al.; Phys. Rev. Lett. **112** (2014) 241101;

arXiv:1403.3985v2
8. PLANCK Collaboration; arXiv:1409.5738
9. G. Mangano et al.; Phys. Lett. **B534** (2002) 8-16
10. Th. A. Mueller et al.; Phys. Rev.; **C 83** (2011) 054615
11. 4. G. Mention et al. Phys. Rev.; **D 83** (2011) 073006
12. 5. C. Zhang et al. Phys. Rev. **D 87** (2013) 073018
13. C. Polly. Results from MiniBooNE experiment. Neutrino 2012 Conference. Kyoto
14. M. Antonello et al. (ICARUS) Eur. Phys. J. C (2013) 73:2599; arXiv:1209.0122
15. N. Agafonova et al. (OPERA), JHEP **1307** (2013) 004, arXiv:1303.3953
16. A. Boyarsky, O. Ruchayskiy, M. Shaposhnikov; arXiv 0901:0011
17. E. Balibut et al.; arXiv:1402:2301 v2
18. R. D. Peccei and H. R. Quinn; Phys Rev Lett **38** (1977) 1440
19. J. E. Kim; Phys Rev Lett. **43** (1979) 103 and M. Shifman, A. Vainstein, V. Zacharov; Nucl Phys **B 166** (1980) 493
20. M. Dine, W. Fischler, M. Srednicki; Phys Lett **B 104** (1981) 199 and A. R. Zhitniski; Sov. J. Nucll. Phys. **31** (1980) 260
21. L. Visinelli and P. Gondolo; arXiv: 1403.4594
22. P. Sikivie; Phys. Rev. Lett. **51** (1983) 1415
23. Astzalos et al.; Phys. Rev. Lett. **104** (2010) 041301
24. G. Rybka; Proceedings of TAUP 2013 to be published
25. J. Jaeckel and J. Redondo, arXiv:1308.1103
26. B.D. Blout et al.; Astrophys. J. **546** (2001) 825
27. E. Aprile e al.; Phys. Rev. Lett. **109** (2012) 181301
28. G. Angloher et al.; Eur. Phys. J.; **C 72** (2012), arXiv:1109.0702
29. C. E. Aalseth et al.; Phys. Rev. **D 88** (2013) 012002, arXiv:1208.5737
30. R. Agnese et al.; Phys. Rev. Lett. **112** (2014) 241302
31. G. Angloher et al.; arXiv:1407.3146
32. L. Bernabei et al. NIM A **592** (2008) 297
33. L. Bernabei et al.; Eur. Phys. J.; **C56** (2008) 333, arXiv:0804.2641
34. M. Fairbairn and T. Schwetz; arXiv: 0808.0704
35. P. Benetti et al.; Nuclear Instruments and Methods **A327** (1993) 203
36. L. Onsager; Phys. Rev. **54** (1938) 544
37. G. Jaffe; Ann. der Phys. **42** (1913) 303
38. E. Aprile et al.; Phys. Rev. **B 76** (2007) 014115, arXiv:0704.1118
39. G. Plante et al.; Phys. Rev. **C 84** (2011) 045805
40. Hitachi et al. Phys Rev **B** (1983) 5279
41. D. Cline et al.; Astrop. Phys. **12** (2000) 377
42. M. Yamashita et al.; Astrop. Phys. **20** (2003) 79
43. http://darwin.physik.uzh.ch
44. D. S. Akerib et al.; Phys Rev **D72** (2005) 052009
45. K. D. Irwin et al.; Rev. Sci. Instrum. **66** (1995) 5322
46. Z. Ahmed et al.; Science **327** (2010); Phys. Rev. Lett. **106**, 131302
47. D. Bauer et al.; arXiv:1310.8327

LECTURER: A. BETTINI

Dark Matter Searches

Scientific Secretaries: P. Kakin, V. Syrotenko, S. Raspopov, O. Iarygina

DISCUSSION

\- *Millington:*

My question is in relation to the heavy neutrino mass spectra that you discussed. So aside from the high GUT scale mass spectra and low scale Shaposhnikov scenarios one can also construct models with electroweak scale quasi-degenerate mass spectra that as far as I know fit well in all current experimental constraints. Would you be able to comment on such scenarios and potential for models of asymmetric dark matter?

\- *Bettini:*

You mean right-handed neutrinos at electroweak scale? To evade the experimental and the observational facts, you must assume that such neutrinos should be almost completely decoupled from normal matter. In particular they do not thermalize. But under such assumptions almost anything can be imagined. This clearly does not prove they do not exist.

\- *Zennamo:*

Can you comment on the proposed solid noble elements detectors i.e. the idea of growing Xenon crystals that can be used for dark matter search?

\- *Bettini:*

Why use crystals when the liquid is so easy to use? It seems like a really complicated thing to do. In addition, crystals have surfaces which are sources of background.

\- *Zennamo:*

Do you have any idea about what will happen when we push below the neutrino floor?

\- *Bettini:*

It might take many years before we get there. In addition, that is not necessarily an absolute floor; there might be new ideas to go further, if really still interesting.

\- *Zennamo:*

Why in a double phase TPC the read out wires are in the gas rather then in the liquid?

\- *Bettini:*

In the gas, not in a liquid, the number of electrons can be amplified. Near a wire the electric field is high and accelerates the primary electrons. They get enough energy to break the gas molecules and produce more free electrons. These are accelerated in turn, etc. and you amplify the charge. In a liquid, the electrons mean free path between collisions is too short to have this effect.

\- *'t Hooft:*

My question is about that mass matrix in the see-saw, namely about the zero on the top left corner. It is needed to have the see-saw mechanism. But, even if you have a zero in the Lagrangian you start with, any interaction or any perturbation, will make the zero going away. Isn't that zero an unnatural zero?

- *Bettini:*

I do not know, I have to look.

- *'t Hooft:*

Maybe someone else knows? It is a very important zero. There has to be some principle that says you have a zero there.

- *Bettini:*

If you put something different from zero, you don't get the see-saw.

- *Minkowski:*

The zero mass of left neutrinos, which are isodoublets, is protected by the symmetries of the Standard Model. A non-zero value would be an isotriplet.

- *Leontsinis:*

On slide 22 you see this 3.5 keV line and I want to make a comment that there was another publication, which came in a week after this one, which confirms that finding. It would be great to see more details on this measurement.

- *Bettini:*

Yes, you are right, there are two publications. They look roughly at the same data and give the same result. They collected spectra from several observatories at different red-shifts, then they blue shifted to the present era each of them and superposed them. They calculated the overall acceptance of the superposition. This shows a wide minimum just in the “signal” region. The "signal" itself is statistically very marginal, about 3σ. Notice that the probability to find a fluctuation of the background at this level, on all the spectrum is not negligible. In addition, it results from subtracting a linear spectrum from an assumed concave background. This is a very dangerous procedure. Both claims appear to be neither systematically nor statistically significant. You have to wait for better measurements.

- *Miyazaki:*

My question is about axions. There are two types of the axions detectors: one is helioscope and the other is haloscope. The former is sensitive to the heavier region, right? And the latter is sensitive to the lighter region. But the assumption is slightly different: the helioscope assumes only the existence of the axions, and the haloscope experiment assumes both the existence and that the axion is cold dark matter. So, suppose, if the IAXO will detect or observe axion in the future, in such case, how to justify which is related to the dark matter, where we need the haloscope experiment? Is there any other ideas, how to extrapolate the ADMX type measurement?

- *Bettini:*

The so called helioscopes search for axions produced in the Sun, the haloscopes search for axions in the Galaxy. They may be the same particle. To detect solar axions experiments use the Primakoff effect, converting axions to photons in a strong magnetic field a detecting photons. For the Peccei Quinn axions there is a strong correlation between mass and coupling constant, a diagonal band in the plane of these two parameters. The CAST experiment excludes a very small region in the axion band at masses > 100 meV. IAXO, if it will be approved, it can reach about 10 meV mass, only if the coupling is on the high side. This is marginally compatible with the model dependent limit coming from white dwarf cooling time measurements. However, if something is there it can be well a component of dark matter. There are already thoughts on how to explore the axion band at masses larger than those reachable of ADMX and smaller than CAST, which is the most interesting region. However, more effort would be needed here.

- *Kolevatov:*

Can you tell once more, where does the lower limit of the axion mass come from?

- *Bettini:*

It depends on the epoch of the U(1) symmetry breakdown. If it happens before the inflation, then there is practically no theoretical lower limit to the axion mass. If it is after inflation, there are both a lower and an upper limit that come directly from the theory.

- *Kolevatov:*

Another question is about the experiment with 3.5 keV line. Some time ago it was shown there is no such line in the Milky Way. If we believe that this line is present in some clusters of galaxies, does it mean that the dark matter could be the mixture of the different particles in the different areas of space?

- *Bettini:*

There is not real evidence for that, you can not trust 3σ effects. Given the number of observations that are done, we can expect such fluctuation rather frequently. We have to wait for better measurements.

- *Ponomarev:*

First, I should warn that I am not working in the field so maybe my question is a bit stupid. So am I right that you discussed a constraint put by CMB on number of neutrino species?

- *Bettini:*

You are right. CMB gives information on the effective number of light neutrinos.

- *Ponomarev:*

So then I was recently present at the talk where it was reported on extraction on the number of neutrino species from primordial deuterium abundance measurement. Results are more or less as precise as from CMB. And the experiment is much cheaper. So can you comment on that?

- *Bettini:*

To what do you refer?

- *Ponomarev:*

To the observations of absorption of light from first stars in deuterium. The resulting evaluation of the number of neutrino species and results are more or less the same than that found from CMB: $N_{eff}= 3.3\pm 0.3$.

- *Bettini:*

Thanks for the information. [Note added after the discussion. The quoted $N_{\rm eff}$ is the result of combining the primordial deuterium abundance with the Planck CMB map]

- *'t Hooft:*

Thank you very much. Are there any other questions? Or comments?

- *Ragazzi:*

To the question on the neutrino background floor and what to do when it hits that level. This neutrino background, solar, atmospheric and relic supernova neutrinos, is certainly very interesting by itself. It would be great to be able to look at that.

- *Ferroni:*

Some comments on BICEP2 results. What have you understood after the collection of information you have got. I mean where are we really?

- *Bettini:*

To me it looks like a serious piece of work. The main problem seems to be the contribution to the observed B-mode polarization by any polarized powder between the CMB and ourselves. A powder map cannot be measured from the surface, but only from outside the atmosphere. Planck did not publish the map yet, but only a mean value. BICEP evaluated the foreground contribution with different models, as I showed this morning. All of them give contributions that are much smaller than the claimed signal. We need to wait for the Planck results. Planck is less sensitive than BICEP2, but looks in a much wider part of the sky. The two effects tend to compensate each other, but we obviously cannot say anything now. It is pretty sure however that a powder map will be produced. May be this will happen pretty soon.

- *Millington:*

I just have a comment in relation to the earlier question about the zero in the mass matrix. There are papers looking for radiative inverse see-saw models. And as far as I can recall that zero is also protected from radiative corrections in the see-saw matrix.

- *'t Hooft:*

Yes, that is what Peter Minkowski also said. I think it is what he implied. So, any further remarks, questions? Then let's thank Bettini for a lovely discussion.

Highlights from ALICE

Paolo Giubellino
INFN Torino and CERN Geneve
E-mail: paolo.giubellino@CERN.ch

The LHC has delivered for the first time collisions of Nuclei in November 2010, at an energy of 2.76 TeV per nucleon pair, which represents a jump of more than an order of magnitude over the highest energy nuclear collisions ever studied before. The high energy, the quality of the state-of-the art detectors, and the readiness of the experimental collaborations at the LHC have allowed a rich harvest of important scientific results. In this lecture a short overview will be given of how the results from the LHC, and in particular from the ALICE experiment, have provided new insight on the properties of matter under extreme conditions of temperature and pressure, analogous to the conditions present in the early phases of the evolution of the Universe. Results from the proton-nucleus run, which took place in early 2013, will also be presented. Finally, a very short outlook to the future will be given.

1.Why high-energy Heavy-Ion Collisions

Collisions between very high-energy nuclei, normally referred to as Heavy-Ions since they are atoms stripped of their electrons to allow acceleration, produce each a huge number of particles, over ten thousand in a single event at the LHC. A typical event is shown in Fig. 1. The study of such extremely complex events poses a formidable challenge both to the detections systems and to the software and analysis methods to be used to treat and eventually understand the data. So, it is legitimate to ask first of all the question of why experimenters are willing to face such extreme problems, what are the underlying questions which we aim to answer.

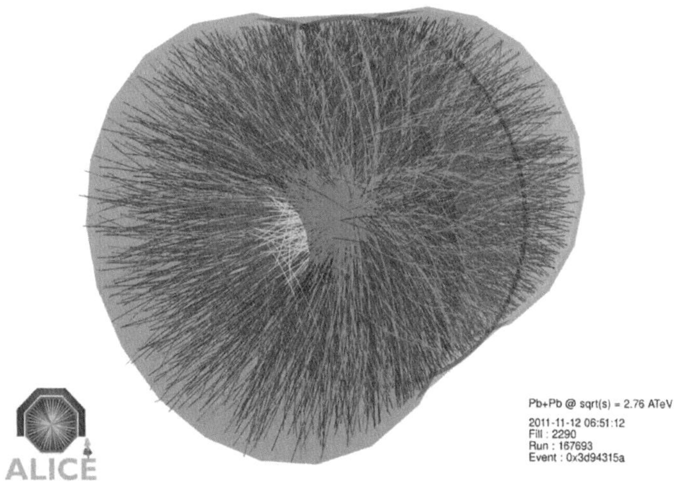

Fig. 1 A collision of two Lead Nuclei as seen by the ALICE detector at the LHC. (CERN-EX-1111290 04)

The strong force, which holds together the nucleons, protons and neutrons, in nuclei and the quarks within the nucleons, is described to great precision by a very successful theory, called Quantum Chromo Dynamics (QCD), in which the quarks possess a color charge and interact exchanging force carriers, the gluons, which are also colored. A fundamental feature of the interaction is the fact that it grows with the distance between the quarks, much like a spring connecting two balls. Therefore, any attempt to isolate a quark are in vain, since the energy stored in the spring will eventually be more than the mass of a quart-antiquark pair: the spring will break generating a pair, and the quarks will not be isolated. This property is called "confinement" and has major consequences. In particular, most of the mass of the hadrons is to be attributed to this feature: more technically, one can say that the mass is acquired dynamically because of the interaction between the quarks, whose Higgs mass would account for just a small fraction, about 1%, of the mass of the light hadrons as shown in Fig. 2. Therefore, since the mass of the matter around us is essentially the mass of the nuclei of the atoms, it can be said that, while the Higgs mechanism is at the base of the masses of the quarks, strong interaction is at the origin of most of the mass of ordinary matter.

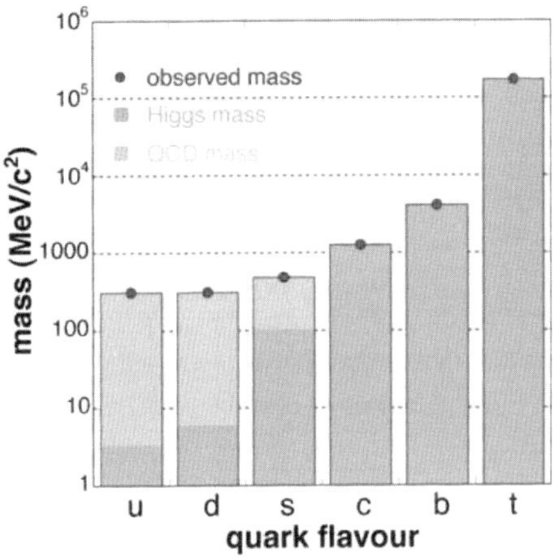

Fig. 2 QCD and Higgs mass components for the light and heay quarks from Ref [1]

The methodof colliding high-energy nuclei to heat and compress hadronic matter dates from the early eighties, but the path had been indicated already several years before. It was in 1975 that T.D. Lee pronounced the famous sentence "it would be interesting to explore new phenomena by distributing a high amount of energy or high nuclear density over a relatively large volume", and the idea of deconfinement had been around for some time. Already in 1965 Hagedornhad observed that the mass spectrum of hadronic states led naturally to the concept of a "limiting" or "critical" temperature beyond which the concept of hadron would cease to make sense. In 1973 the development of QCD as the theory describing strong interactions introduced the concept of asymptotic freedom[3][4]: at small distances quarks and gluons would be loosely interacting, or quasi-free. This fundamental result earned D.J.Gross and F.Wilczek, H.D. Politzer the Nobel Prize for Physics in 2004. The motivation reads: "for the discovery of asymptotic freedom in the theory of the strong interaction".

In 1975, N. Cabibbo and G. Parisi[5]proposed for the first time the existence of a ``different phase of the vacuum in which quarks are not confined", and drew the first schematic diagram of hadronic matter with temperature andbaryonic density as axes. Two regions were defined: at low-T and low baryonic density, ordinary hadrons, at high-T, high-μ_B a deconfined state: a new field of research was born!In the same year, J.C. Collins and M.J. Perry[6]developedthe idea that matter under extreme density conditions, as in neutron star cores and in the early universe, is made of quarks instead of hadrons, which overlap and lose their identity. Their argument was based on the factthat quarks interact weakly when they are close together, and therefore at high densities could bedescribed as a gas of free massless quarks. Finally, in 1980 E. Shuryak introduced [7]the term quark gluon plasma (QGP) to describe the state of nuclear matter in which quarks are deconfined, stressing the analogy between a classical plasma made by ionized atoms and the state made of colored strongly interacting objects.

In a nucleus-nucleus collision the energy of the incoming projectiles is dissipated in the relatively large volume defined by the overlap region of the two colliding nuclei,thus creating the conditions for a phase transition to deconfined quark matter by heating.

So, by colliding nuclei at high energies we compress and heat a system with a large number of quarks, to form a "bubble" of deconfined strongly-interacting matter, which in a very short time, of the order of 10^{-23}s, expands and cools until ordinary hadrons reconstitute, just as they did in the evolution of primordial Universe, some 10 millionth of a second after the Big Bang.

Over the last three decades, a vigorous experimental program, carried both in Europe and in the United States, has dramatically increased our understanding of nuclear matter under extreme conditions of temperature and density, allowing a real exploration of the phase diagram of strongly interacting matter. Our current understanding is graphically summarized in Fig. 3, with the trajectories allowed by the most important accelerators in operation or under construction: the two colliders LHC (in Europe) and RHIC (in the USA) and the fixed-target high-luminosity facility FAIR under construction in Germany.

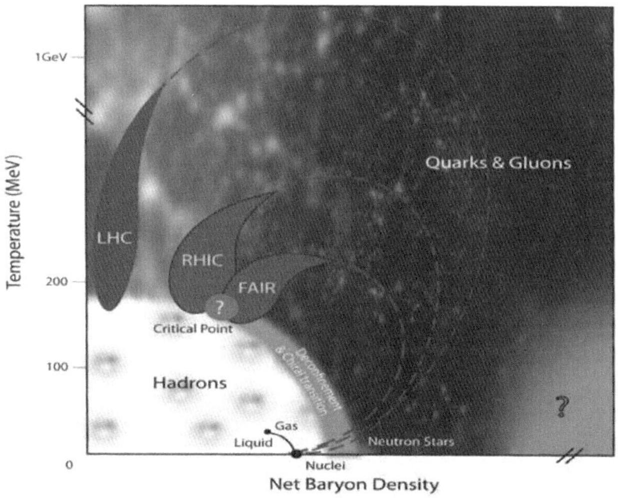

Fig.3 The phase diagram of nuclear matter, from Ref[1]

Together, these programs aim at providing answers to questions such as:

What are the properties of matter at the highest temperaturesand densities?

What is the QCD equation of state? How can we test it?

What are the dominant microscopic mechanisms ofQCD non-equilibrium dynamics and thermalization?

How does hadronization proceed dynamically? How is it changed in dense QCD matter?

How can the QCD phase diagram be efficiently explored?

Clearly, a comparison of the experimental results with theory is not a trivial task. Since the transition to the deconfined phase is outside the limits of validity of the perturbative QCD, a number of different approaches has been developed to derive the equation of state and the transition temperature. Only at the LHC some aspects of the interactions will be accessible with perturbative methods. Effective lagrangians have been extensively used, and also potential models, and even calculations based on the AdS/CFT correspondence have been providing a fresh look and new insight. Still, the most important and widely used are QCD calculations on a space--time lattice. They generally predict a phase transition to occur, and are able to determine its temperature. Current results indicate a critical temperature around T_c~170 MeV, or about 10^{12} K, 100000 times higher than at the center of the Sun, corresponding to a critical energy density of ε_c~ 700 MeV/fm^3

In the following, I will give a brief overview of the main experimental results obtained so far at the LHC, just little over one year after the very first collision of Lead Nuclei was recorded in the experiments. Three experiments joined in this pioneering work: ALICE[9], a dedicated experiment specifically designed to study nuclear collisions, and both ATLAS [10]and CMS[11], the two general-purpose experiments of the LHC, primarily designed for the study of high-p_T processes in pp collisions. I will focus mostly on the ALICE results.

While I will not describe the experimental apparata, I would like to remind here that ALICE has a very different optimization compared to the other LHC experiments. For the understanding of nuclear collisions it is essential that as many of the produced particle as possible are measured and identified, for the widest possible range of transverse momenta (typically from 0.1 to 100 GeV/*c*) and for all particle types: electrons, muons, photons and hadrons. So, the two guiding principles in the ALICE design have been robust tracking, able to handle the very large number of tracks which characterize Heavy Ion Collisions, and Particle Identification capability. The first is provided by very high granularity tracking detectors providing 3-Dimensional space points (several hundred millions of points) operating in a moderate magnetic field and with very low material. To identify leptons, photons and hadrons, including short-lived particles, such as hyperons, D and B mesons, ALICE features essentially all known techniques: dE/dx, Cherenkov & Transition radiation detectors, Time Of Flight, calorimeters, muon filter and topology (secondary vertices, kinks).

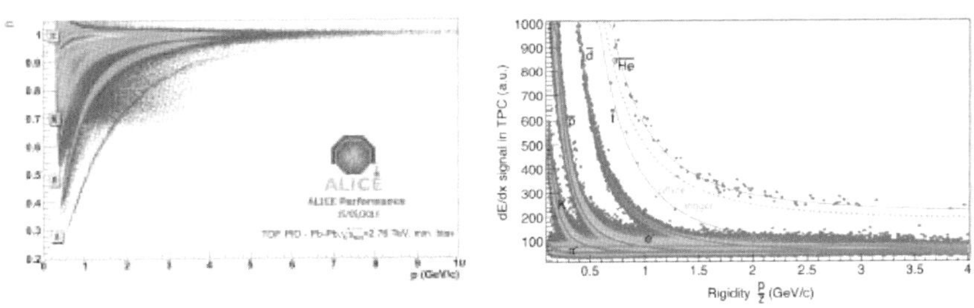

Fig. 4: PID in ALICE with TOF (left panel) and TPC (right panel) from ref.[12]

As an example of the ALICE PID capability, Fig. 4 shows the of the performance of the Time Of Flight (TOF) and of the Time Projection Chamber (TPC)'s energy loss measurement. Both are performing better than originally specified, a remarkable result in detector systems of such complexity observing events with thousands of tracks! The TOF has provided a time resolution below 100 picoseconds, and the TPC an energy loss measurement with a precision better than 6%.

In figure 5 (left panel) the momentum coverage of the various ALICE PID detectorsis shown, demonstrating the extent of the coverage achieved. In addition, short-lived particles are identified through their decay topologies: in the right panel of figure 5 the impact parameter resolution expected in ALICEis shown, which is adequate for the measurement of D and B meson decays. K and Λ decays can be efficiently reconstructed for transverse momenta up to more than 10 GeV/*c*. The very fine granularity of the tracking detectors, providing a large number of space points per track, provides high efficiency tracking for charged particles even at the highest particle multiplicities foreseen at the LHCwith Pb beams, with a lower limit in momentum lower than 200 MeV/*c*. The very long track length measured allows ALICE to achieve an excellent momentum resolution (δp/p < 5% at 100 GeV/*c*). Several smaller detectors (ZDC, PMD, FMD, T0 andV0) used for global event characterization and triggering are located at small angles. The combination of the central and forward detectors provides coverage for the measurement of charged multiplicity which extends from -3.4 to +5 in pseudorapidity. An Electromagnetic Calorimeter, EMCal, covers about a third of the azimuth and provides improved energy resolution and triggering for jets. A high-resolution crystal photon spectrometer, PHOS, is placed at a very large radial distance from the interaction point (4.6 m). The muon spectrometer covers the forward rapidities, between-2.5 and -4, and features a complex arrangement of absorbers, starting very close to the interaction point (90cm) to minimize muon background from weak hadronic decays, and fourteen planes of tracking and triggering chambers, using 2-D pad readout to handle the high multiplicities.

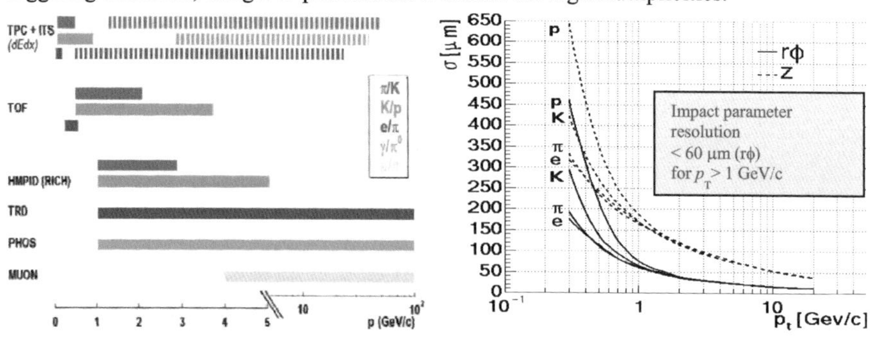

Figure 5: Left: particle identification in ALICE: momentum range and detectors used. Topological identification, not included in this plot, can reach over 10 GeV/*c* for K and Λ. Right: ALICE impact parameter resolution as a function of transverse momentum.

The hardware trigger in ALICE combines the input from detectors with fast trigger capability (T0, V0, ZDC, SPD, TOF, PHOS, EMCal, Muons, ACORDE). It operates at several levels (pretrigger, L0, L1 and L2) to satisfy the individual timing requirements of the different detectors, as the ALICE electronics is in general not pipelined. In addition, a software based High-Level Trigger (HLT) using a farm of up to 1000 multiprocessor PCs performs a preprocessing of the data allowing for a compression of the data volume by a factor of 3.5. It also allows real timedata quality monitoring.

2. ALICE results

The first LHC heavy-ion results were already published during the 2010 lead–lead run, lasting for a month. In fact, there were three papers submitted almost simultaneously: the multiplicity measurement [13] and the measurement of v_2 coefficient of the azimuthal anisotropy [14] by ALICE, and by the ATLAS collaboration a paper on jet-energy imbalance [15]; these appeared together in one issue of Phys. Rev. Lett. One of the most spectacular findings at RHIC was that the matter generated in heavy-ion collisions flows like a liquid with very low viscosity, almost at the limit of what is allowed for any material in nature. This tells us that the constituents of this QGP are quite different from freely interacting quarks and gluons, and led to the definition of the matter under study as sQGP, for strongly interacting Quark Gluon Plasma, a liquid very different from the expected gas of quasi-free quarks and gluons. The first LHC azimuthal anisotropy measurement [14] confirms the RHIC results: elliptic flow of particles with the same transverse momenta is almost identical at the two energies. The nearly-perfect fluid has been found to be opaque to even the most energetic partons (quarks and gluons), which appear as jets of particles from the collisions, an effect known as jet quenching. This is an interpretation of the reported strong jet-energy imbalance [15]. However, the physical mechanisms underlying these phenomena are not well understood. Another manifestation of jet quenching manifests is a reduced yield of high-transverse-momentum-particle in central collisions compared to that expected from the measurements in proton–proton reactions. To express such reduction a properly normalized ratio of yields in heavy-ion and in proton–proton interactions – the nuclear modification factor R_{AA} – is used. Since proton-proton collisions are rescaled according to the number of binary collisions, R_{AA} should be well below one at low transverse momentum, where particle production is rather proportional to the number of participant nucleons, and grow toward one for increasing transverse momenta. The suppression pattern observed by ALICE [16] gives factor about 7 (= $1/R_{AA}$) lower charged-particle production in lead–lead collisions at transverse momentum around 6GeV/c, a striking evidence of the energy loss in the extreme density matter.

The size of the pion-emitting source in central lead-ion collisions is deduced from the shape of the Bose-Einstein peak in the two-pion correlation functions [17]. The collective flow makes the size of the system appear smaller with increasing momentum of the pair. This behaviour is clearly visible for the radii measured in the ALICE experiment. The results for measurements of the radius of the pion source in three dimensions indicate a short duration for the emission, hence an "explosive" emission. Time when the emission reaches its maximum is 10–11fm/c, significantly longer than it is at RHIC, see fig. 6 (left side). Moreover, the product of the three components at low pair-momentum, an estimate of the homogeneity volume of the system at decoupling, is twice as large as at RHIC, see fig. 6 (right side).

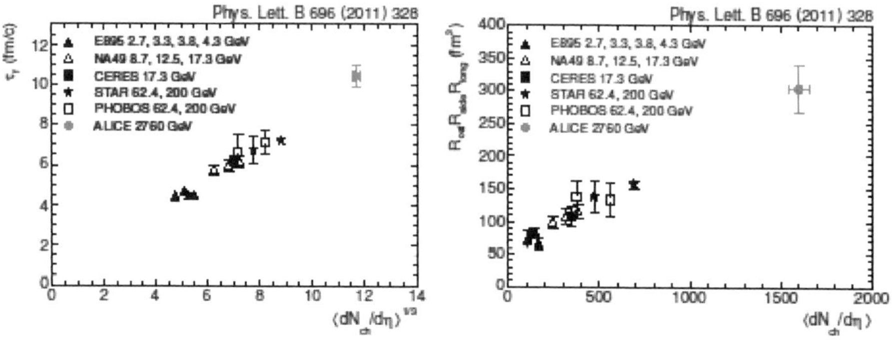

Figure 6. Bose–Einstein pion-interferometry results from different experiments [17]. Estimate of lifetime of emitting source as a function of cube root of particle density (left side). Volume of homogeneity region as a function of particle density (right side).

One of the crucial measurements for the characterization of the fireball produced in heavy-ion collisions are the spectra of identified hadrons, which encode the collective expansion velocity developed in the QGP and in hadronic stages. Moreover, the overall abundances of identified hadrons are believed to be fixed at hadronization, thus they indicate the temperature when chemical composition was established. The measured spectra [18] show an increase of about 10% in the radial-flow velocity when compared to RHIC results. At present, however, the yield ratios observed by ALICE seem to challenge both previous experiments and theory. While the K/π, Ξ/π and Ω/π ratios are compatible with the expectations from the thermal model with a temperature for the "chemical freeze-out" of about 165MeV, as in previous observations, the p/π ratio points to a significantly lower temperature. On the experimental side, there are indications of a similar effect at lower energies, which call for further investigations. On the theoretical side, a number of different possibilities are being investigated, none of them conclusive at the moment.

As mentioned above, the first results on elliptic flow were published [14] during the initial Pb–Pb run. Flow is a fundamental observable since it carries information on the equation of state and the transport properties of matter created in a heavy-ion collision. This observable relates final state anisotropies with features of the initial one, thus allowing study of the medium response and characteristics. The azimuthal anisotropy in particle production is a clear experimental signature of collective flow. It is caused by multiple interactions between the constituents of the created matter and the initial asymmetries in the spatial geometry of a non-central collision. The amount of elliptic flow measured by the v_2 coefficient, integrated over transverse momentum, increases by 30% compared to RHIC energies. However, this increase is entirely a consequence of the growth in transverse momenta. The overall azimuthal distribution of particles emitted in a heavy-ion collision can be described with a Fourier expansion, whose coefficients relate to different aspects of the initial state. ALICE measured also the higher harmonic coefficients of azimuthal distribution [19]; they exhibit much shallower centrality dependence than v_2, and the symmetry planes of v_2 and v_3 coefficients are uncorrelated (see fig. 7, left side). These observations point to fluctuations in the initial geometry as the origin of higher-order azimuthal asymmetries. The flow coefficients can also be studied with a different approach, using two-particle correlations in azimuthal angle [20]. A Fourier decomposition of the correlation function gives squares of different v coefficients, and describes well the structures seen in azimuthal correlations, see fig. 7 (right side), such as the "long-range ridge" (a small angle correlation between particles distanced in longitudinal direction) and the "Mach cone" (a correlation on two sides of back-to-back direction). The collective response to initial spatial anisotropy that causes elliptic flow economically explains these puzzling features, once event-by-event initial-state density fluctuations are considered. The ALICE experiment has analyzed azimuthal asymmetry also separately for different particle species, which constrains models incorporating a realistic initial state and hydrodynamic evolution.

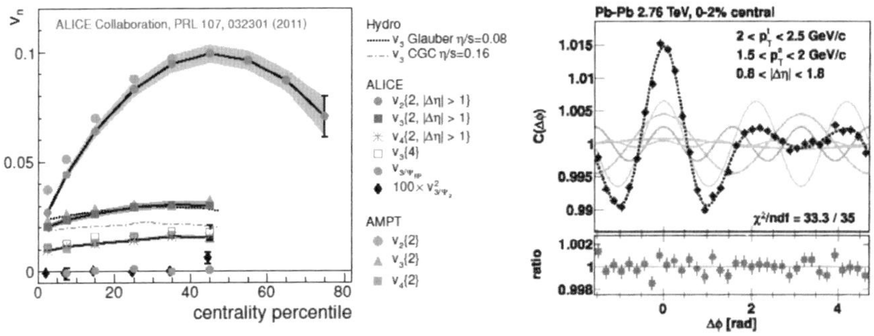

Figure 7. Measurements of azimuthal anisotropy in lead–lead collisions at LHC. Dependence of different v_n coefficients (see legend) on collision centrality [19] (left). Fourier decomposition of two-particle azimuthal correlation function in 2% most central collisions [20] (right).

The two-particle azimuthal correlations at higher transverse momenta measure particle-yield modifications in jet-like structures [21]. When compared to the expectation from proton–proton interactions, the observed yield around high-transverse-momentum particle is slightly higher, while the yield on the opposite side is reduced. This measurement assists the understanding of jet-quenching phenomena. The ALICE collaboration continues to contribute to this topic by studying reconstructed jets, underlying event fluctuations, modifications of jet-fragmentation function, jet-particle composition and providing detailed studies of the dependence of the energy loss on the flavor and mass of theparton traversing the QGP.

Heavy-flavour particles are particularly effective probes of a very dense and hot medium formed in nucleus–nucleus collisions, sensitive to its energy density, through the mechanism of in-medium energy loss. At LHC energies, heavy-flavour particles are copiously produced and thus provide ideal tools for QGP studies. The nuclear modification factor R_{AA} is well established as a sensitive observable for the study of the interaction of hard partons with the medium. Parton energy-loss is caused by the strong interaction, hence the amount of energy loss depends on the colour charge of parton. Therefore, quarks are predicted to lose less energy than gluons.

In addition, heavy quarks (up to some high momentum), which are slower than light, cannot emit gluons within the so-called "dead-cone" around their trajectory; this effect is expected to reduce the energy loss of heavy quarks with respect to light ones. Thus, a pattern of gradually decreasing suppression (i.e. increasing R_{AA}) should emerge when going from the light-flavour hadrons (e.g. pions), which mainly come from gluons, to the heavier D and B mesons: $R_{AA}(\pi) < R_{AA}(D) < R_{AA}(B)$. The measurement and comparison of these different probes provides, therefore, a test of the colour-charge and mass dependence of parton energy-loss. The ALICE collaboration has measured the production of the charmed mesons D^0 and D^+, detecting their hadronic decays in lead–lead collisions [22]. In central collisions a large suppression with respect to expectations at large transverse momentum was found, indicating that charm quarks undergo a strong energy loss in the hot and dense state of strongly-interacting matter formed at the LHC. This is the first time that D meson suppression has been measured directly in central nucleus–nucleus collisions. The results show a suppression by a factor 4–5, almost as large as for charged pions, above 5 GeV/c (see fig. 8). At lower momenta, there is an indication of smaller suppression for D than for π mesons.

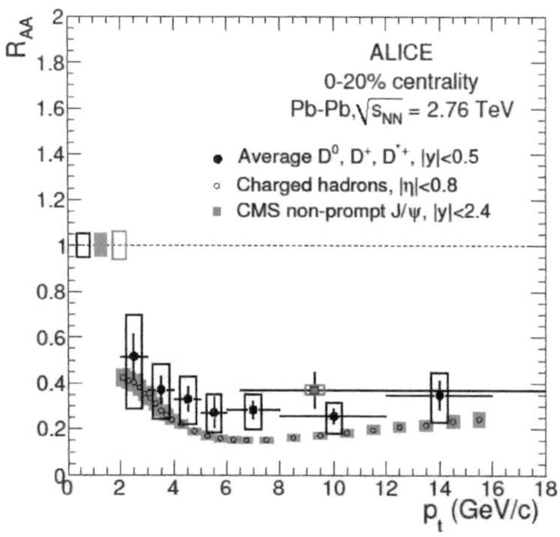

Figure 8. Charmed-meson nuclear modification factor as a function of transverse momentum compared to that of charged hadrons and of non-prompt (mostly from B-decays) J/ψ mesons (as measured by CMS).

Suppression of charmonium production was for a long time considered as one of the main probes for a deconfined medium. At large enough temperatures bound charm–anticharm states are supposed to be dissolved due to Debye screening. However, at LHC energies, new mechanisms of charmonium production in the QGP could occur because of a large number of charm quarks. Around one hundred charm–anticharm pairs are expected to be produced in a central lead–lead collision. Several dynamical transport models predict that charm and anticharm quarks could combine at later stages of the interaction, leading to an enhancement of charmonium production in the most central collisions. ALICE detects charmonium down to very low transverse momentum in two different regions: the central barrel in the dielectron

channel and the forward muon arm in the dimuon channel. The detection at low transverse momentum is crucial because the recombination of the charm and anticharm quarks is expected to be the main production mechanism for charmonium below 3GeV/*c*. The different regions allow for the study of QGP with different charm densities. ALICE has studied the nuclear modification factor R_{AA} for J/ψ mesons as a function of collision centrality [23], [26]. The results indicate that the J/ψ nuclear modification factor shows little dependence on centrality (see fig. 9, left), a trend that is different from that observed at lower energies. For central and mid-central collisions the J/ψ R_{AA} is larger at the LHC than that measured at RHIC. In a complementary study, the CMS collaborations at the LHC have measured a smaller value for the J/ψ nuclear modification factor at transverse momenta above 6.5 GeV/*c*. The nuclear modification factor exhibits a marked dependence on transverse momentum (see fig. 9, right), with a strong increase for values below 3GeV/*c*.
These observations hint at the recombination of charm and anticharm quarks in the QGP as the main mechanism for J/ψ production in central lead–lead collisions at LHC energies. As shown in fig. 9, the effect of suppression followed by recombination accounts for both the shape and the magnitude of the effect.

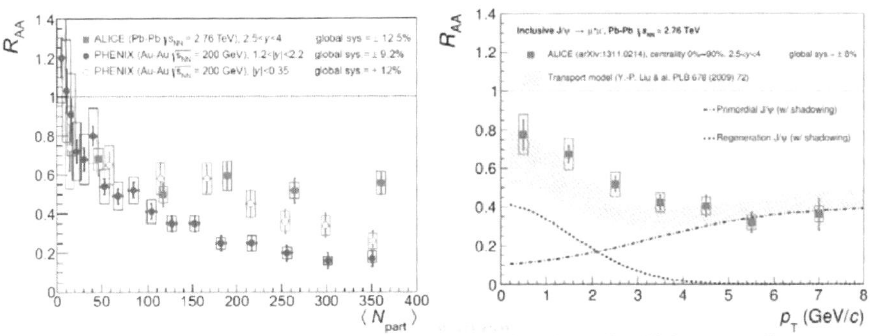

Figure 9. Left: Nuclear modification factor for J/ψ as a function of collision centrality, expressed as number of participants. ALICE data are compared to those from the PHENIX experiment at RHIC. Right: Dependence of the J/ψ nuclear modification factor on transverse momentum, showing the contributions of regenerated and primordial J/ψ (from [26])

It is clear from these first results that a coherent picture of the characteristics of the extreme density matter of which our Universe was made of in the first microseconds of its existence. Even more, the tool chosen to study it, heavy-ion collisions at the world's most powerful accelerator, has proven to be the right one, allowing for a continuous improvement of our understanding. The results shown here are but a small subset of the wealth of results obtained so far, and a was a mere hint to the ones coming. It is indeed a very exciting time for the physics of the Quark-Gluon Plasma.

3. Results from the pA run

When studying the difference between pp and PbPb collisions it is essential to understand the initial wave function of the colliding nuclei, and thus disentangle 'initial state effects" from those due to the hot and dense matter created in PbPb collisions. Understanding the structure of the initial state is very interesting in its own right, because at the LHC energy, experiments probe the structure of the nucleus in a novel and unexplored QCD regime of very low values of the longitudinal parton-momentum fraction (x <10–3). In this kinematic region, the extremely high gluon density is expected to saturate by means of strongly non-linear coherent processes, leading theorists to predict the existence of yet another pre-collision state of matter – the so-called colour glass condensate (CGC).Moreover, pPb collisions at the LHC open a new window for the study of high density strongly interacting matter, allowing a study as a function of the volume of the system.

The results from the basic control measurements, the charged-particle multiplicity density and transverse-momentum spectrum, were in agreement with expectations from theoretical models that incorporate present knowledge of the function describing the longitudinal distribution of partons inside the nucleus [24]. In addition, as shown in the left panel of fig 10, when compared with the results obtained for pp collisions at 2.76 and 7 TeV, on one hand the charged-particle multiplicity in pPb collisions was found to scale roughly with the mean number of nucleons participating in the collision, as expected for particle production through soft processes. On the other hand, the particle spectrum at large transverse momenta – beyond 3–4 GeV/c – was found to scale with the mean number of binary nucleon–nucleon collisions, as expected for particle production through hard processes. Also charmed mesons showed no significant deviation from binary nucleon–nucleon scaling, as shown in Fig. 10. The above observations confirm that the suppression of high transverse-momenta hadrons and jets observed in central PbPb collisions can be attributed to a final-state effect, namely, the energy loss experienced by partons traversing the medium created in these collisions.

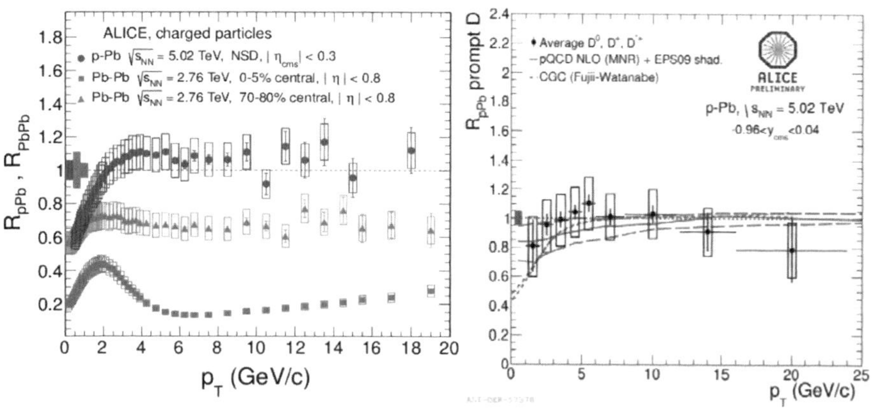

Figure 10. Left (from ref [24]): modification factor for charged particles. Right: Ratio of charmed mesons production in pPb collisions to the one measured in pp scaled by the number of collisions R_{pPb}. $R_{pPb}$is plotted as a function of pT for prompt D mesons (average of D0, D+ and D*+) in p-Pb collisions at √sNN=5.02 TeV. Results are compared to model calculations of initial-state effects.

The study of two-particle correlations in high-multiplicity pPb events led instead to a surprising result: a near-side, long-range (elongated in pseudorapidity) correlation, forming a ridge-like structure observed in high-multiplicity pp collisions, accompanied by an essentially symmetrical away-side ridge, opposite in azimuth ,was observed [25]. This double ridge was revealed after the short-range correlations arising from jet fragmentation and resonance decays were suppressed by subtracting the correlation distribution measured for low-multiplicity events from the one for high-multiplicity events. Similar long-range structures in heavy-ion collisions have been attributed to the collective flow of particles emitted from a thermalized system undergoing a collective hydrodynamic expansion. The origin and nature of this structure are currently subject of debate and research. In the subsequent months, and more evidence of features which in heavy-ion collisions have been attributed to collective behavior have emerged, in particular a characteristic mass scaling in the measurement of v_2 harmonic coefficients of the azimuthal anisotropy, and in the baryon-to-meson ratios. So the questions of how collectivity and even hydrodynamic behaviour could take place in such a small system, and of the compatibility with the energy loss results are currently hotly debated. The whole community interested in the study of strongly interacting matter at extreme conditions is eagerly waiting for the further developments of these intriguing findings.

4. The future of ALICE

The LHC operates in a quite different mode than previous CERN accelerators. During running, only a very short stop of operation takes place around the end of the year, while most of the time is devoted to quasi-continuous running. After O(3) years of operation, a long shutdown of more than a year allows for major maintenance and improvements of both the accelerator and the detectors.

Data taking for ALICE in the so-called RUN1 has included, apart from data taking with pp collisions, three periods of data taking with nuclear beams. In 2010 the ALICE detector collected with a minimum-bias trigger about 30 millions of lead–lead collisions at centre-of-mass energy of 2.76TeV per nucleon pair. In 2011 there was a short period of proton–proton running at the same energy at the start of the LHC operation, to gather comparison data; these were used for normalization in practically all measurements described above. The rest of the proton running proceeded at (a standard) energy of 7TeV. At the end of the year, the LHC switched again to the heavy-ion mode. This time the instant luminosity grew to above $10^{26}cm^{-2}s^{-1}$, and was above the design value for energy 2.76TeV per nucleon pair. ALICE used triggers for different types of lead–lead collisions: central, semi-central, dimuons, photons, jets, ultraperipheral, and others, and collected about 100 million events, inspecting over $0.1nb^{-1}$ of integrated luminosity. At the beginning of 2013, the Heavy-Ionrun has been dedicated to proton–lead collisions, to improve the baseline comparison, taking into account modifications of structure functions in nuclei. ALICE collected $0.03nb^{-1}$ of luminosity in the four weeks of pA running, half of which in pA mode and half in Ap mode, to take into account the asymmetry of the apparatus.

Since then the LHC operation has been paused for the upgrade necessary to increase the collision energy. The Long Shutdown (LS1), will last until early 2015. During this period, ALICE is carrying a very intense program of detector consolidation, installation of new detectors (DCAL, a second arm of the ALICE electromagnetic calorimeter, a set of forward scintillator counters to improve the coverage at large rapidities, the last five supermodules of TRD and a fourth supermodule of PHOS).

During RUN2, after the restart in 2015, the ALICE collaboration aims to complete its originalprogramme, collecting $1nb^{-1}$ of heavy-ion collisions at the higher collision energy (5.5TeV per nucleon pair in the centre of mass, being the design value). This will provide a 10-fold increase in statistics, on top of a better quality of data and the higher energy. In addition, a second proton-nucleus run is planned, in order to further explore the very exciting features discovered in RUN1. This agenda will be completed before the second long shutdown of the LHC, planned to start in 2018.

A majorupgrade of the ALICE detector allowing for high-luminosity heavy-ion running after this period, to reach a tenfold increase in integrated luminosity in RUN3 and RUN4 (extending to the mid 2020's) has been proposed by the ALICE Collaboration. The extended physics program justifying the LHC operation in heavy-ion mode beyond 2020, which would imply collecting over $10nb^{-1}$ of data, has been included among the top priority goals of the European Strategy for High Energy Physics.The Letter of Intent [27], defining the scientific goals of the Upgrade, has been approved in late 2012, and the five Technical Design Reports discussing the technical implementation of the upgrade are being progressively submitted for approval in 2013 and 2014. The one for the new Inner Tracking System, based on monolithic silicon pixel detectors, and the one dedicated to the new electronics and trigger have already been approved.

After having confirmed the main discoveries obtained at RHIC, ALICE has entered an exciting phase of new measurements, allowing a much broader and deeper study of the QGP. At the same time, ALICE is already preparing, on the basis of what has been learnt so far, a next step in more detailed characterization of the extreme state of matter produced at LHC; new, unexpected discoveries may come!

References

[1] NUPECC Long Range Plan 2010: Perspectives of Nuclear Physics in Europe, published by the European Science Foundation

[2] R. Hagedorn, NuovoCim. Suppl. 3 (1965) 147-186

[3] D.J. Gross and F. Wilczek Phys. Rev. D 8(1973) 3633—3652.

[4] H. D. Politzer Phys. Rev. Lett. 30 (1973) 1346-1349.

[5] N. Cabibbo and G. Parisi Phys. Lett. B 59 (1975) 67-69

[6] J.C. Collins and M.C. Perry~M.C. Phys. Rev. Lett.34(1975) 1352—1356.

[7] E.V. Shuryak Phys. Rept. 61 (1980) 71—158

[8] B. Muller and J. Nagle, Ann. Rev. Nucl. and Part. Phys. **56** (2006) 93

[9] ALICE Collaboration, Aamodt K, et al., JINST 3:S08002 (2008).

[10] ATLAS Collaboration, Aad G, et al., JINST 3:S08003 (2008).

[11] CMS Collaboration, Adolphi R, et al., JINST 3:S08004 (2008).

[12] A. Kalweit for the ALICE Collaboration, J.Phys.G G38 (2011) 124073.

[13] K. Aamodt et al. (ALICE collaboration), Phys. Rev. Lett. **105** 252301 (2010).

[14] K. Aamodt et al. (ALICE collaboration), Phys. Rev. Lett. **105** 252302 (2010).

[15] G. Aad et al. (ATLAS collaboration), Phys. Rev. Lett. **105** 252303 (2010).

[16] K. Aamodt et al. (ALICE collaboration), Phys. Lett. **B696** 30 (2011).

[17] K. Aamodt et al. (ALICE collaboration), Phys. Lett. **B696** 328 (2011).

[18] M. Florisfor the ALICE collaboration, J. Phys. **G38** 124025 (2011).

[19] K. Aamodt et al. (ALICE collaboration), Phys. Rev. Lett. **107** 032301 (2011).

[20] K. Aamodt et al. (ALICE collaboration), Phys. Lett. **B708** 249 (2012).

[21] K. Aamodt et al. (ALICE collaboration), Phys. Rev. Lett. **108** 092301 (2011).

[22] B. Abelev et al. (ALICE collaboration), JHEP **1209** 112 **(2012)**

[23] B. Abelev et al. (ALICE collaboration), Phys.Rev.Lett. **109** 072301(2012)

[24] B. Abelev et al. (ALICE collaboration), Phys.Rev.Lett. **110**082302 (2013)

[25] B. Abelev et al. (ALICE collaboration), Phys. Lett. **B 719** 29(2013)

[26] B. Abelev et al. (ALICE collaboration), Phys. Lett. **B 743** (2014) 314-327

[27] B. Abelev et al. (ALICE collaboration), J. Phys. G 41 (2014) 087001

LECTURER: P. GIUBELLINO

Highlights from ALICE

Scientific Secretaries: D.V. Perepelitsa, M. Ajaz

DISCUSSION

- *Csanad:*

There are some very nice PID (identified particle) flow results from ALICE in Pb+Pb collisions. You showed part of them, and commented that since the phi lies on the proton curve, there is mass ordering instead of constituent quark scaling.

- *Giubellino:*

On the other hand, previous talks showed that both explanations were plausible. So even though the scaling is worse than RHIC, there is some constituent quark scaling. In light of a previous speaker's appeal to Occam's Razor, we should consider the simplest explanation before introducing a more complicated one. Radial expansion is an undisputed ingredient in interpreting these results. If it alone can account for the particle species dependence of the flow results, one doesn't need to invoke additional factors linked to recombination and the number of quarks. Thus, rather than saying whether or not constituent quark scaling is present in the data, we should say it is not a fundamental factor that we are forced to invoke to explain the measurement. The RHIC data were more prone to requiring to an additional ingredient to explain the mass ordering. There, the observed baryon anomaly is likely a combination of two effects: radial expansion and something related to the quark counting rules. With the current ALICE phi data, the constituent quark dependence does not seem to be an essential feature.

- *Boren:*

What will ALICE be doing in the near future to push measurements of heavy flavor suppression to lower transverse momentum (pT)?

- *Giubellino:*

ALICE has quantified the expected performance of the detector upgrades in 2019 for measuring the RAA of identified heavy flavor hadrons with two figures of merit [P. Giubellino's slide 79]: the minimum pT reached, and the statistical uncertainty on the measurement. In general, since a large fraction of the cross-section is near pT = 0, it is important to achieve as low a pT as possible. With the post-LS2 [Long Shutdown 2] upgrades, ALICE will have substantially thinner inner detectors, reaching one third of a percent of a radiation length per layer. This will improve the capability of reconstructing the secondary vertex, and the efficiency for low pT charged particles. For many observables, this will move the minimum pT either to zero or much closer to it and improve the statistical uncertainty substantially. This improved capability is one of the core goals of the upgrade programs, expected to be completed around 2020.

- *Esposito:*

Are there prospects for measuring exotic mesons (such as tetraquarks or meson-molecules) in heavy ion collisions?

- *Giubellino:*

There has been some work on measurements of exotic nuclei, but less progress on exotic mesons. The data is available, especially in low-multiplicity proton-proton collisions where these states could be produced diffractively, and would serve as a good project for future students. In Pb+Pb collisions, these types of spectroscopy measurements become more challenging due to the much larger multiplicity. In an analysis focusing on a specific meson, the cuts can be optimized to extract the signal, but a blind search across the mass range is much more challenging.

- *Zennamo:*

Question n. 1: Is there any ongoing effort to use heavy ion data at RHIC and the LHC to inform our understanding of high-energy neutrino - large nucleus scattering, or of models of the structure of large nuclei?

Question n. 2: During high luminosity data-taking, what are there effects due to space charge build-up in the ALICE time projection chamber [TPC], and how are they mitigated?

- *Giubellino:*

Question n.1. Due to the large differences in the relevant beam energies and scattering kinematics, the present data will probably not be able to provide significant information for those purposes.

Question n.2. Like most modern TPCs, the ALICE TPC is gated, which limits the amount of ion backflow and tames the extent of the space charge problem to some extent. The remainder of the problem is treated with a detailed analysis of fake track rates induced by a laser system. This is approach is no trivial, but is straightforward. For future data-taking, during which ALICE will have a much higher data taking rate, no gating will be possible. One of the challenges for the ALICE collaboration in the near future is how to handle the space charge corrections in situations with substantial ion backflow. Part of the solution can come by minimizing the rate of ion backflow by, for example, replacing the wires with GEMs or MicroMegas. Ultimately, it will be important for the physics program to keep 100-micron precision in situations where the space charge can be 10 or 15 centimeter. This is a challenge ALICE is working hard to address for 2022. Thus, for the present, the gating is a good solution. For the future, the GEM technology (also being used at CMS) is under development and looking promising.

- *Minkowski:*

How do we know that the same plasma, which we identify as being produced in heavy ion collisions, is not also present in proton-proton collisions?

- *Giubellino:*

As recently as two years ago, the conventional belief in the field would have been to dismiss this possibility. Since the results produced in the recent proton-lead LHC run, we have had to rethink our assumptions. It is clear that there are some signs of collectivity or of hydrodynamic expansion in smaller systems. An interesting question for future study is what is the smallest system for which such collective behavior can happen. This has inspired renewed interest in proton-proton collisions, for example when investigated at the same multiplicity as proton-lead collisions. For example, a recent ALICE measurement of three pion correlations (which has several advantages over more traditional two pion correlation measurements) shows a division between proton-proton and proton-lead collisions on one side which have similar properties, but lead-lead collisions on the other. In general, this can be experimentally challenging to study in proton-proton collisions, because unless we are analyzing a special minimum bias dataset with a small number of mean interactions for crossing, selections on high multiplicity tend to select only high in time pileup events.

- *Millington:*

This question relates to transport properties of the QGP. What are the current limits on the formation or thermalization times (or any other transport coefficients) of the QGP?

- *Giubellino:*

The main transport parameter that we measure is the shear viscosity (divided by the entropy density, since we are only sensitive to the ratio of the two), which is very close to the quantum limit. The formation time cannot be measured directly but must rather be inferred in a model-dependent way. The lifetime can be accessed experimentally through interferometry measurements and has been done so by ALICE.

- *Kofarago:*

We heard about the high luminosity (HL) of LHC a few days ago and I would like to ask, what are the plans for ALICE for the HL-LHC?

- *Giubellino:*

The increase in the luminosity of the proton-proton running does not have a direct impact on ALICE. The increase in the luminosity for the heavy ion running will be implemented after the long shutdown 2 (LS2) which will be around 2019. The program for which ALICE has been approved after that which is when the steps in the luminosity of heavy ion will take place stands about 6-7 years. So, yes ALICE will be working during the high luminosity LHC, but it will not profit much from the higher luminosity of proton-proton. The bigger change will be after long shutdown 2 when the luminosity of the lead beams will increase.

- *t' Hooft:*

You noticed that there is a lot of thermalization going on, nevertheless your system is needed where in a temporary phase thermal equilibrium, so that should mean in principal you should be able to figure out all that is needed to determine a phase diagram of the theory to some extent or well, ALICE will enable to figure out the picture of the whole phase diagram with ALICE domain so we have to put all these data together and what we know about neutron stars and so on. It looks like we mean have everything to figure out what we have in phase diagram of QCD will be?

- *Giubellino:*

Yes, there are a few critical points in that, that we are not skilled so far. We do have a typical understanding of the area near the zero net baryon density. So we understand type of smooth transition, which is a smooth one and is not a sharp one and with the RHIC data we have done a lot to fill this part. So this is fairly well known. Let me know that there are a number of points at the AGS and the SPS that do not cross the line and there is a search on going from a number of experiments that is really a huge conversion experimental efforts to work on finding this critical point. I mean the images can at RHIC, there is a dedicated experiment at CERN in 1961, there is an accelerator been built at Dubna, NICA, there is FAIR under construction in Germany and now the HERA program at JPark in Japan. All going on essentially this point which we do not have. Yes we have a number of points on the phase diagram. So we know quit a bit. We are missing a very essential point whether there is a, this should actually have a very dramatic feature I mean if we look at the critical point in the phase diagram of other materials they show a very dramatic fluctuation so on and so forth.

- *Savirskyi:*

In Pb+Pb collisions, the ALICE detector sees a charged particle multiplicity of up to 8000. Which algorithms you are using for jet identification, in particular, in order to separate jets from the particles not associated with them?

- *Giubellino:*

Well, first of all, ALICE was designed for big track multiplicities. But actual amount of tracks are close to 3000. Usually jet reconstruction at ALICE is very complex procedure and based on several long-term studies, which are described in a recent publication. The algorithms used by ALICE are based on algorithms which subtract or suppress the role of the underlying event, for example by using special definitions of jets such as a particular cone size.

- *Russo:*

Will it be possible during LHC Run 2 to have a pp data sample matching the PbPb center of mass energy?

- *Giubellino:*

I cannot predict this now. It will mainly depend on how the LHC machine is operating during Run 2.

- *Miyazaki:*

Is it possible to measure pure the Yukawa mass of quarks free from confinement mass in QGP?

- *Giubellino:*

Not directly. There is indirect approach to do so. One of the points that is always being considered as the most promising and the most difficult is still are/R. There are dedicated experiments but the measurement of Yukawa mass is extremely difficult.

Planck Highlights

Alain Riazuelo

13 January 2015

1 Introduction

The Cosmic Microwave Background (hereafter CMB) can be defined in simple terms as the afterglow of the Big Bang. The early Universe was very dense, very hot and therefore filled by intense radiation. This radiation was diluted and cooled down as the Universe expanded, but has survived till then. In the early phases of the Universe, the high density of matter associated to the fact that it is ionized made it completely opaque to radiation. As a consequence the free mean path of photons was very short. This situation changed drastically when the Universe reached the age of ~380 000 years, the so-called recombination era. Its temperature dropped below 3 000 Kelvins, which corresponds to the threshold where free electrons can combine with hydrogen nuclei. At that temperature, opacity is mostly determined by Thomson scattering of photons on free electrons, so that the disappearance of the latter made rather suddenly the free mean path of the former become larger than the size of the observable Universe. As a consequence, the CMB photons we observe today have travelled mostly unaffected since this epoch and give us a very clean picture of the state of the Universe at that time. More importantly, some events that occured significantly earlier than 380 000 years after the Big Bang, and hence at much higher energies have left an imprint on it, which allows us to recover some information on such events. For these reasons, since around two decades, the CMB has become the most useful tool to study cosmology. The reason for this to have become possible comes from a series of fortunate coincidences that allows us to extract a very large number of information from a single object.

2 A short introduction to CMB physics

Even today, the CMB photons are overdominant as compared to those produced by any other astrophysical processes. The electromagnetic energy that is stored within the CMB corresponds to around 95% of the electromagnetic energy of the Universe, the five remaining percents being mostly split in more or less equal parts between starlight (in the visible part of the spectrum) and the thermal radiation of interstellar dust (in the far infrared domain). All the other processes at other wavelengths (radio, X-ray, gamma) bring an astrophysically interesting but numerically negligible contribution to this total, as shown in Fig. 1. Since the Universe is homogeneous and isotropic on large scales, the CMB is also expected to show a very high degree of isotropy. However, structure formation is a rather slow process and therefore necessitates that even at the time of CMB emission, the matter distribution was not exactly homogeneous. This inhomogeneity in the matter distribution should be somehow reflected in the radiation distribution, which is expected to show some degrees of inhomogeneity, either in its photon number or temperature. Observing these inhomogeneities is at the core of modern studies of CMB. The observability of these inhomogeneities comes from a very fortunate coincidence: not only the CMB dominates within the microwave part of the spectrum over any other sources, but it dominates in a limited frequency interval by more than 10^5. Therefore, even if the homogeneities we are interested in are rather small (10^{-5} in relative amplitude) they are still dominant over the other sources (hereafter named as "foregrounds" since they originate from later epochs and therefore from nearer to us than the CMB). This situation is shown on Figure 2 below.

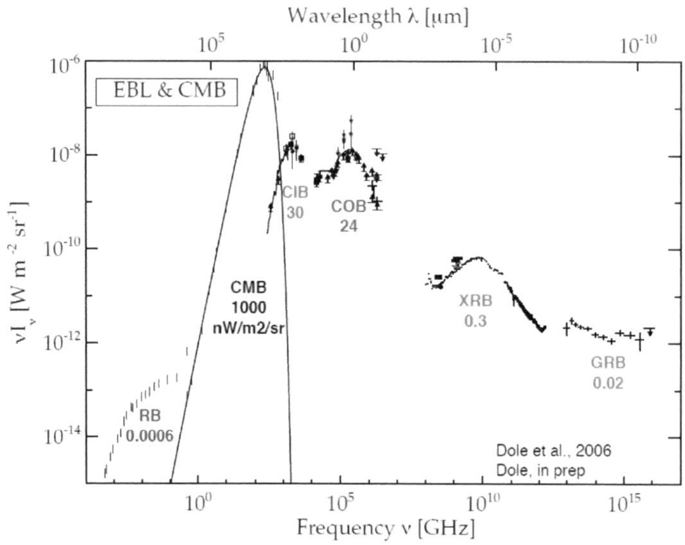

Figure 1: The overall integrated electromagnetic spectrum of the Universe, as seen by a fictious observer far from any source of radiation. The CMB completely dominates any other sources, the others being at most 30 and 40 times smaller (for the Cosmic Infrared Background and the Cosmic Optical Background, respectively). However, what will make the CMB really interesting to study is the relative amplitude of its fluctuations with respect to the other sources as shown in the next Figure.

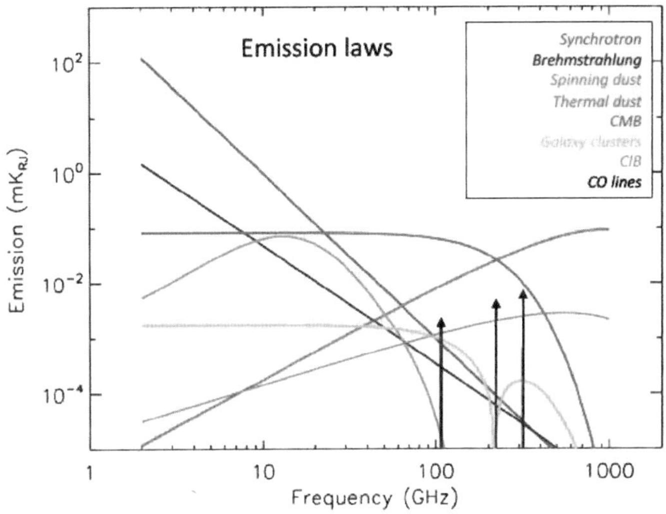

Figure 2: The relative amplitude of CMB fluctuations as compared to various foreground emissions. The amplitude is expressed in term of antenna temperature (therefore a constant value for CMB at low frequency since it is a thermal radiation) as a function of frequency. Some foregrounds dominate at low frequencies such as synchrotron and free-free emission, whereas some others dominate at higher frequencies (thermal emission from dust and the Cosmic Infrared Background). However, and somehow by chance, CMB fluctuations are still dominant in a narrow range of frequencies. In order to perform the most accurate extraction of the CMB map, one observes the sky in a variety of channels, from 30 to 857 GHz, most of which are dominated by foregrounds. This allows to obtain a very accurate map of these, which can then be substracted in the frequencies where the CMB dominates in order to obtain the cleanest CMB map.

Where does this 10^{-5} amplitude come from? Before recombination, Thomson scattering between photons and electrons, combined with Coulomb interaction between electrons and atomic nuclei make ordinary matter and raditation behave as a single fluid. Therefore the temperature fluctuations in the photon fluid reflect the density fluctuations in the baryon fluid. After recombination, baryons and photons cease to interact, and baryon then behave as an almost pressureless fluid.

It is well-known that some gravitational instability occurs on large scales when one considers the propagation of sound waves within a medium, the critical wavelength associated to this process being the so-called Jean's length, λ_J, whose value is given by $\lambda_J = c_s\sqrt{\frac{\pi}{G\mu}}$, where c_s is the sound speed within the medium and μ is mass density. A simple interpretation of this result is that, up to some numerical constants, a spherical object of average density μ will have equal content of (positive) thermal energy and (negative) gravitational binding energy if its radius is of order λ_J. Above this radius, it will collapse under its own weight because pressure force are not able to stabilize it. In a non expanding medium, the instability associated to the Jeans length makes overdensities grow exponentially with time, but in an expanding Univers, overdensities can still grow (but only under some conditions), and only linearly with respect to the scale factor a, i.e. the overall relative distance increase between objects because of the expansion of the Universe. Today's CMB temperature is 2.73 K only, a factor 1000 smaller than the temperature at CMB emission[1]. Therefore, one expects that today's fluctuations in the baryon fluid will be 1000 times larger than at recombination. However, when confronting this ansatz to the observations, we find that this is not the case: many lengthscales today have reached the nonlinear regime (i.e., density contrast larger than 1) and therefore, following the above reasonning, were expected to be at least at the 10^{-3} level at recombination, which is not the case since the observed fluctuations are at level 10^{-5}. This result proves that the smallness of the CMB fluctuations necessitates the adjunction of some new physics, the most convincing candidate of which being dark matter. The reason for this is that since (by definition) dark matter does not interact with radiation, its pressure is negligible and consequently it is much more rapidly subject to the Jeans instability than ordinary matter and can have much larger fluctuations than the observed 10^{-5} level of the photon + baryon fluid at recombination. Since the presence of dark matter is, independently, very strongly suggested by the analysis of the galactic rotation curves and even more by the dynamics of the galaxy clusters[2], the existence of dark matter arises as a very natural solution to many astrophyscial problems.

Once this background is settled, let us go back to the physics of CMB before recombination. The combined baryon + radiation fluid has some energy density that is dominated either by the radiation part (at earliest epochs) or matter (at later epochs), as well as some pressure (only contributed by radiation but transferred to baryons as well through Thomson scattering and Coulomb ilteractions). Any inhomogeneities in such a fluid will lead to the propagation of density waves. Overdense regions also contain more photons and therefore are hotter, whereas underdense regions are cooler, which leads to observable patterns in the CMB map. To this simple framework, one has to add the influence of gravitational effects on the wave propagation, which range from the expansion of the universe itself to the presence of dark matter which evolves rather independently from matter and radiation but whose inhomogeneities influence the latter through their gravita-

[1] The CMB was emitted when free electrons combined with hydrogen and helium nuclei. A naive estimate of the corresponding temperature would therefore be $T_{\rm rec} \sim 13.6$ eV$/k_{\rm B}$ ($k_{\rm B}$ being the Bolzmann constant), however, the Universe as a whole is a system with a very large photon-to-baryon ration, which is larger than 10^9. Consequently, what matters at recombination is not the average energy of photons, but the average enery of the tiny (10^{-9}) fraction of photons which still have a chance to interact significantly with atomic nuclei even though the temperature is lower than 13.6 eV$/k_{\rm B}$. On can easily show that when this energy is around 13.6 eV, then the temperature is only 3000 K.

[2] In both cases, Kepler third law links the velocity of stars (within galaxies) of galaxies (within clusters) to the total mass of their host. This gravitational mass can then be compared to the baryonic mass deduced from the radiation it emits, once one can modelize the mass-to-light ratio of the system. Whatever structure one observes, there is always an important gap between the luminous, baryonic mass and the gravitational one, the former being systematically lower to the latter. Even if one assumes that some baryons do not radiate much, it does not seem possible to explain the observations without invoking the existence of some new form of matter which does not show any amount of non gravitational interaction with the rest of the Universe (see, e.g., [1].

tional potential. Since the fluctuations are small, all this can be computed by using a linearized version of the General Relativty equations around a homogeneous and isotropic solution, a task that is not very complicated technically, but that depends on a moderately large parameter space that has to be explored through Monte-Carlo simulations in order to find the set of parameters that fits the data best.

What is the CMB angular spectrum supposed to look like? Since we are in an expanding universe, physical lengths (and, hence, wavelengths) grow with time. Moreover, there is an important timescale associated with the expansion. It is the Hubble radius, that is the inverse of the expansion rate, which is (in general) proportional to the age of the Universe at the epoch of interest, as well as the size of the observable Universe. General Relativity tells us that the expansion rate is determined by the energy content of the Universe which during most of its history is dominated by radiation (at early times) or matter (at late times). In these two situations, one can show that physical lengths grow more slowly than the Hubble radius. This means that waves are, at sufficiently early time, larger than the then observable Universe before becoming smaller than it. Cosmologist often summarize this by saying that perturbations "enter into the Hubble radius" when their (growing) wavelength becomes smaller than the (growing) Hubble radius. As long as this has not occured, we are in a situation where the perturbations evolve on a timescale larger than the age of the Universe, and therefore have not yet begun to evolve in practice. One therefore may say that at sufficiently early times, any cosmological perturbation is frozen once it has been formed by some yet-to-determine process. Once density waves begin to evolve, they are expected to oscillate with a frequency that is all the more large that the wavelength is small. Therefore, different wavelengths will show a progressive phase shift with time. Finally, since CMB shows a picture of the Universe at recombination, any angular scale in the CMB map can be converted into a physical scale at recombination, and propagating sound waves should translate into an oscillating angular power spectrum, the detailed features of which depends on the detailed physics that is at work there. An outlook of this, as well as a sketch of the approximate dependence of the spectrum with respect to some cosmological parameters is given in Fig. 3.

3 Temperature spectrum and cosmological parameter estimation

After a long and tedious processing of the raw data (which, of course included the foreground emissions), the final processed CMB fluctuation map was obtained from the Planck data. It is shown in Fig. 4 The angular power spectrum can also be extracted from the data (altough in a more subtle way than extracting it from the map itself). The result is shown in Fig. 5. From the temperature data alone, it is possible to constrain the cosmological parameters. The main difficulty we have to face is that we do not know for sure which underlying model has to be confronted with the data. The first step is therefore to determine which minimal model can possibly fit the data. Since the advent of the WMAP data ten years ago, we know that we need at least six parameters in order to interpret the CMB data. These six parameters are:

- **Baryon density.** This is the most obvious parameter. At present, no particle physics theory is able to predict the baryon asymetry which settled the baryon-to-photon ratio at around $n_{\rm b}/n_\gamma \sim 6 \times 10^{-10}$, so that we need to extract this parameter from the data. In practice, we extract the present day average baryon density $\rho_{\rm b}^0$, and, following the usual conventions, it is expressed in term of the baryon density parameter times the squared normalized Hubble constant, i.e. the quantity $\omega_{\rm b} \equiv \Omega_{\rm b} h^2$ where h corresponds to the normalized Hubble constant, i.e. $h = H_0/(100\ {\rm km\,s^{-1}\,Mpc^{-1}})$. The conversion factor is then $\rho_{\rm b}^0/(1\ {\rm GeV\,m^{-3}}) \simeq 10.48\omega_{\rm b}$.

- **Dark matter density.** The exact nature of dark matter (axions, WIMPs, or whatever) does not matter from a cosmological point of view. The only property which plays a role in structure formation is the epoch at which it begins to behave as a pressureless fluid. In case

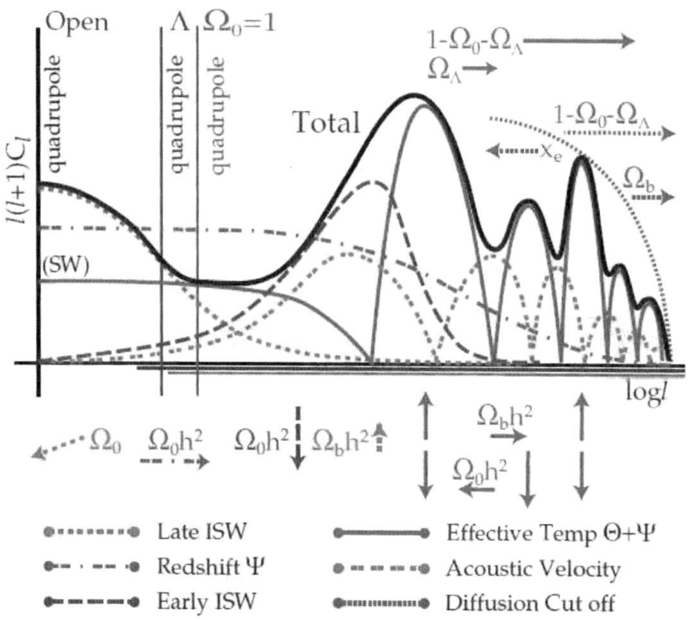

Figure 3: A sketch of what the CMB angular spectrum is expected to look like, and its dependence on some cosmological parameters. To each angular scale corresponds (roughly) a physical scale whose evolution mostly depends on the ratio of its period to the age of the Universe at recombination. The series of peaks we see here corresponds to mode which did not have time to evolve (left of the first peak), which experienced half of an oscillation (first peak; from overdensity to underdensity, or vice-versa), one full oscillation (second peak) and so on. The relative amplitude of the peaks mostly depends on the baryon density ($\Omega_b h^2$) and at a lesser extent the total (dark + baryonic) matter density ($\Omega_0 h^2$). Some damping occurs because of viscosity and heat conduction at small scales (right of the spectrum) and its amplitude is determined by the baryon density as well. Finally, the overall physical to angular scale conversion factor mostly depends on the history of the expansion of the universe and therefore dark energy as well as its geometry (Ω_Λ and $1 - \Omega_0 - \Omega_\Lambda$, respectively). For more details, see Ref. [2] from which this picture was borrowed.

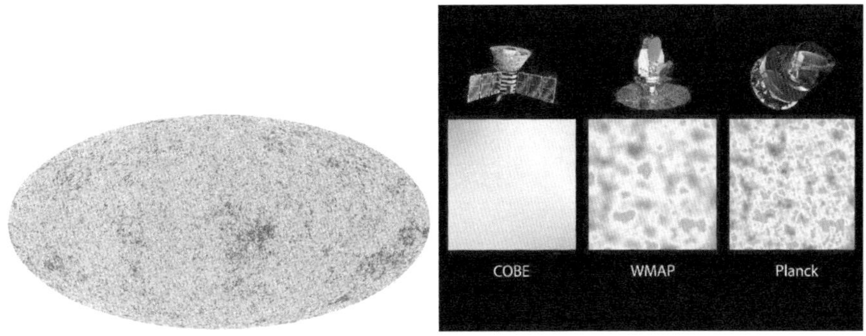

Figure 4: The Planck CMB temperature fluctuation map (left). The Planck resolution is of course much better than the pixel size of the picture shown here, so that there is in fact more information in the real map. As a comparison, the right picture shows the increase in resolution between the historical COBE satellite map (1992), the WMAP data (2003-2009) and Planck. In addition to the obviously much better resolution, the increase in sensitivity and foreground removal is also crucial.

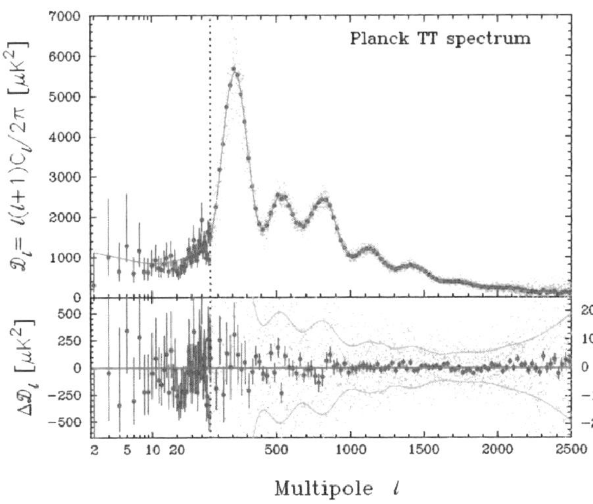

Figure 5: The corresponding Planck CMB temperature fluctuation spectrum. The angular two point correlation function of the map $C(\theta)$ is expanded in term of Legendre polynomial, $C(\theta) = \sum_\ell C_\ell P_\ell(\cos\theta)$, the coefficient of which are shown here (grey dots) up to some factor. The spread in individual measured C_ℓ can be reduced by binning the data (blue circles). The red curve shows the best fit to the data for the six parameter cosmological model described in the text. The break in the curve comes from the fact that the graph shifts from logarithmic to linear horizontal scale at $\ell = 45$. The bottom part of the graph shows the residuals once the best fit is sbstracted from the data. Only the small scale ($\ell > 45$) data are used for parameter extraction as explained later in the text. Figure taken from Ref. [3].

this happens when any scale of cosmological interest has not yet begun to evolve, one talks about cold dark matter (in the other case, the conventional name is "hot" or "warm" dark matter depending on the relativistic to non relativistic epoch). Just as for baryonic matter, its present day energy density $\rho_{\rm m}^0$ is parametrized by $\omega_{\rm m} h^2$, with the same conversion factor as above.

- **Dark energy.** In addition to normal and dark matter, the study of the history of the expansion of the Universe in the recent billions of years reveals that an extra component of negative pressure has to be present in the Universe. The reason is that we measure an acceleration of the scale factor with time, i.e. of the distance between any pair of far away galaxies. General Relativity predicts that the second derivative of the scale factor a has to obey an equation like $\ddot{a} \propto -(3P+\rho)$, where P and ρ respectively stand for the total pressure and energy density of the matter-energy content of the Universe. A positive $\ddot{a}$ therefore implies a negative total pressure ($P < -\rho/3$, actually), something which cannot be achieved with radiation, neutrinos, baryons or even dark matter. The extra component responsible for this negative pressure is conventionally named "dark energy". It could be vacuum energy (constant energy density with exactly opposite pressure) or something more complicated.

- **Reionization.** Today, no star whose composition can be determined through spectroscopy shows a total absence of metal, i.e., anything else that hydrogen and helium. This is at odds with the expected content of the first stars which originated from matter that had been unprocessed since the Big Bang Nucleosynthesis and that should therefore be devoid of any metal. This is not a surprise, however, since it has been known for long that zero metallicity stars are far less opaque to radiation and therefore much more short lived than their present-day counterparts. Even though these short lived objects have never been seen, their presence is known indirectly by the fact that spectroscopy reveals that most matter is in ionized form one billion years after the Big Bang, a result that can be explained only if one assumes that some intense UV radiation has been emitted before then, just as one expects to be the case from these primordial stars which, being less opaque, could burn their fuel at a much higher rate and hence had a much larger surface temperature. If matter was reionized sufficiently early, which therefore amounts to say that the first generation of stars formed quite early as well, then the free electron density increase can possibly rescatter once or twice a fraction of the CMB photons and alter the primordial CMB map.

- **Initial power spectrum of density fluctuations (2 parameters).** Cosmological perturbations naturally evolve from a set of initial conditions, but cannot arise from nothing. For example, Poissonian fluctuations in the baryon or photon number are far, far too small to explain the amplitude of the observed temperature fluctuations at recombination. We therefore need to a starting point for the conventional evolution of the cosmological perturbations that already contains fluctuations, and we therefore need to assume some yet unknown phenomenon that produced them. Observations suggest however that, at least approximatively, this "initial" spectrum is close to be a power law, that need (at least) two parameters to be described: an amplitude at some fixed scale and a spectral index. These are the parameters that can be extracted from the data. Then, having a class of early Universe scenarios that could produce cosmological perturbation during a very early epoch, these parameters can possibly be related to the true physical parameters of the underlying theory that was at work then.

A fit to the Planck data was made in the framework of this six parameter model, the results of which are given in Fig. 6.

Then, some extension of this parameter space was attempted by extending the model in various ways such as:

- **Non zero curvature.** Any early Universe scenario has to address the question of why is the Universe so homogeneous and isotropic. Whatever the detailed explanation, it usually

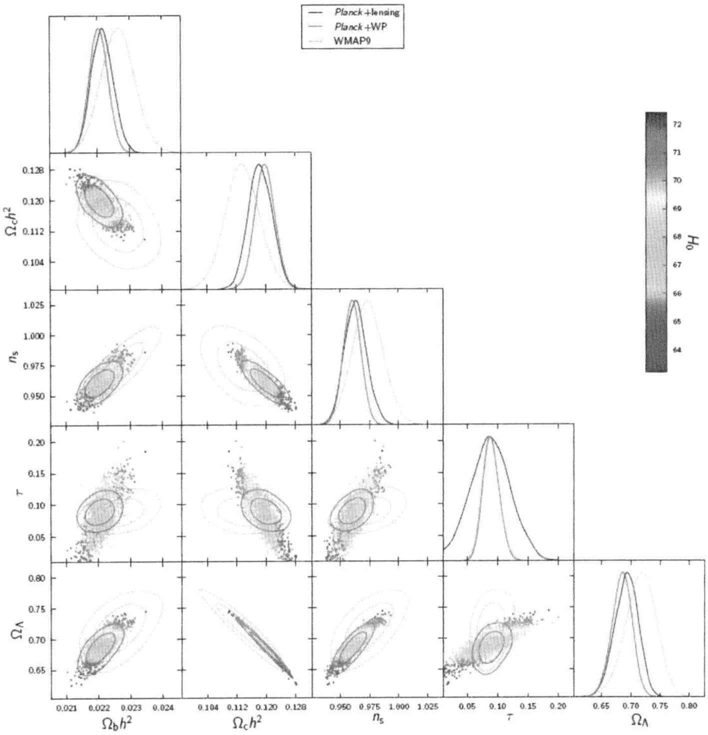

Figure 6: The one- and two-dimensional likelihoods for five of the six parameters of the so-called "concordance model" outlined by the WMAP data and more accurately constrained by the Planck data. From top to bottom are shown the estimation for the baryon density ($\Omega_b h^2$), the dark matter density ($\Omega_m h^2$), the initial spectrum spectral index (n_S), the reionization optical depth (τ) and the dark energy fraction (Ω_Λ). The one dimensional likelihood are on the diagonal and the two-dimensional likelihood under it. The constraints from Planck data alone (including lensing) are in black, those from Planck (without lensing) + WMAP polarization in red. The previous WMAP data are in grey. In most case, the Planck alone constraints are significantly better than the WMAP data, at the exception of the reionization constraints, an unsurprising result since reionization is mostly constrained by the large scale polarization data which are not yet public for Planck. The expansion rate of the Universe, the so-called Hubble constant is not estimated directly from the process, but comes trivially from the formula $H \propto \sqrt{(\Omega_b h^2 + \Omega_c h^2)/(1 - \Omega_\Lambda)}$. The corresponding value of the Hubble constant is shown in the two-dimensional likelihood plots through color codes. Adding some external constraints on it can possibly shift the best fit. The two dimensional plots show some of the degeneracies between parameters. Some parameters are not very well estimated because of these, whereas some combinations of them are exquisitely well constrained. The only parameter that is not shown is the amplitude of the perturbations at some fixed scale, which is not really important in practice. Figure taken from Ref. [3].

predicts that the spacial curvature of the Universe (i.e. the inverse square of the scale beyond which the Euclidean laws of egometry cease to be valid) had to be negligible. This can be checked within the data, mostly for consistency purpose.

- **Non trivial dark energy properties.** Dark energy is usually supposed to behave as a cosmological constant, or, equivalently, as vacuum energy within constant energy density and opposite pressure. However, more complicated dark energy models can be considered, where dark energy is made of some scalar field with negative but dynamical pressure, which is not exactly opposite to the energy density.

- **More complicated description of initial conditions.** It is clear that an amplitude (at some scale) and a spectral index are the minimum ingredients to the description of initial conditions, but more realistic models usually predict deviation from this simple law. Most notably, the spectral index might depend on the scale, and in addition to density perturbations, there may exist primordial gravitational waves which may leave an imprint on the CMB.

- **Variation of fundamental constants.** Laboratory experiments can put very stringent constraints on the variation of fundamental constants, but only during a limited time span (rarely more than one year – at most a decade). On the contrary, astrophysical data can scan larger timescales, although with less precision in the constraints (or model-dependent constraints). Recombination era is mostly determined by Thomson scattering which depends on the fine structure constant. Also, the efficiency of the baryon drag by photons depends on the electron-to-proton mass ratio. Moreover, any epoch of structure formation depends on the formulation of the gravitation theory, which can be more complicated that General Relativity. All these possibilities can, at some extent, be tested or constrained by the CMB data.

The initial six parameter model was not really in difficulty when confronted to the new Planck data, so that adding one extra parameter did not improve anything. Only upper limits one these extra parameters could be put with Planck, however, this situation might evolve when the polarization data are included in the analysis next year.

4 Some crosschecks

The Planck data very strongly strengthen the already popular six parameter "concordance model". However, the now standard model can be even more strengthened by including other data coming from the Planck satellite.

4.1 Polarization

It is known that Thomson scattering generates some polarization. If one considers the scattering of a plane wave, then the outgoing radiation is linearly polarized perpendicularily to the scattering plane. Consequently, if one considers scattering at right angle from two orthogonal direction (horizontal and vertical, see Fig. 7), then if there is more radiation coming along the two horizontal directions than along the two vertical directions, then the scattering radiation seen by the observer will acquire a net vertical linear polarization. In other word a quadrupole anisotropy in the photon distribution function at scattering translates into polarized scattered radiation. Now, if one starts from only density fluctuations in the photon distribution, then it is pretty obvious that density gradients will generate dipolar fluctuations in the distribution function and that subsequently dipole gradients will generate quadrupolar fluctuations (and so on). This shows that in addition to the temperature anisotropy, one expects to have polarization anisotropies which will be partially correlated with the temperature ones. In the end, one should observe two extra spectra that have to be confronted to the cosmological models one wants to test: the polarization angular spectrum and the cross correlation between temperature and polarization. At present, the polarization

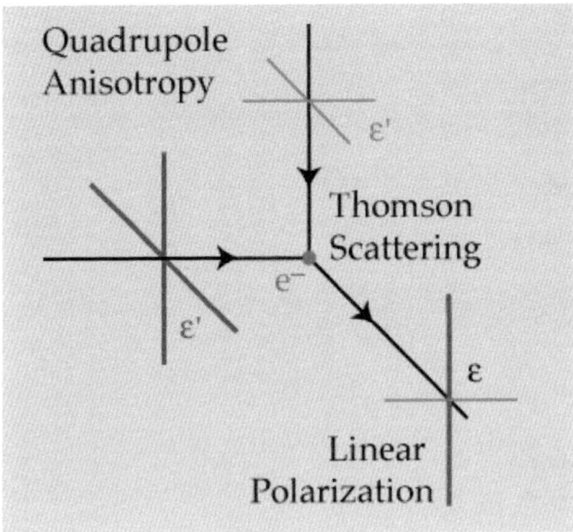

Figure 7: A simple sketch showing how an anisotropic radiation becomes linearly polarized because of Thomson scatteing. Considering radiation that is only scattered at 90 degrees, if the incoming radiation along the horizontal direction of graph has a different flux than the one that travels along the vertical direction before scattering, then the observed radiation gets more linear polarization in one direction than in the others and therefore exhibits some net linear polarization. The polarization can be mapped throughout the microwave sky and its power spectrum can be computed, just as for temperature. The cross-correlation between temperature and polarization can also be measured.

data are not ready, but the Planck collaboration have made public the present state of the data processing, which is given in Fig. 8. The blue points show the likely polarization data (on small scales, since the large scale polarization data are more foregound dependent and more difficult to process), and the red curve show the best fit obtained *from the temperature data only.* The impressive agreement between the two is a very convincing proof that no serious problem is likely to occur in the interpretation of the data. However, it has to be noted that it is very difficult to find distinct models that share the same temperature spectra but different polarization spectra, so that the agreement shown on Fig. 8 is mostly a check that the foreground removal procedure (and hence much of the foreground modelling) is very satisfactory.

4.2 Gravitational lensing

Probably the most impressinve crosscheck that can be done in the Planck data comes from gravitational lensing, i.e. the deflection of light produced by the foreground matter distribution fluctuations. The main effect of gravitational lensing on CMB (at the scale observed by Planck) is to distort the "true" map of the CMB fluctuations.This distortion does not induce loss of information as in the case of turbulence, because the distortion does not increase when one goes to smaller scales. However, an initially Gaussian distribution will exhibit non Gaussian features because of this distortion, a very fortunate thing since those can be identified through the connected four-point correlation function of the CMB fluctuations. Although the lensing signature had already been found at more than 5σ in previous CMB observations starting from the Atacama Cosmology Telescope in 2011 [4], it is the first time that the data are sufficiently precise so that the observed CMB map can be delensed: the lensing is detected here at more than 25σ. This makes possible to deduced the distortion field produced by lensing and hence map the projected gravitational potential of all the structures between the last scattering surface and us. In practice,

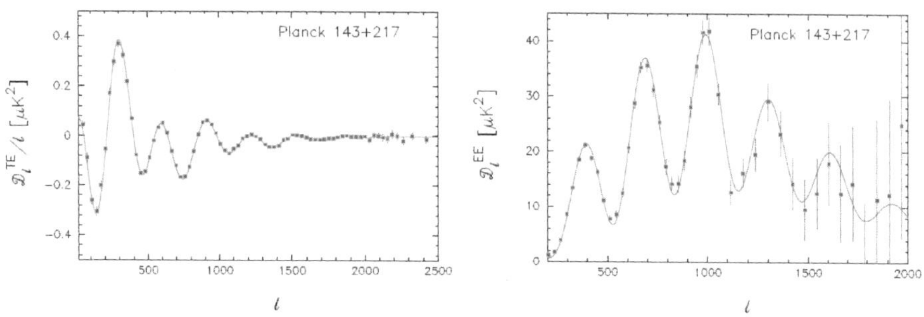

Figure 8: Present (March 2013) polarization anisotropy spectrum and the temperature-polarization cross spectrum. The dots represents still-to-be improved data points, whereas the solid curve corresponds to the best-fit cosmological model extracted from the temperature data only. The agreement is excellent, but at present, this mostly tells that the polarization foreground removal technique is accurate. How much the polarization data will improve the parameter estimation remains undetermined at the moment.

this reconstruction is difficult to do very accurately, and the gravitational potential maps that are shown in Fig. 10 have in fact a rather low signal-to-noise ratio (around 0.6 per individual mode), however, what one is in fact interested in is the correlation functions associated to the map, and it is possible to construct the projected potential two-point correlation function. Such a function can then be compared to the one that is predicted by the best-fit model of the temperature data. This is shown in Fig. 9, and even if noisy, the agreement between the data and the predicted curve is excellent. More stringent cross checks can be done with lensing as the projected gravitational potential maps are expected to show correlation with galaxy catalogs in the regions where they overlap. This is indeed what is found between 7 and 20σ level when considering the NVSS quasar catalogue (at 20σ), the SDSS luminous red galaxies catalogue (at 10σ), and the MaxBCG SDSS cluster catalogue and the WISE infrared satellite catalogue (at 7σ).

4.3 Baryon content and nucleosynthesis

One of the most impressive result of CMB study is that it allows to recover the baryonic content of the Universe independently of the predictions of the Big Bang Nucleosynthesis (BBN). During the Big Bang, weak equilibrium breaks down as the Universe cools down. This fixes the amount of available neutrons when the temperature then drops to 0.1 MeV so that nucleosynthesis can begin. With the CMB, we see a coupled baryon + photon fluid whose hydrodynamical properties allow to infer the relative amount of baryons with respect to photons. In other words, cosmology allows to measure the baryon content of the Universe either by doing some nuclear physics at $T \sim 0.1$ MeV or by studying the propagation of sound waves at a temperature of a few thousand Kelvins only. These measurements are made by completely different observations, instruments, teams, etc. The systematics, uncertainties are also completely unrelated, but still both methods give compatible results, with a nucleus-to-photon ratio of order of 6×10^{-10} and an uncertainty that is now of only a few percents.

However, the CMB parameter extraction necessitates to know the separate amounts of hydrogen and helium since recombination history is a three-step process, with a first electron combining to helium nuclei, then the second one and soon after hydrogen recombination. Usually parameter estimation is made by fitting the total baryon content of the Universe to CMB data assuming that the hydrogen-to-helium ratio is given by the predictions of standard nucleosynthesis as a function of the total baryon density. But the Planck CMB data are so precise that the *separate* imprint of

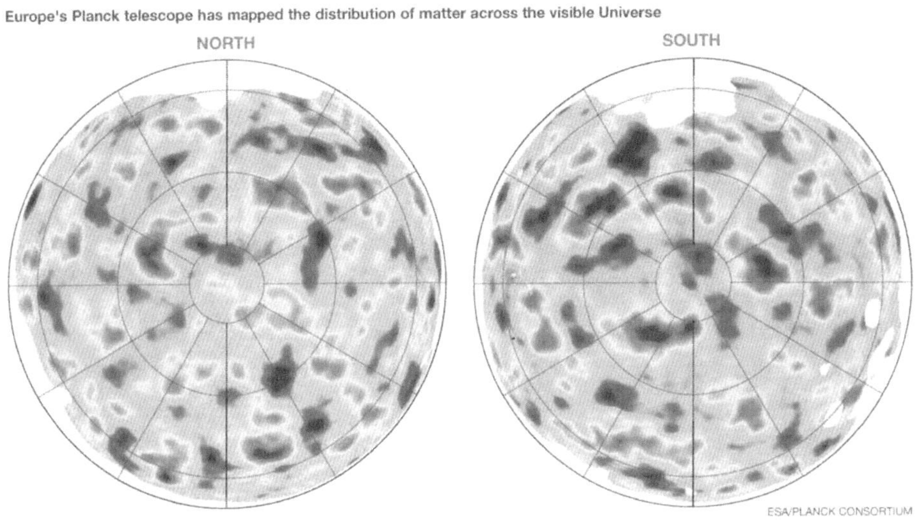

Figure 9: From the non Gaussinaities induced by gravitational lensing, one can attempt to reconstruct the projected gravitational potential on the celestial sphere, as shown here for the north and south Galactic pole regions. The lensing reconstruction suffers from large uncertainties, so that the maps shown here are compatible with the observed lensing signal, but very noisy. The SNR for individual mode is 0.6, so that there is in fact more noise than signal here.

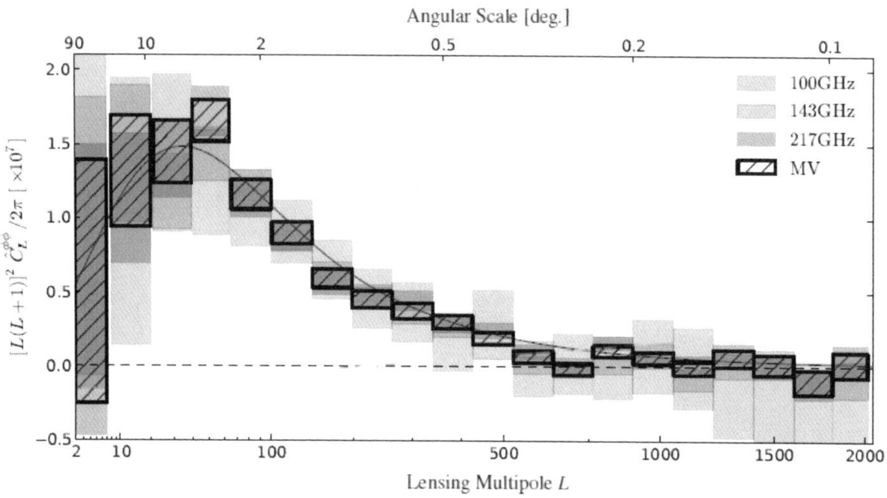

Figure 10: From the (noisy) reconstructed projected gravitational potential, one can reconsctruct the associated two-point correlation function. The procedure can be done in several ways (using different frequency maps, or a combination of them. The hashed bins correspond the best combination. The solid curve corresponds to the expectation of the best-fit model found without using the lensing data, so that the agreement between the curve and the bins is in fact a very powerful crosscheck of both the relevance of the cosmological model and the data processing pipeline.

hydrogen and helium can now be recovered directly from the data, so that the hydrogen-to-helium ratio can be estimated independently of the BBN prediction. And, again, we recover a perfect agreement between CMB and nucleosynthesis predictions.

4.4 Curvature

In the six parameter model estimation, we have explicitely assumed that the spacelike sections of the Universe were Euclidean, i.e., "flat" according to cosmologists' ambiguous vocabulary. This is justified by the fact that most early universe models that generate large scale coherent fluctuations such as the ones that are observed also contribute to erase any departure from Euclidean spacelike section, and this is indeed one of the first prediction that was made from the inflationary paradigm [5], but this has to be checked at some point. From the temperature spectrum alone this difficult as spatial curvature almost only affects the angular scale under which the CMB features are seen, just as dark energy does, so that the two parameters are degenerate and only a combination of the two can be extracted from the data. In order to remove this degeneracy, one has to combine the temperature spectrum with some other data such as the Hubble parameter constraints of the supernova data, however both are shown to be in (very) slight tension with the CMB data. It is therefore better to perform some internal checks of the flatness of space, which can be (again!) done with lensing. Indeed the amplitude of the distortion map is directly linked to and angular scale at which a given physical length appears when seen from some redshift, and this angular size vs. redshift relation depends on the dark energy + curvature content of the universe, which allows to break the above mentioned degeneracy.

From the Friedmann equation, one can define the density parameter Ω_X of any form of matter X by its ratio with respect to the critical density, $\rho_{\rm crit} = 3H^2c^2/8\pi G$. Then, the departure from 1 of the sum of all the Ω's, $\Omega_K \equiv \sum_X \Omega_x - 1$ can be put in correspondance with the local curvature radius $R_{\rm curv}$ through the formula $R_{\rm curv} = c/H\sqrt{|\Omega_K|}$. The Planck 1σ constraint on Ω_K here transform from $\Omega_K = -0.042^{+0.027}_{-0.018}$ without lensing data to $\Omega_K = -0.0096^{+0.010}_{-0.0082}$ including them. Taking a realistic value of 65 $\mathrm{km\,s^{-1}\,Mpc^{-1}}$ for the Hubble parameter, the curvature radius is bound to be larger than more than 100 billion lightyears. Restraining ourselves from considering Planck data alone, the curvature can be further constrained by adding the baryon acoustic oscilation (BAO; they correspond to featured that are observed in the present three-dimensional distribution of galaxies) data, which also allow to extract some even more accurate information on the angular size vs. redshift relation at lower redshift. When doing so, the 2σ (instead of previously 1σ) constraint on Ω_K reaches $\Omega_K = -0.001^{+0.0062}_{-0.0065}$, so that the curvature radius is then bound to be at least 3.7 times larger than the radius of the observable Universe at the 2σ level.

5 Search for neutrino properties

One of the hottest topic in CMB study was what could be deduced about the energy density of neutrinos. The motivation for ths originates from the so-called reactor anomaly, which is a series of observations which suggests a third neutrino oscillation that is not compatible with he two other, well-established, oscillations, the atmospheric neutrino oscillation and the Solar neutrino oscillation. The inconsistency of the reactor anomaly comes from the fact that the squared mass difference it predicts is neither compatible with the sum or the diffrence of the two other squared mass diffences found in the two well-established oscillations, hence cannot corespond to a situation with only the three known neutrinos. The reactor anomaly remains somewhat controversial and needs future experiments to be settled in either way. Before these, cosmology can give some answers by measuring independently the ultra relativistic matter energy density at recombination. In standard cosmology, there are only three species of neutrinos. As long as the temperature is above 1 MeV, weak equilibrium is established and both neutrinos and photons share the same temperature. Below this temperature, the coupling does no longer occur and neutrinos and photons evolve independently from each others. The neutrinos temperature keeps the $1/a$ scaling due to the expansion of the Universe, just as the photons do, except during the electron positron annihilation that occur around 0.5 MeV where some energy injection occurs in the photon fluid. In order to estimate this increase of temperature with respect to the usual $1/a$ behaviour, we remind that since there are (almost) an equal number of electrons and positrons, their chemical potentiel is 0, so that according to the first law of thermodynamics, the photon fluid evolves according to $\mathrm{d}U = -P\mathrm{d}V + T\mathrm{d}S$, just as for the photon fluid. However the "conservation law" for the common stress energy tensor of the photons + electrons + positrons fluid implies that for this system, one has $\dot{\rho} = -3H(P+\rho)$, where ρ is the energy density, which corresponds to the thermodynamic constraint $\mathrm{d}U = -P\mathrm{d}V$ once we note that $U = \rho V$ and that $3H$ is the relative time variation of a comoving volume, i.e. $3H = \dot{V}/V$. Consequently, the energy injection of the electron positron annihilation into the photon fluid is adiabatic, and the entropy S is conserved, or, equivalently, the entropy density $s \equiv S/V$ scales as $1/a^3 \propto T_\nu^3$, so that the quantity s/T_ν^3 is constant. Now, since U, V and S are extensive variables, the law $\mathrm{d}U = -P\mathrm{d}V + T\mathrm{d}S$ translates into $U = -PV + TS$ or, when dividing by V, $s = (\rho + P)/T$. As well known, the energy density for a relativistic fluid writes $\rho = 2g\sigma T^4/c$, where g is the number of fermionic or bosonic degrees of freedom and σ the Stefan-Boltzmann constant. Long before the electron positron annihilation, the quantity s/T_ν^3 is therefore given by $s/T_\nu^3 = (g_{e^+e^-} + g_\gamma)2\sigma T_\gamma^3/cT_\nu^3 = (g_{e^+e^-} + g_\gamma)2\sigma/c$. Long after the annihilation, only the photon fluid exists, so that one has $s/T_\nu^3 = g_\gamma 2\sigma T_\gamma^3/cT_\nu^3$, so that the temperature ratio between photons and neutrinos is given by

$$\left(\frac{T_\gamma}{T_\nu}\right)^3 = \frac{g_{e^+e^-} + g_\gamma}{g_\gamma} = \frac{\frac{7}{8}\times 2\times 2 + 2}{2} = \frac{11}{4}. \tag{1}$$

The energy density ratio between photons and neutrinos is therefore

$$\frac{\rho_\gamma}{\rho_\nu} = \frac{g_\gamma}{g_\nu}\left(\frac{T_\gamma}{T_\nu}\right)^4 = \frac{2}{3\times\frac{7}{8}\times 1\times 2}\left(\frac{11}{4}\right)^{\frac{4}{3}} = \frac{22}{21}\left(\frac{11}{4}\right)^{\frac{1}{3}} \simeq \frac{1}{0.68132...}. \tag{2}$$

In practice however, the electron positron occurs too close to weak equilibrium breaking so that part of the e^+e^- annihilations end up in injecting energy in the neutrino fluid, which, for cosmological purposes, amount to claim that the neutrino to photon temperature is still given by the $(4/11)^{\frac{1}{3}}$ factor, but that the "number" of neutrinos species is larger than 3. In the cosmological literature, one thus parametrizes the neutrino energy density by the "effective number" N_{eff} given by

$$\frac{\rho_\nu}{\rho_\gamma} = \frac{N_{\text{eff}}}{3} \times 0.68132... \tag{3}$$

Detailed weak interaction calculations then show that in the framework of the standard model, one has $N_{\text{eff}} \simeq 3.046$ [6].

Finally, CMB data are weakly dependent on neutrino number or masses. Since the well established squared mass differences that have been found are fairly small, it is safe to consider them as equal for CMB studies, so that in what follows, we shall focus on three equal mass neutrinos (the Standard Model ones), plus possibly something else. This explains the choice of the four analyses that were performed beyond the standard three massless (or negligible mass) neutrino hypothesis:

1. An arbitrary N_{eff} number of massless neutrinos
2. Three neutrinos ($N_{\text{eff}} = 3.046$) of arbitrary equal mass
3. The mixture of these two cases, with an arbitrary number of equal mass neutrinos
4. $N_{\text{eff}} = 3.046$ massless neutrinos, plus another set of massive neutrinos of arbitrary (non integer N_{eff}) amount.

The effect of neutrinos on CMB spectrum is well-known and has deserved significant scrutiny in the recent year because of the insights that CMB could bring to the reacor anomaly puzzle. What CMB constrains well is the matter to radiation ratio at recombination. Therefore, adding extra radiation necessitates to have more matter at recombination, and from Friedmann equations, higher Hubble parameter. This reduces the age of the universe at recombination, which mostly affects the small scale integrated effect of photon + baryon fluid evolution, therefore it is important to add the small scale CMB data beyond Planck in order to address this issue as accurately as possible. Those data come from the South Pole Telescope (SPT) and the Atacama Cosmology Telescope (ACT). Also, polarization helps in reducing the degeneracies between the effect of N_{eff} and other cosmoogical parameters, therefore the minimum combination of data that has to be consider is Planck + WMAP polarization + other ground based CMB observations at higher angular resolultion (hereafter Planck+WP+HighL).

The first case already leads to mixed results. If one takes the Planck data only, then the likelihood function peaks somewhere between $N_{\text{eff}} = 3$ and $N_{\text{eff}} = 4$, with the 2σ limits at

$$N_{\text{eff}} = 3.36^{+0.68}_{-0.64} \qquad \text{Planck} + \text{WP} + \text{HighL}$$

This constraint puts the standard $N_{\text{eff}} = 3.046$ at around 1σ to the preferred value, but the $N_{\text{eff}} = 4$ is disfavoured at almost 2σ. Constraining better the Hubble parameter should allow to decrease these bounds. Adding the BAO data allows to reduce this uncertainty as the baryon acoustic oscillations are sensitive to some angular scale that depend of the late time value of the Hubble constant. Since BAO data are consistent with Planck data, one expects them of contribute to narrow the constraints on N_{eff}, and this is indeed the case, as one obtains $N_{\text{eff}} = 3.30^{+0.54}_{-0.51}$, which prefers a little bit more significantly the $N_{\text{eff}} = 3.046$ case over the $N_{\text{eff}} = 4$ one, which is now exluded at more than 2σ. However this conclusion differs significantly if one instead adds the external Hubble parameter measurements. The reason is that those prefer higher value of this parameter, they accomodate with a higher total radiation + matter content at recombination with a fixed ratio between the two. One obtains here $N_{\text{eff}} = 3.62^{+0.50}_{-0.48}$, which now prefer the $N_{\text{eff}} = 4$ case over the $N_{\text{eff}} = 3.046$ one. Adding the BAO constraints to the Hubble parameter puts the two competing cases on almost equal grounds, with $N_{\text{eff}} = 3.52^{+0.48}_{-0.45}$.

Considering now the combined mass of the neutrinos in a $N_{\text{eff}} = 3.046$ situation (case 2 above), one observes that the constraints somewhat depend on the datasets one considers, although in a far less dramatic way. By taking the already mention Planck+WP+highL combination, one obtains the constraint

$$\sum_{m_\nu} < 0.66\,\text{eV} \qquad \text{Planck} + \text{WP} + \text{HighL}$$

at the 2σ level. Such constraint are found by using various other combinations, that are more on less stringent, ranging from 0.23 eV if one includes the BAO measurements to more than 1 eV. Although impressive, these results remain inconclusive in order to get some insight on the neutrino masses, since this figure is more than one order of magnitude above the neutrino oscillation constraints (which say that the total neutrino mass has to be larger than 0.06 eV).

Combining the two former assumptions (case 3 above) allow to put simultaneous constraints on N_{eff} and $\sum m_\nu$. The effect of massive neutrinos (which behave as non relativistic matter at late times) is quite different from that of increasing the neutrino number (which only increases the radiation content at early times), so that we do not expect both effects to be degenerate and hence that the simultaneous contraints will differ much from the seperate one. Indeed, by taking the base combination of data, the simultaneous constraints are

$$\left.\begin{array}{l} N_{\text{eff}} = 3.29^{+0.67}_{-0.64} \\ \sum m_\nu < 0.60\,\text{eV} \end{array}\right\} \quad \text{Planck} + \text{WP} + \text{HighL}$$

Adding the BAO data slightly improve the constraint to $N_{\text{eff}} = 3.32^{+0.54}_{-0.52}$ and $\sum m_\nu < 0.28$ eV at 2σ.

Considering now a massive sterile neutrino of arbitrary abundance in addition to the three ordinary ones (for which one takes $\sum m_\nu = 0.06$ eV in order to realistically account for the oscillation data), one has to distinguish between the actual sterile neutrino mass and what can be called its effective mass. The sterile neutrino becomes non relativistic at the time its temperature

T_s drops below its mass. But since we deal with a sterile neutrino that might be have decoupled since a long time from the rest of the known particles, T_s can possibly be different (and lower) from that of the known neutrinos, just as the normal neutrino have a lower temperature than the CMB. If this sterile neutrino has a thermal distribution, we therefore define the effective sterile neutrino mass $m_s^{\rm eff}$ as a function of its real mass in the thermal hypothesis, $m_s^{\rm ther}$ by $m_s^{\rm eff} = m_s^{\rm ther}(T_s/T_\nu)^3$. In practice, it is the effective mass that is contrained, the real mass being somewhat arbitrary, at least as long it is smaller than 10 eV since above this mass, a particle behaves like ordinary cold dark matter and therefore cannot be distinguished from it. The constraint on obtains here is therefore split into two domains. One first allows low $N_{\rm eff} - 3$ with almost arbitrary $m_s^{\rm eff}$ since such particles would behave (from CMB parameter extraction point of view) like cold dark matter. Second, and more interestingly, one can put the set of constraint

$$\left.\begin{array}{l} N_{\rm eff} < 3.91 \\ m_s^{\rm eff} < 0.59 \text{ eV} \end{array}\right\} \quad \text{Planck + WP + HighL}$$

and, as earlier, adding the BAO data only slightly improves the constraint to $N_{\rm eff} < 3.80$ and $m_s^{\rm eff} < 0.42$ eV at 2σ. In this respect the most naive, straightforward interpretation of the reactor anomaly would probably be $N_{\rm eff} \sim 4$ and $m_s^{\rm eff} = m_s^{\rm ther} \sim 1$ eV, which is significantly away from the 2σ region.

6 What can Planck could say about inflation

Presumably, the most interesting part of the Planck results, at least from a particle physicist point of view, deals with the contraints on the initial spectrum of cosmological perturbations, from which one can infer constraints on the physical process that produced them. At present, the most favored model for generating the first seeds of cosmological perturbations is (by far) inflation. It relies on the hypothesis that one (or more) scalar fields drove the expansion of the early Universe some time during which the field kinetic term was negligible as compared to the potential term so that it behaved close to a cosmological constant, resulting by an almost exponential expansion phase. One can show that this process naturally explains the homogeneity of the Universe, but also that it freezes toward a classical level the quantum fluctuations of the field itself, which are therefore the seeds of the cosmological pertubations that evolve later on. From this rather general setup, one can build a very large number of inflationary scenario, by varying the field potential and/or by adding some extra fields. If one focuses on single field inflation, then the physical parameter is the inflation potential $V(\phi)$, and the three main predictions of inflation (which only marginally constrain the shape of $V(\phi)$ are:

- Negligibly small spacelike curvature;
- Almost "scale invariant" spectrum of the fluctuations, i.e., with the astrophycists' definition, the spectral index of the fluctuations, n_S is expected to be very close but smaller than 1
- (Almost) Gaussian fluctuations.

To this, one may add that one may have primordial gravitational waves, but at a level lower than the density fluctuations. The Gaussianity of the fluctuation is strongly related to single field inflation. With Planck sensitivity, non Gaussianities coming from multi-field inflation were expected to be detectable. Finally, the order of magnitude of the energy density during inflation is related to the amplitude of the density perturbation, and since the latter are of order of 10^{-5}, the energy scale is very likely to be within the range 10^{14}–10^{16} GeV (the precise value depending on the amount of gravitational waves, i.e. the value of the r parameter — see below.

The results from the Planck data analysis confirmed these three predictions. As already said, no spacelike curvature is hinted by the data, which makes the inflationary ansatz relevant to the CMB interpretation. In this framework, no non Gaussianities are seen in the data, which strongly favors single field inflation among the complete inflationary zoology. Finally, and for

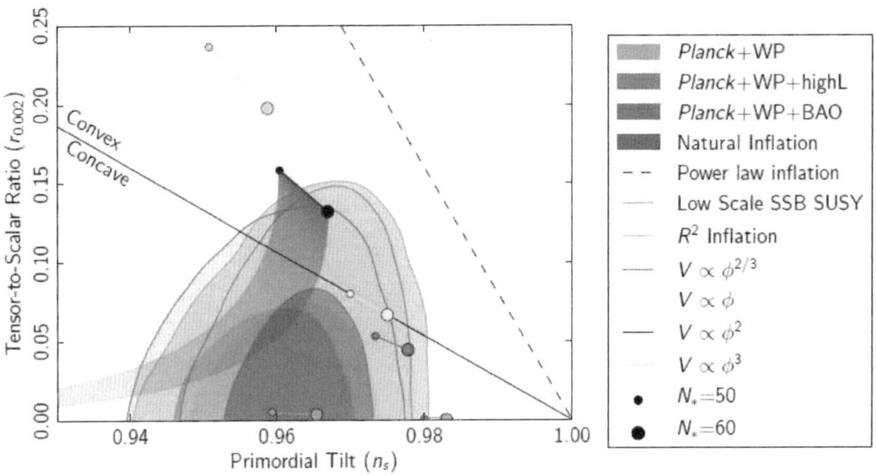

Figure 11: A sketch of the constraint derived from Planck plus other data on the two most important inflationary parameters, the density perturbations spectral index $n_{\rm s}$ and the gravitational wave to density perturbation power spectrum ratio r. In the six parameter concordance model does not include r as a parameter, and adding it results in interesting upper limits on it. On the contrary, the concordance model already detects a 5σ deviation of $n_{\rm s}$ from 1, a result that persists when one includes the r parameter in the data. As explained in the text, a given inflationary model can be put somewhere in the $n_{\rm s} - r$ plane, the exact position depending on the epoch of end of inflation, a given model therefore lying along some short segment. Some simple inflationary models are ruled out, such as the $V \propto \phi^3$ potential (several σ's) and the $V \propto \phi^4$ one (not shown as it is outside the graph). Only models with a sufficently low r are of interest now. Note that in practice, this procedure does not recover/constrain the full inflationary potential $V(\phi)$, but rather its shape on the limited range of ϕ which correspond to the values of the inflaton field when the cosmological perturbations we observe were generated.

the first time, deviation from the $n_{\rm s} = 1$ value is detected for the first time at 5σ level. This conclusion remains robust even is we allow some non zero amount of gravitational waves, and the constraints in the $n_{\rm s}$–r plane (where r characterizes the gravitational waves-to density fluctuation power spectrum ratio at some scale) are summarized in Fig. 11. In this graph, the predictions associated to some specific inflationary scenario can be drawn, and their consistency with respect to the Planck data can be checked. For example, an important result is that a power law $V \propto \phi^n$ potential for the inflationary field is excluded at 2σ for $n = 2$ and many more sigmas for larger values of n. Actually, locally concave rather than convex potentials are preferred by the data, which means that the potential is more complicated than a simple power law form since it must have some minimum and be locally convex for some value of ϕ (different from the one it was when it produced the fluctations we oserve in the CMB sky). Whether or not it is possible to perform a full reconstruction of the inflationary potential is something that cannot be decided today. In any case, even if the Planck data are not precise enough to pinpoint a very limited region of the $n_{\rm s}$–r plane, they are already able to say something about the inflationary potential, and therefore to say something about some process that probably occurred when the temperature of the Universe was around 10^{14} or even 10^{16} GeV. More details on the selection that Planck performs among inflationary models can be found in Ref. [7].

7 Various other results

7.1 Did Planck do some indirect detection of dark matter?

A poorly advertised result of the Planck mission is the discovery/confirmation of the so-called "microwave haze". This radiation was hinted from the WMAP observations back to 2004 [9] and was then received with skepticism[3]. It consists in a synchrotron-like emission that is seen on low frequencies (mostly 30 and 44 GHz), but whose spectrum is harder[4] than that of a typical synchrotron radition, and softer than free-free emission. The WMAP claim became more seriously considered in 2009 when the Fermi satellite discovered the so-called "Fermi-bubbles", which are a large area of high energy (GeV range) emission arising around the Galactic center. The microwave haze could then be naturally explained as the synchrotron radiation emitted by charged particles associated with the high energy photons detected by Fermi, however the association remained to be proven, and this is what Planck could do, by benefiting from its better sensitivity and foreground control [10]. The detailed analysis of the Planck data did justify some of the hidden assumption made in the WMAP analysis, and exclude some of possible the caveats that had been put forward against the existence of the microwave haze. The importance of the haze-Fermi bubble association from a particle physics point of view arises from the fact that it could be the consequence of dark matter coannihilation in the Galactic center region. In their WIMP interpretation, dark matter particles are likely to coannihilate, produing high energy photons, as well as particle-antiparticle pairs which will then travel through our Galaxy magnetic field and then radiate synchrotron radiation. As an exemple, TeV range dark matter particle coannihilation could possibly produce the desired Fermi radiation as well as the observed microwave haze, provided, for the latter, that the Galactic magnetic field has a value of order 5 μG [5]. However, the dark matter coannihilation hypothesis is only a possibility among others as more conventional astrophysical processes could produce it [11]. These include enhanced star formation/supernova rate in the Galactic center, second order Fermi acceleration, galactic wind and active galatic nucleus-like activity from the cental black hole. The Fermi emission remains somewhat puzzling at this stage. First, the emission flux is very flat in the sense that it is very uniform along the whole area of the sky where it is seen, and it has also a sharp cutoff at its edges. This is a surprising result, because if the GeV photons arise from a volume emission, then there should be a brightening in the center and a limb darkening, the latter not being present in the data. If, on the contrary, they are due to a surface emission (related to some shock-wave, for example), then one should see some limb-brightening, something which is not seen either. From what we understand of the different astrophysical explanations, some seem already ruled out because they should be accompanied by various other counterparts that are not seen: for example, galactic winds should produce an Hα emission which is not seen here. Still, none of the proposed explanation seems able to explained the flatnes of the emission combined with the sharpness of its cutoff. For example, AGN hypothesis should produce volume emissivity and therefore limb darkening. This leaves, at the moment of writing of the se proceedings, the possibility that the dark matter cannihilation is valid, and in any case the Planck data regarding the synchrotron counterpart will play a role in setting the answer.

7.2 Negative results of various exotic searches

We gave a detailed account of what Planck could say about an extension of the six parameter concordance model in the neutrino sector, but this is far from being the only extension that has been considered and for which one obtained upper limits.

For example, the detection of cosmic strings in the Planck temperature map has been considered. Cosmic strings are objects belonging to the general class of topological defects, that can be formed during a symmetry breaking phase transition. String-like objects are known to exist in

[3] An instructive personal account of this can be found in a blog post [8] by the author of Ref. [9]

[4] i.e., the ratio of the former to the latter increases with frequency.

[5] We apologize for spreading the astrophyscist obsolete habit of using Gauss and other weird esu units instead of Teslas and the rest...

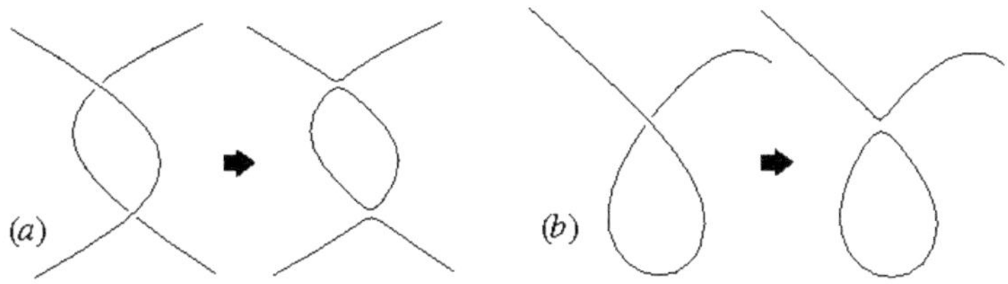

Figure 12: The process of intercommuting (a), possibly leading to the formation of cosmic string loops (b). Picture taken from [15].

condensed matter, such as the magnetic flux tube in a superconductor of the vorticity tubes in a superfluid. Regarding the cosmological context, if during some phase transition one breaks an U(1) symmetry, then one can form extended string-like objects, which are either closed loops or infinite strings. Under quite general conditions, these objects can intercommute when they cross, that is, they can exchange their ends (see Figure 12). This important property makes that the number of long strings (those that are either infinite or which form closed loops larger than the size of the observable Universe) per unit of Hubble volume remains more or less constant with time, so that these object constantly generate density fluctuations of order of the mass per unit of length, $G\mu$, whose scale is given by the symmetry breaking scale by $G\mu \sim (E_{\rm SB}/E_{\rm Pl})^2$. Such regime is the so-called scaling regime and is known to be established both during the matter dominated and the radiation dominated eras. It has also been known for a long time that cosmic strings cannot generate a wiggly anisotropy spectrum such as the one that is observed. The reason is that since generate continuously density perturbations, what one sees can be interpreted as the incoherent sum of wiggly spectra which are progressively shifted with respect from each others [12]. The wiggly shape of the observed CMB spectrum therefore rules out cosmic strings as the main contributor to cosmological perturbations, however, it might happen that a symmetry breaking occurs sufficiently soon after an inflationary phase, so that a subdominant string network leaves a small but observable imprint in the CMB anisotropies. In such case, as it is fairly complicated to compute the two-point correlation function produced by a string network, it is better to look for some simple characteristic signature left by sch object. One of these is due to the fact that spacetime around a string is conical, i.e., it is locally Minkowskian, but there is a deficit angle when one turns around a string. The perimeter of a circle of radius R is not $2\pi R$, but $2\pi(1-4G\mu)R$. Consequently, two photons propagating parallel to each other and passing on different side of the string will be focused after passing close to the string, as shown in Fig. 13. This string wake effect will translate into a redshift difference of light rays passing on different side of a string as seen from an observer looking at the string along a direction that is orthogonal to its motion. One therefore expects to see curves of temperature discontinuities in the CMB sky, the so-called Kaiser-Stebbins effect [13], which is proportional to $8\pi G\mu$. The total length of the discontinuities does not depend on $G\mu$ since the scaling regime does not depend on it, but the amplitude of the discontinuities does. The Planck data do not show any hint of discontinuities (or smoothed version of them because of the finite resolution of the telescope), nor any other signature associated with cosmic strings[6], which translates into an upper bound on $G\mu$ of 1.5×10^{-7} [14].

7.3 The intriguing large scale anomalies

Although the data impressively fit the six parameter concordance model, some analyses also reveal some intriguing features in the data.

[6]Actually, the cosmic strings anisotropy spectrum peaks at rather high multipoles, so that it is the direct impact of cosmic strings on the C_ℓ spectrum that constrains them better than the specific non Gaussian signatures, see [14] for the detailed constraints depending on the methods that are used.

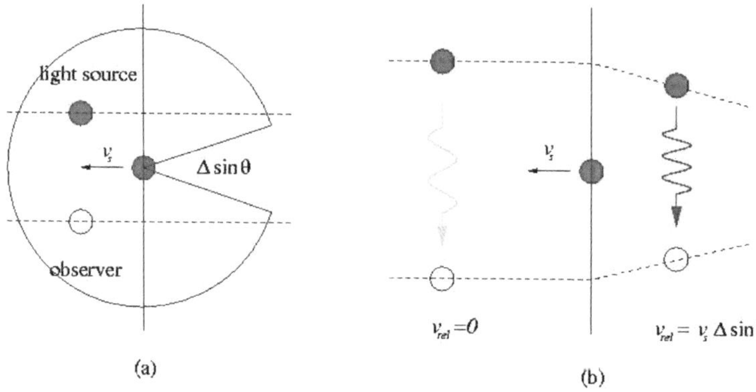

Figure 13: A sketch of the Kaiser-Stebbins effect, i.e. the fact that initially parallel light rays will be focused in the wake of a cosmic string as a consequence of the conical space that exists around these objects. If, in addition, the cosmic string is moving perpendicularily to the observer, a temperature discontinuity will be observed around the string, a feature that can make its detection possible. The amplitude of the discontinuity is proportional to the deficit angle $\Delta \sin\theta$, which in turn is proportional to the cosmic string mass per unit of length μ expressed in Planck units. Picture taken form [16].

One of the most obvious is that there is a slight lack of power at large angular scales when considering the best fit to the data (see left part of Fig.5). Most of the $\ell < 50$ multipoles are below the best fit curve, which is rather unusual from a statistical point of view. Since this feature is rather weak (a less than 3σ effect), it is tempting to discard it completely, but when trying to extend the six parameter model to something else, the inclusion or not of the low ℓ multipoles somewhat affect the outcome of the analysis. Basically, new physics appears to be a more favored possibility when one includes the large angular scales, although with a weak statistical significance. Also, this feature adds to various "anomalies" that were already hinted by the WMAP data and confirmed by Planck. For example, the $\ell = 3$ multipole is unusually flat and well aligned with the quadrupole. There seems also to be many phase correlations between the different multipoles, whereas there are expected to be uncorrelated in a statistically isotropic Universe. A comprehensive list of these "anomalies" is given in Ref. [17].

Even more unexpected is the so-called power asymmetry. It consists in the fact that two opposite hemispheres on the sky have statistically significant different overall amplitude spectra in the sense that we can model the observed temperature map by the phenomenological formula $\Delta T/T(\hat{n}) = s(\hat{n})(1 + A\hat{p} \cdot \hat{n})$, where $s(\hat{n})$ is a statistical isotropic map, A some amplitude and $\hat{p}$ a direction. If one includes a sufficiently large amount of scales, then there seems to be a more than 3σ detection of a nonzero value of A. This fact was hinted by the WMAP results, and proven to be foreground unrelated by Planck. At present, it seems that no simple inflationary model is able to explain this feature.

None of the effect mentionned here are very significant (rarely above 3σ) but their accumulation definitely makes them intriguing. One could argue that we may naturally inclined to pay attention to the rare anomalies that exist in the data and give them more relevance than they deserve, but one has also to consider that these features are real and hint for some incompleteness of our description of the Universe. A fascinating possibility would be that during inflation, large scales experience less homogeneization than small scales, so that the features that are seen only on large scales could give some hints about the pre-inflationary universe. Although some attempts have aleardy been done, it is too early to decide whether these feature will prove being able to change our view about the early Universe.

8 The overhyped and overoptimistic claim of primordial gravitational waves detection

Whereas 2014 was the year where the new release of Planck data was expected (they will be published in 2015), it will remain in CMB history as the year where the first detection of primordial gravitational waves was claimed by a ground-based mission, BICEP2.

The importance of primordial gravitational waves in cosmology was noticed decades ago. A typical scenario that produces density waves often also produces gravitational waves. Detecting those therefore gives a large amount of extra information about the underlying process that is at work in the early universe. Gravitational wave do not produce overdensities as they travel through spacetime, but a photon can exchange energy with those. Gravitational waves therefore produce an imprint on CMB temperature map. However, such a distortion is difficult to identify as such since many properties of the gravitational wave distorted map are essentially the same as the non distorted one (contrarily to gravitational lensing). However, one can predict the contribution of primordial gravitational waves to the CMB power spectrum, in the same way as we do for density waves. We only need for this the initial power spectrum of gravitational waves, which is usually taken as a power law (this is predicted by most models). This therefore adds new parameters to the cosmological parameter extraction procedure.

It was already known prior to the Planck data that detecting those gravitational waves would be difficult because the procedure outlined above was compatible with no gravitational waves at all, and, even worse, the best fit was actually no gravitational waves at all. One had, in practice, only upper limits on the amount of gravitational waves. Such upper limits are usually parametrized by a single number, r, which fixes the ratio in the initial power spectra of gravitational and density waves at a given scale, usually $0.002\,h\,\mathrm{Mpc}^{-1}$, which corresponds to the low-ℓ/large angular scales flat part of the CMB spectrum. The constraints on r are of course model-dependent : they depend on the number of parameters that are contrained. Unsurprisingly, the more parameters, the less constraining are the data, but the typical constraint was around $r < 0.12$ prior to Planck data, including all large (WMAP) and small scale (ACT and SPT) CMB data, plus galaxy catalog constrainst (baryonic acoustic oscillations). Such a constraint was found considering the base ΛCDM model, and was sloppier when including a larger set of parameters, such as running of the spectral index, that is, the fact that the density perturbation spectral index is allowed to (slightly) vary as a function of the wavenumber.

In order to perform a clearer detection of primordial gravitational waves, one has to turn to the polarized part of the CMB light. The reason is that Thomson scattering naturally produces linearily polarized light in a direction that is orthogonal to the scattering plane (see Figure 7 above). If, in addition, light is anisotropically distributed, then a net linear polarization shall remain when taking account all contributions.

Now, whereas both density and gravitational waves produce a slight amount of polarization in the CMB, it happens that at linear order, some polarization patterns are exclusive to gravitational waves. Basically, only linear polarization is produced by scattering of CMB light on free electrons. The observed polarization can therefore be decomposed into an arrowless vector field. Such a vector field can further be decomposed into parity invariant patterns (radial-like or circular pattern), which are called E-modes. Non parity invariant patterns (curls) are then called B-modes, following a very formal analogy with the way electric and magnetic fields evolve under parity transformation. Figure 14 summarizes this definition.

Unfortunately, life is more complicated because in addition to primordial gravitational waves, B-modes can be produced by two other processes:

- First, some non linear effect can produce B-modes from density waves only. The main contributor to this is lensing, which causes a distortion of the "true" CMB map because of deflection of light. Such deflection can transform a small part of the intrinsic E-modes into B-modes.
- Second, the more complicated physics of foregrounds (for example Galatic dust) produces

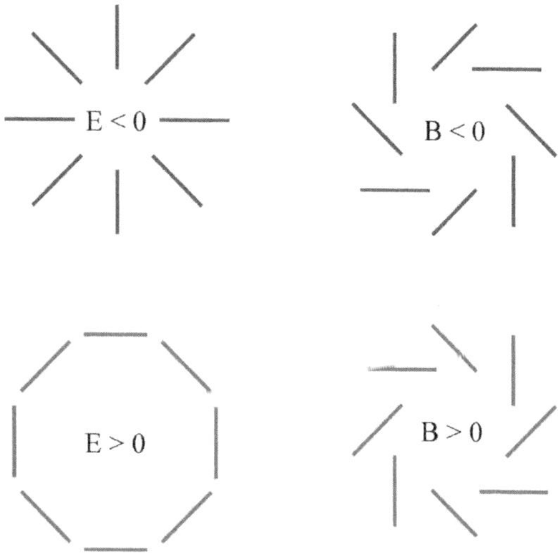

Figure 14: A sketch of the definition of E-modes and B-modes. These patterns can be constructed from the more common Skokes Q and U parameters, but are more convenient in the sense that both Q and U depend on the somewhat arbitrary choice of orientation of a set of axes. A rotation of the system of axes transforms both Q and U into each other. On the contrary, the E and B mode are rotation invariant. The crucial point here is that at linear order in perturbation theory, only gravitational waves can produce a significant amount of B-modes. Detecting those, and insuring that no other (non linear or foreground related) process produces them is therefore a smoking gun of primordial gravitational waves.

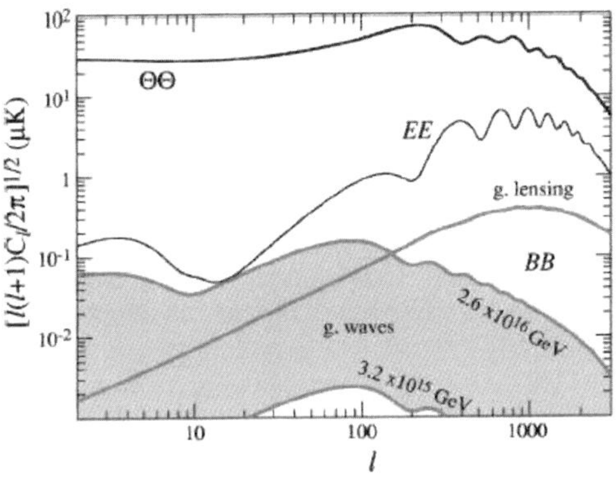

Figure 15: A sketch of the expected temperature and polarization power spectra coming from the density waves. The $\Theta\Theta$ power spectrum is that of temperature. The EE spectrum E-mode polarization generated at linear order by density waves. The upper BB spectrum is then the non linear distortion of the EE spectrum arising from gravitational lensing. Its amplitude and shape are very closely related to that of the EE spectrum, with a smoothing arising from non linear effects (in principle, the observed EE spectrum is also slightly distorted, but this is a small effect here). Under these three spectra, the shaded region correspond to the possible location of the spectrum arising from gravitational waves, the amplitude of which is not predicted theoretically but rather depends on the specific early universe scenario one considers. However, a general feature is that the primordial gravitational wave signal may dominate at large angular scales if its initial amplitude is large enough. In the case of some model of inflation considered here, the amplitude is directly related to the energy scale at which these perturbations were produced during inflation.

some amount of polarization. This polarization is expected to produce comparable amount of E and B modes.

Detecting the imprint of primordial gravitational waves therefore necessitates to be able to distinguish between these three contributions. The lensing contribution is the easiest to handle with because the amount of B-modes produced by lensing is entirely determined by the initial amount of E-modes (see Fig. 15). Removing this contribution leaves only the foreground plus primordial B-mode contribution. In March 2014, the BICEP2 experiment reported for the first time the detection of B mode on not-so-small angular scales [19]. Being a ground experiment, BICEP2 did not perform a full-sky scan coverage but rather focused on a limited patch of the sky (approximatively 1% of the whose celestial sphere) which was observed during several months. The small optical system (a 30 cm refractor) and the atmospheric emission were compensated by exquisite state-of-the-art, new generation detectors and a long exposure time. The observed patch of the sky was of course chosen so that it could be observed continuously from the South Pole (where the experiment was set) and because of its low foreground content, relying dust maps performed by WMAP and other instruments.

The detection of B-modes by BICEP2 was estimated to be a 7.0σ effect, and the amplitude of this signal at some of the scales it was detected was larger than the lensing contribution. Furthermore, the estimates of the amount of polarized foreground emission at these angular scales and in the observed patch of the sky were also smaller than what was measured. This led the BICEP2 team to conclude that the "long search for tensor B-modes [was] apparently over". Although this announcement led to a flurry of press articles, many cosmologists remained somewhat skeptical about the conclusion drawn by the BICEP2 field. The reason was that the only justification for the observed signal to be of primordial origin relied on estimates of the amount of polarized emission rather on actual measurements of those. No such measurement was publicly available at that time. The reason is that dust is most easily detected at high frequencies (typically above 250 GHz, see Fig. 2), but measurements at those frequencies are not possible from the ground because of atmospheric emission. The only actual measurements that were made at that time were those by the Planck spacecraft but were not available then. Therefore the BICEP2 people had to trust foreground models whose reliability had not been checked at that time. They also tried to strengthen these estimates by using some slides taken from official seminars from the Planck in which was advertised the quality of the (yet) unpublished polarization data. But of course those such slides were not meant to be usable to do any science, they were rather teasers showing that they had good data. Another source of concern is that the amount of gravitational wave that was claimed to be detected was of order of $r = 0.2$, a value that was significantly higher than the upper limits derived from Planck temperature data.

The importance of the BICEP2 claim naturally led to a joint effort between the Planck and BICEP2 groups in order to obtain a reliable estimate of the the polarized dust emission, which led to a disappointing result [20]. The main conclusion was that, although the dust content of the BICEP2 patch was indeed small, its polarization fraction was larger than average. Worse still, the average polarization of the foreground was significantly larger that what was expected. The combination of those two factors made the BICEP2 signal perfectly compatible with foreground emission only. This does not mean that no primordial gravitational wave is present, but at the moment it is not possible to see it in the data.

The BICEP2 measurements, although far less interesting for cosmology as initially claimed, remain however an important step in polarized CMB studies. The improvement in detector technology made by the BICEP2 team made possible, for the first time, to detect a signal of potential cosmological origin and opens the door to more ambitious projects that could end in actually detecting these long-awaited gravitational waves.

9 Reflection on the future of CMB studies

The Planck temperature anisotropy data can hardly be improved by future missions because the main limitation we face is the so-called cosmic variance, i.e., the fact that we observe a finite volume of the Universe, which therefore limits, from a purely statistical point of view, the robustness of our results. Regarding polarization, the situation can be very significantly improved since we are (very) far from being in reach of the ultimate sensitivity achievable, since we are here limited by the diffraction limit and photon noise. Some reflection have begun since a few years regarding a possible successor to Planck. A recently proposed project (from European side) was called PRISM (Polarized Radiation Imaging and Spectroscopy Mission) [21]. It was based on a larger telescope (3.5 m instead of 1.5 m of equivalent diameter) and more importantly, a much larger number of detectors (7600 vs. 74 for Planck). This huge leap in sensitivity could allow a number of new studies to be done. For example, the CMB frequency spectrum could be measured to such a precision that some tiny deviations from thermalization due to exotic phenomena as early as $z \sim 10^6$ could be detected. Among these, let us mention dark matter annihilation, dark matter decay, or primordial black hole evaporation.

A considerable amount of technical challenges had to be addressed in order to make this project possible, since, for example, the amount of data collected by the instrument would have been too large to be downloaded to Earth so that a heavy amount of real time processing had to be done onboard. All this strongly impacted the overall cost of the mission and put it in the category of the Large ESA mission (overall cost around 1 billion euros) whose first available launching slots would not occur before 2027. Very recently, ESA chose to attribute the two presently planned L-class mission slots to other fields (spaceborne gravitational waves observatory and X-ray telescope), so that the PRISM project will probably have to be rescaled down to a medium class mission, but still keeping the large increase in detector number of the first L-class proposal.

10 Conclusion

The results presented here come from the first release of the Planck data. They confirm the overwhelming consistency of the concordance model, but point out to some anomalies whose interpretation (and existence) are unclear. The situation regarding these anomalies should improve in 2015 with the release of the polarisation data. Whether or not they will disappear or point toward the necessity to improve the concordance model is impossible to tell now.

References

[1] K. Freese, B. Fields, D. Graff, Proceedings of the 19th Texas Symposium on Relativistic Astrophysics and Cosmology (CD-ROM), arXiv:astro-ph/9904401 .

[2] W. Hu, Lect. Notes Phys., **470**, 207 (1996).

[3] P. A. R. Ade, *et al.*, *Astronomy and Astrophysics*, **571**, A16 (2014).

[4] S. Das, *et al.*, *Physical Review Letters*, **207**, 021301 (2011).

[5] A. H. Guth, *Physical Review D*, **23**, 347–356 (1981).

[6] G. Mangano, *et al.*, *Nuclear Physics*, **B729** 221–234 (2005).

[7] J. Martin, *et al.*, arXiv:1312.3529.

[8] `http://blogs.discovermagazine.com/cosmicvariance/2012/09/04/ /guest-post-doug-finkbeiner-on-fermi-bubbles-and-microwave-haze/`

[9] D. P. Finkbeiner, *Astrophysical Journal*, **614**, 186–193 (2004).

[10] P. A. R. Ade, *et al.*, *Astronomy and Astrophysics*, **554**, A139 (2012).

[11] G. Dobler, *Astrophysical Journal*, **750**, 17 (2012).

[12] A. Riazuelo, & N. Deruelle, *Annalen der Physik*, **9**, 288–298 (2000).

[13] N. Kaiser, & A. Stebbins, *Nature*, **310**, 391–393 (1984).

[14] P. A. R. Ade, *et al.*, *Astronomy and Astrophysics*, **571**, A25 (2014).

[15] `http://www.damtp.cam.ac.uk/research/gr/public/cs_home.html`

[16] `http://aether.lbl.gov/eunhwa_webpage_2/coverpage.html`

[17] , P. A. R. Ade, *et al.*, *Astronomy and Astrophysics*, **571**, A23 (2014).

[18] W. Hu, & M. White, *Phys. Rev.* D**56**, 596–615 (1997).

[19] P. A. R. Ade, *et al.*, *Phys. Rev. Lett.*, **112**, 241101 (2014).

[20] P. A. R. Ade, *et al.*, *Phys. Rev. Lett.*, **114**, 101301 (2015).

[21] `http://www.prism-mission.org/`

LECTURER: A. RIAZUELO

Planck Highlights

Scientific Secretaries: I. Yeletskikh, K. Kornieieva

DISCUSSION I

- *Zennamo:*

I have questions concerning picture on slide 16. First, I did not really know what 'spinning dust' was, could you please explain that? Second, in the 'galaxy cluster' curve you have a sharp dip before it turns back on, so I was wondering, what was that? Is it just a coincidence, that you see all spectrum lines lined up with all features – the end of the 'spinned dust' curve and the spike in the 'galaxy clusters' curve?

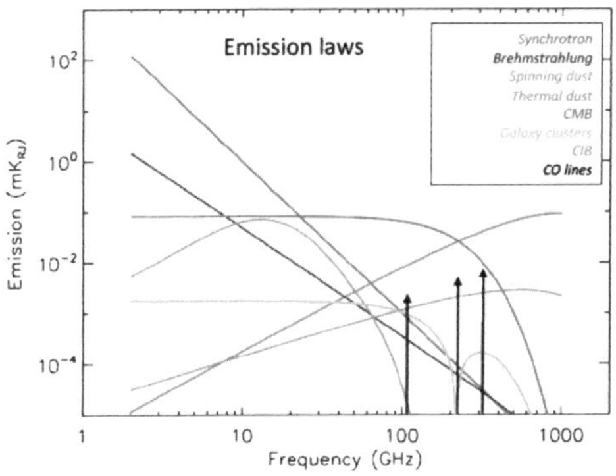

- *Riazuelo:*

The 'spinning dust' is something that was discovered while processing the Planck data. It was discovered that the part of low energy emission was probably explained by the fact that dust was made of small, spinning grains. Regarding CMB (Cosmological Microwave Background) – the regions, where CMB is analyzed, are those, where the backgrounds are small.Though we do not understand all the emissions in the galactic plane (say, at 800 GHz), it does not affect the CMB analysis. Concerning the 'galaxy clusters', this is the so-called Sunyaev-Zel'dovich Effect. When some cold radiation crosses some hot gas in the inter-galactic medium – it undergoes a distorsion, with a depletion of low energy photons and a small increase of higher energy photons, without net energy exchange between the hot gas and the cold radiation. One can show that at some specific frequency (217 GHz for a 2.73 K blackbody), no distorsion occurs, hence the dip in this contribution at that frequency. This effect is completely unrelated to the carbon monoxide lines which are the lowest rotational excitations of the carbon monoxyde molecules.

- *Esposito:*

Is that true that BICEP-2 never published any paper with a detailed description of the analysis of systematics? If so, could you comment on that?

- *Riazuelo:*

Yes, it's true. They removed all comments regarding the interpretation of their data. In the first arXiv publication they claimed they saw a signal with 7σ significance, which was not due to polarized dust. In the published version of their paper they removed their claim of having dust effects under control. It was actually shown that a proper removal of the background (especially dust) was probably not performed. We will have to wait a bit before knowing more about this. There will probably be a joint paper of Planck and BICEP-2 – after Planck will publish paper with polarized dust data.

- *Boren:*

Why does the Plank satellite use L2 Lagrange point, instead of more stable L4 and L5 points?

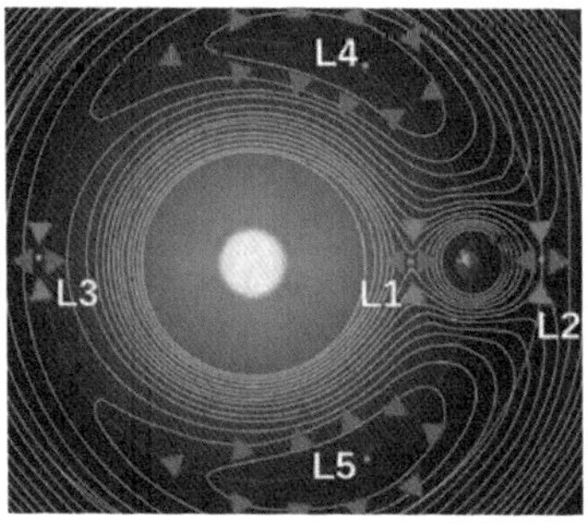

- *Riazuelo:*

Main issue is expense, and I don't think you would gain more, going to L4 or L5. At L4 or L5 the telemetry would be more complicated, since much larger distance from the Earth (150 million kilometers instead of 1.5 million). The L2 Lagrange point is unstable, as you may know, but the time scale at which unstability develops is about 6 months, so one needs to do some orbit correction maneuvre less than once per month. Moreover, cruising to L4 would take several months, and since satellite contains He-3 in it's tanks, which are not perfectly isolated, all the way it would leak a little bit into space. It also needs to be stressed that Planck is said to use three years of He-3 world production, and this is not a negligible part of the cost of the satellite. Thus, it is useless to spend months to go to L4, wasting He-3.

- *Miyazaki:*

How do you calibrate detectors at different frequencies? It is very important to know the response because you use data at high/low frequency to estimate background in CMB-dominant region.

- *Riazuelo:*

If you want to detect the fluctuations with the amplitude of 10^{-5}, you need the perfect thermal stability and the detector that is much cooler than the radiation to be detected. The temperature of 0.1 K is the compromise of what could be achieved in space, giving the amount of He-3 and the stability people want. For calibration we use several methods. One of these is the CMB dipole. Earth is not at rest with respect to CMB, so that it shows a dipolar modulation throughout the sky, whose amplitude had been very precisely measured by the COBE satellite. Moreover, we observe an annual modulation of this CMB dipole, because of the motion of the Earth around the Sun. We again know the absolute amplitude of this modulation, which allows to calibrate the detectors. Then, the scanning strategy of the satellite allows to periodically observe all the planets of the Solar system. Their spectra are sufficiently well known, and this also allows doing calibration, which is of course a crucial issue.

- *Miyazaki:*

After the result of BICEP-2, do you have any tricks to eliminate human bids in analysis of Plank data?

- *Riazuelo:*

That is a very good question. I'm not sure how biased the analysis is. There are many complicated phenomena (like the backgrounds) and it is very difficult to account for all the human biases. But we definitely do all what we can to put this more or less under control.

LECTURER: A. RIAZUELO

Planck Highlights

Scientific Secretaries: V. Kovalenko, J. Zennamo

DISCUSSION II

- *Kofarago:*

With the Planck data we cannot decide between the inflation theories. Is it only a lack of statistics or do we need new quantities to be measured? If we need new quantities, what are they? Both with the Plank data and beyond.

- *Riazuelo:*

The most important unknown quantity at the moment is the tensor-to-scalar ratio, r. If it is below 0.1, it would be difficult to detect it with Planck temperature data only because the temperature power spectrum is the sum of these two contributions (scalar modes, i.e., density waves, and tensor modes, i.e., gravitational waves) without possibility to distinguish between them explicitely. If one adds polarization data, one can hope to detect directly this tensor-to-scalar ratio because some polarization patterns, the so-called B-modes are specific to gravitational waves, but Planck experiment has not enough sensitivity to detect them if r is below 0.1 If this ratio is indeed 0.2, as BICEP-2 people claim, then Planck could possibly detect it as well. The difference between the Planck and BICEP-2 experiment is that BICEP-2 observes small part of the sky with higher sensitivity and Planck observes the full sky, but with lower sensitivity; it is hard to say which experiment is best. Concerning future missions, some people in Europe proposed to ESA a next generation space telescope, a bigger satellite than Planck with detectors better than BICEP2 with good covering of a focal plane (with Planck, only 1% of the light hitting the focal plane reaches the detectors). The experiment will be designed to observe CMB polarization data. Instead of 9 channels with 70 detectors like for Planck it could have 8000 detectors operating in 30 different channels, which is, of course, a huge difference in term of efficiency. We will have such a precise measurement of the black body spectrum of CMB, so we will be able to measure the deviations from it, for example, the Lyman-alpha lines during the recombination (when electrons combine with atomic nuclei). More importantly, such a satellite will be able to measure r down to 10^{-3} – $3{\cdot}10^{-4}$. The limit of what can be measured for this ratio comes from the fact that E-mode polarization produced by density waves are partially converted map into B-modes due to gravitational lensing. Those lensed B-modes will dominate over the B-modes produced by gravitational waves we are interested in if the gravitational waves amplitude is too low. One can partially get rid of those spurious lensed B-modes by performing some delensing procedure, however, the process cannot be 100% efficient. Current estimates are that without delensing, the upper bound on r can be decreased by two order of magnitude with respect to current ones (that is, one could go from 10^{-1} to 10^{-3}), and one could gain an extra half order of magnitude with delensing. Said in a different way, in order to extract CMB temperature power spectra, we do not need a better telescope than Planck. In term of temperature, Plank is the ultimate thing. The new project is a kind of equivalent for the polarisation data. The very preliminary design of the new telescope, PRISM, was proposed as a Large ESA mission (1 billion euros – quite expensive). Recently it was not approved by ESA and was recommended to be rescaled as an M-class mission (twice cheaper), for which there are more slots available in the future. It is possible that such telescope will be built some day. It will depend on whether the observation of the gravitational waves motivate more detailed study.

- *Nanopoulos:*

Can you comment on a paper which says that the BICEP signal is coming from dust inside the galaxy?

- *Riazuelo:*

This is something that is investigated by Planck, but no result has been released yet. It is pretty clear that everyone is working on all the issues, especially on the BICEP-2 patch. Basically, BICEP-2 scientists claim that they have detected primordial gravitational waves because the signal they observe is compatible with this and that the patch of the sky they are observing is free from any significant amount of foreground emision which could mimic the gravitational waves signal. This patch was selected by the BICEP-2 team for this very reason but this now deserves better scrutiny. For example, people found recently that there is supernova remnant bubble in this region, and this could have an impact on the observed signal. The Planck investigation regarding the amount of foreground emission in the BICEP-2 field should be released in October.

Highlights from BNL and RHIC 2014

M. J. Tannenbaum *
Physics Department, 510c,
Brookhaven National Laboratory,
Upton, NY 11973-5000, USA
mjt@bnl.gov

1 Introduction

The Relativistic Heavy Ion Collider (RHIC) at Brookhaven National Laboratory (BNL) is one of the two remaining operating hadron colliders, the other being the LHC at CERN. Unlike the LHC, which is buried in a deep underground tunnel, RHIC is built in an enclosure on the surface which is covered by an earth berm for shielding which can be seen from outer space (Fig. 1).

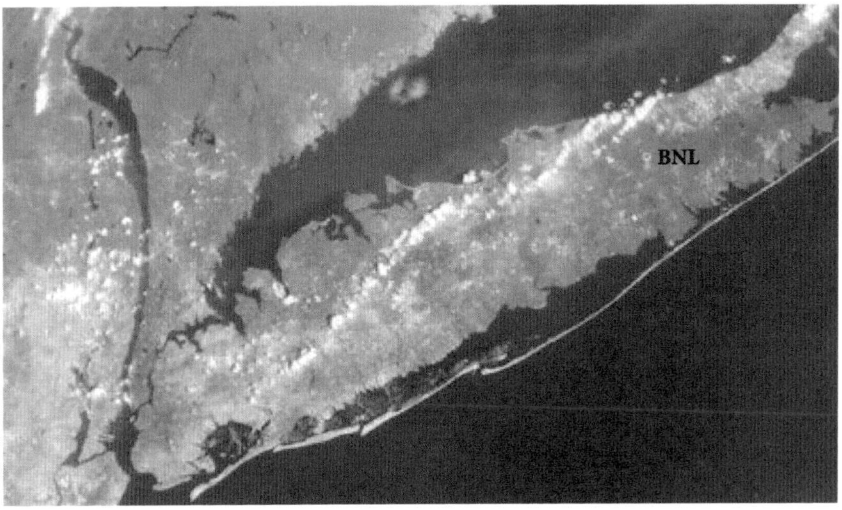

Figure 1: NASA infra-red photo of Long Island and the New York Metro Region from space. RHIC is the white circle to the left of the word BNL. Manhattan Island in New York City, ~100 km west of BNL, is also clearly visible on the left side of the photo.

*Research supported by U. S. Department of Energy, DE-AC02-98CH10886.

BNL is a multipurpose laboratory, quite different in scope from Fermilab and CERN, with many "cutting edge" major research facilities in addition to RHIC. Figure 2 shows the two newest facilities: the National Synchrotron Light Source II (NSLS II) to come on-line October 1, 2014; and the Long Island Solar Farm which has an experimental section as well as supplying 32MW peak power to nearby homes in partnership with the local electric company.

Figure 2: Aerial view of BNL with NSLS II, RHIC and the Solar Farm indicated.

2 News from BNL since ISSP2013

Although Fiscal Year 2014 started on October 1, 2013 with the U. S. Government shut down for the first 16 days due to the lack of an approved budget, the rest of the FY2014 turned out very well for BNL and RHIC. At the administrative level, Professor Robert Tribble of Texas A&M University, well known for both his physics research and leadership of two important U. S. Department of Energy (DoE) Panels in 2005 and 2013, was appointed Deputy Director for Science & Technology, effective February 2014. Not long after that, in March, the DoE issued a Request For Proposals for a "management and operating (M&O) contractor" for BNL, which is owned by the U.S. Government, but run by an M&O contractor. The present contractor, BSA, is a partnership of Battelle Memorial Institute, a private non-profit science and technology development company, headquartered in Columbus, Ohio, and Stony Brook University. BSA has been the M&O contractor at BNL for the past 15 years (out of the BNL's 67 year existence), with the "engagement" of six of the world's leading research universities (Columbia, Cornell, Harvard, MIT, Princeton and Yale) who were among the universities that formed the founding M&O contractor, Associated Universities Incorporated, along with Johns Hopkins, and the Universities of Pennsylvania and Rochester. The new contract starts on January 1, 2015, preceded by a maximum 2 month transition phase-in period so should be awarded near or soon after November 1, 2014 [1].

The U. S. High Energy Physics bureaucracy was not idle during this period, with the release of the "Particle Physics Project Prioritization Panel (P5)" Report to the High energy Physics Advisory Panel (HEPAP) on May 21, 2014. The charge of the panel was "to develop an updated strategic plan for U.S. high energy physics that can be executed over a 10 year timescale, in the context of a 20 year global vision for the field." Their reasonable top priority for constrained budget scenarios was to "Use the Higgs boson as a new tool for discovery" which is good news for the U. S. HEP groups working at the LHC at CERN; but lots of internal U. S. activities were "redirected" [2]. For the unconstrained budget scenario, all they could come up with was:

- Develop a greatly expanded accelerator R&D program that would emphasize the ability to build very high-energy accelerators beyond the High-Luminosity LHC (HL-LHC) and ILC at dramatically lower cost.
- Play a world-leading role in the ILC experimental program and provide critical expertise and components to the accelerator, *should this exciting scientific opportunity be realized in Japan.*

which IMHO lacks the imagination and drive of previous generations of U.S High Energy Physicists who had proposed and were constructing a 40 TeV p+p collider for completion in 1995 if not for ... [3].

Not to be outdone, the new Long Range Planning exercise for U. S. Nuclear Physics was initiated in April 2014.

3 RHIC Operations and accelerator future plans

Since beginning operation in the year 2000, RHIC, which can collide any species with any other species including polarized protons, has provided collisions at 14 different values of nucleon-nucleon c.m. energy, $\sqrt{s_{NN}}$, and ten different species combinations including Au+Au, d+Au, Cu+Cu, Cu+Au, U+U, and in 2014 He^3+Au, if differently polarized protons are counted as different species. The performance history of RHIC with A+A and

RHIC Run	Year	Species	Energy	Ldt
Run-1	2000	Au+Au	130 GeV	1 μb-1
Run-2	2001-2	Au+Au	200 GeV	24 μb-1
Run-2		Au+Au	19 GeV	0.4 μb-1
		p+p	200 Gev	150 nb-1
Run-3	2002/3	d+Au	200 GeV	2.74 nb-1
		p+p (L)	200 GeV	0.35 nb-1
Run-4	2003/4	Au+Au	200 GeV	241 μb-1
		Au+Au	62.4 GeV	9 μb-1
Run-5	2005	Cu+Cu	200 GeV	3 nb-1
		Cu+Cu	62.4 GeV	0.19 nb-1
		Cu+Cu	22.4 GeV	2.7 μb-1
		p+p (L)	200 GeV	3.8 pb-1
Run-6	2006	p+p (T+L)	200 GeV	10.7 pb-1
		p+p	62.4 GeV	100 nb-1
Run-7	2007	Au+Au	200 GeV	813 μb-1
Run-8	2007/2008	d+Au	200 GeV	80 nb-1
		p+p (T)	200 GeV	5.2 pb-1
		Au+Au	9.2 GeV	
Run-9	2009	p+p (L)	200 GeV	16 pb-1
		p+p (L)	500 GeV	14 pb-1
Run-10	2010	Au+Au	200 GeV	1.3 nb-1
		Au+Au	62.4 GeV	100 μb-1
		Au+Au	39 GeV	40 μb-1
		Au+Au	7.7 GeV	260 mb-1
Run-11	2011	p+p (L)	500 GeV	27 pb-1
		Au+Au	200 GeV	915 μb-1
		Au+Au	27 GeV	5.2 μb-1
		Au+Au	19.6 GeV	13.7 M events
Run-12	2012	p+p (L)	200 GeV	9.2 pb-1
		p+p (L)	510 GeV	30 pb-1
		U+U	193 GeV	171 μb-1
		Cu+Au	200 GeV	4.96 nb-1
Run-13	2013	p+p (L)	510 GeV	156 pb-1
Run-14	2014	Au+Au	15 GeV	44.2 μb-1
		Au+Au	200 GeV	2.56 nb-1
		He3+Au	200 GeV	134 nb-1

Years	Beam Species and Energies	Science Goals	New Systems Commissioned
2014	15 GeV Au+Au 200 GeV Au+Au	Heavy flavor flow, energy loss, thermalization, etc. Quarkonium studies	Electron lenses 56 MHz SRF STAR HFT, MTD
2015-16	p+p at 200 GeV p+Au, d+Au at 200 GeV	Extract η/s(T) + constrain initial quantum fluctuations More heavy flavor studies	PHENIX MPC-EX Coherent e-cooling test
2017	No Run		Low energy e-cooling upgrade
2018-19	5-20 GeV Au+Au (BES-II)	Search for QCD critical point and onset of deconfinement	STAR iTPC upgrade Partial commissioning of sPHENIX (in 2019)
2020	No Run		Complete sPHENIX installation STAR forward upgrades
2021-22	Long 200 GeV Au+Au with upgraded detectors p+p, p/d+Au at 200 GeV	Jet, di-jet, γ-jet probes of parton transport and energy loss mechanism, color screening for different quarkonia	sPHENIX
2023-24	No Runs		Transition to eRHIC

Figure 3: a)(left) Year, species and proton polarization (**L**ongitudinal or **T**ransverse), $\sqrt{s_{NN}}$, integrated luminosity of RHIC runs. b) (right) Future run schedule and new equipment.

polarized $p+p$ collisions is shown in Fig. 5a; and in Fig. 5b, the plans for future runs are shown.

For this year's run (2014) the full 3 dimensional cooling including electron lenses for partial compensation of the beam-beam tune shift and 56 MhZ storage r.f. for stronger longitudinal focusing were implemented which led to a higher initial luminosity and much longer lifetime of the beam with a more level luminosity load due to the 3d stochatic cooling (Fig. 4). The luminosity performance of RHIC with A+A and polarized $p+p$ collisions is shown in Fig. 5. Notably, the Au+Au $\int \mathcal{L}\, dt$ in 2014 exceeds all previous Au+Au runs combined as did the p+p $\int \mathcal{L}\, dt$ in 2013.

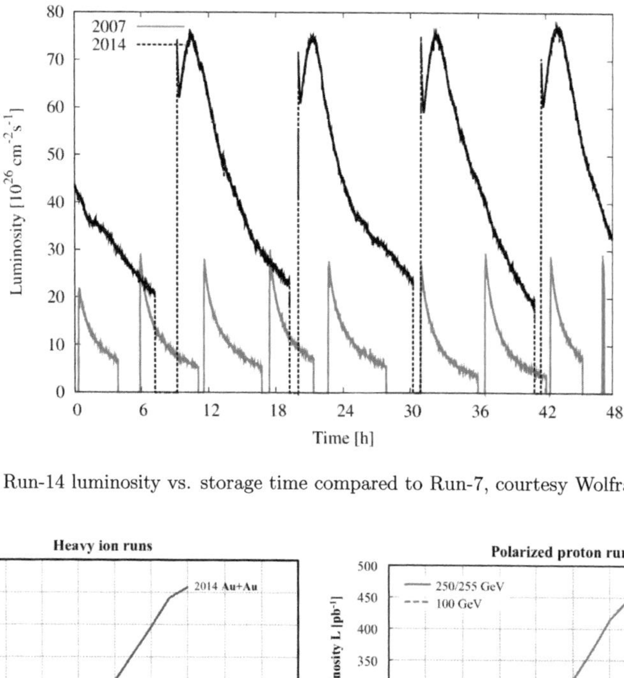

Figure 4: Run-14 luminosity vs. storage time compared to Run-7, courtesy Wolfram Fischer.

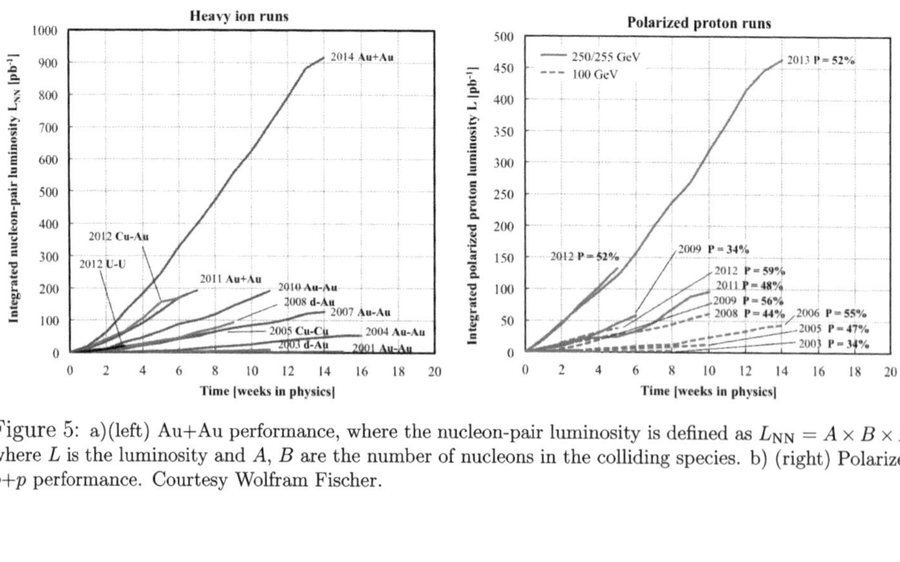

Figure 5: a)(left) Au+Au performance, where the nucleon-pair luminosity is defined as $L_{\mathrm{NN}} = A \times B \times L$, where L is the luminosity and A, B are the number of nucleons in the colliding species. b) (right) Polarized $p+p$ performance. Courtesy Wolfram Fischer.

The major future plan for accelerators in Nuclear Physics concerns an electron-ion collider, which if located at BNL will be called eRHIC. A new highly innovative and cost-effective design of eRHIC was proposed this year based on a Fixed Focus Alternating Gradient (FFAG) electron accelerator and an Energy Recovery Linac (Fig. 6).

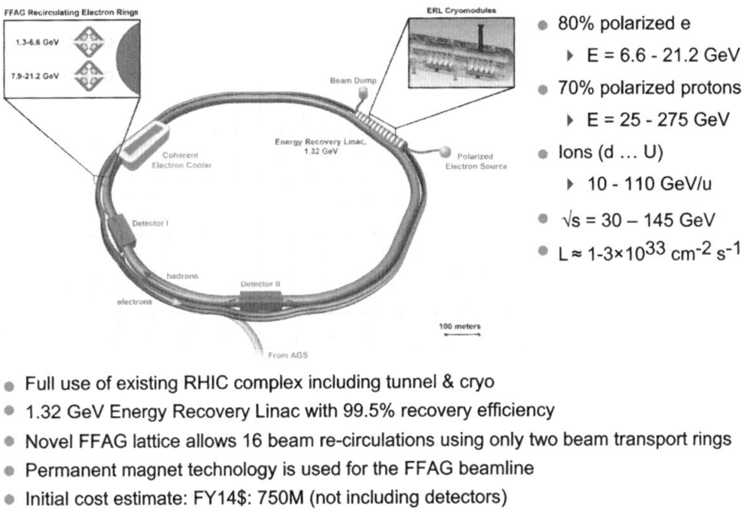

Figure 6: New BNL design for eRHIC with annotations.

4 Detector issues in A+A compared to p+p collisions

A main concern of experimental design in RHI collisions is the huge multiplicity in A+A central collisions compared to p+p collisions. A schematic drawing of a collision of two relativistic Au nuclei is shown in Fig. 7a. In the center of mass system of the nucleus-

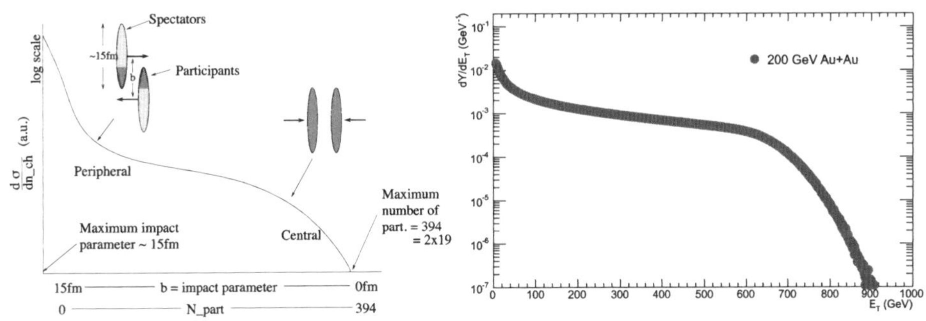

Figure 7: a) (left) Schematic of collision in the N-N c.m. system of two Lorentz contracted nuclei with radius R and impact parameter b. The curve with the ordinate labeled $d\sigma/dn_{\rm ch}$ represents the relative probability of charged particle multiplicity $n_{\rm ch}$ which is directly proportional to the number of participating nucleons, $N_{\rm part}$. b)(right) E$_T$ distribution in Au+Au at $\sqrt{s_{NN}}$ = 200 GeV from PHENIX [4].

nucleus collision, the two Lorentz-contracted nuclei of radius R approach each other with impact parameter b. In the region of overlap, the "participating" nucleons interact with each other, while in the non-overlap region, the "spectator" nucleons simply continue on their original trajectories and can be measured in Zero Degree Calorimeters (ZDC), in fixed target experiments, so that the number of spectators can be measured from which the number of participants ($N_{\rm part}$) can be determined for symmetric A+A collisions. The degree of overlap is called the centrality of the collision, with $b \sim 0$, being the most central and $b \sim 2R$, the most peripheral. The maximum time of overlap is $\tau_\circ = 2R/\gamma\, c$ where γ is the Lorentz factor and c is the speed of light in vacuum. The energy of the inelastic collision is predominantly dissipated by multiple particle production, where $N_{\rm ch}$, the number of charged particles produced, or E_T, the energy emitted transverse to the beam direction, is directly proportional [5] to $N_{\rm part}$ as sketched on Fig. 7a. Thus, $N_{\rm ch}$ and E_T in central Au+Au collisions are roughly A times larger than in a p+p collision, as shown in actual events from the STAR and PHENIX detectors at RHIC (Fig. 8).

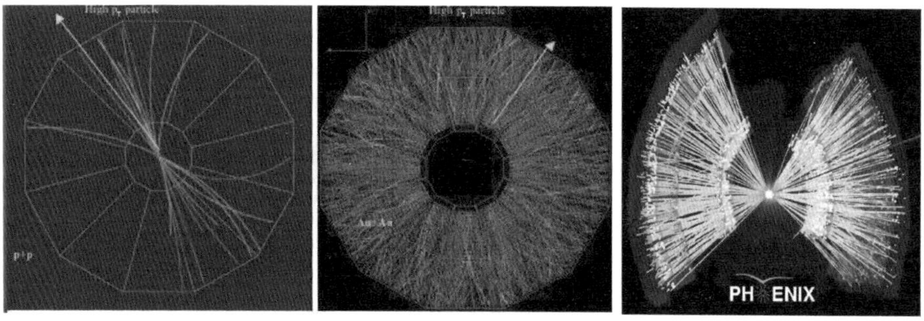

Figure 8: a) (left) A p+p collision in the STAR detector viewed along the collision axis; b) (center) Au+Au central collision at $\sqrt{s_{NN}} = 200$ GeV in STAR; c) (right) Au+Au central collision at $\sqrt{s_{NN}} = 200$ GeV in PHENIX.

At colliders, the impact parameter b can not be measured directly because charged spectators are swept away from zero degrees by the collider magnets. Instead, the centrality of a collision is defined in terms of the upper percentile e.g. top 10%-ile, upper 10−20%-ile, of $N_{\rm ch}$ or E_T distributions as in Fig. 7b. Unfortunately the "upper" and "-ile" are usually not mentioned which sometimes confuses the uninitiated. Also a model is required to derive $N_{\rm part}$ from the measurement so that the derived value of $N_{\rm part}$ at a collider or the number of binary nucleon-nucleon collisions ($N_{\rm coll}$) is model dependent and may have biases.

Since it is a huge task to reconstruct the momenta and identity of all the particles produced in these events, the initial detectors at RHIC [6] concentrated on the measurement of single-particle or multi-particle inclusive variables to analyze RHI collisions, with inspiration from the CERN ISR which emphasized those techniques before the era of jet reconstruction (see, for example, Refs. [7] and [8]). There are at present two major detectors in operation at RHIC, STAR and PHENIX, and there were also two smaller detectors, BRAHMS and PHOBOS, which have completed their program. As may be surmised from Figs. 8a,b and 9a, STAR, which emphasizes hadron physics, is most like a conventional general purpose collider detector, a Time Projection Chamber to detect all charged particles over the full azimuth ($\Delta\phi = 2\pi$) and ± 1 units of pseudo-rapidity (η); while PHENIX (Figs. 8c and 9b),

a) 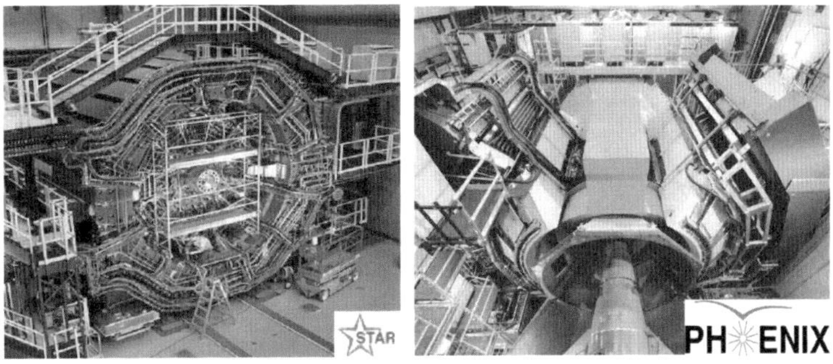

b)

Figure 9: Actual STAR (a) and PHENIX (b) detectors, compare with Figs 8b,c. The direction of the beam is along the axis of the STAR solenoid; and in (b) between the two spectrometer arms (perpendicular to the caption).

is a very high granularity high resolution special purpose detector: a two-arm spectrometer at mid-rapidity, with each arm covering solid angle $|\eta| \leq 0.35, \Delta\phi = 90°$, together with two full-azimuth muon detectors at forward and backward rapidity ($1.1 \leq |\eta| \leq 2.3$).[1] For the present runs, both STAR and PHENIX have excellent particle identification (PID) capability with electromagnetic calorimeters (EMcal) for photon and electron detection and Time of Flight for charged hadrons. PHENIX has a Ring Imaging CHerenkov counter for enhanced electron detection and triggering and small but full azimuth EM calorimeters (MPC) just before each muon arm covering $3.1 \leq |\eta| \leq 3.7$, while STAR obtains enhanced hadron identification using dE/dx in the TPC. For the 2014 run, both PHENIX (VTX, FVTX) and STAR (HFT) are equipped with micro-vertex detectors for tagging Heavy-Flavor c and b quarks via displaced vertices.

The main objectives of buliding RHIC were i) to discover the Quark Gluon Plasma (QGP), which was achieved as I have discussed in detail in review articles based on previous ISSP proceedings [7, 9]; ii) to measure its properties, which were much different than expected, namely a "perfect fluid" of quarks and gluons with their color charges exposed rather than a gas. The latest measurements from RHIC continue to be very interesting.

5 $N_{\rm ch}$, E_T distributions and constituent-quarks as the fundamental elements of particle production

The first experiment specifically designed to measure the dependence of the charged particle multiplicity in high energy p+A collisions as a function of the nuclear size was performed by Wit Busza and collaborators at Fermilab using beams of $\sim 50 - 200$ GeV/c hadrons colliding with various fixed nuclear targets. They found the extraordinary result [10] that the average charged particle multiplicity $\langle N_{\rm ch}\rangle_{hA}$ in hadron+nucleus (h+A) interactions was not simply proportional to the number of collisions (absorption-mean-free-paths), $\langle N_{\rm coll}\rangle = \overline{\nu}$, but increased much more slowly, proportional to the number of participants $\langle N_{\rm part}\rangle$. Thus, relative to h+p collisions (Fig. 10a) [11]:

$$R_A = \langle N_{\rm ch}\rangle_{hA} \,/\, \langle N_{\rm ch}\rangle_{hp} = \langle N_{\rm part}\rangle_{hA} \,/\, \langle N_{\rm part}\rangle_{hp} = (1+\overline{\nu})/2 \quad . \qquad (1)$$

[1]The detector is so non-conventional that it made the cover of Physics Today, October 2003.

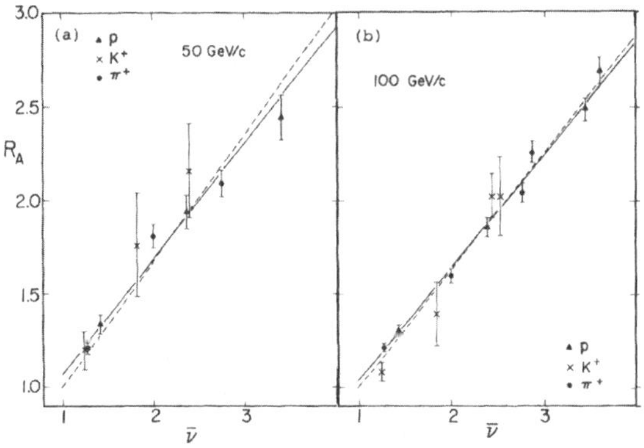

a)

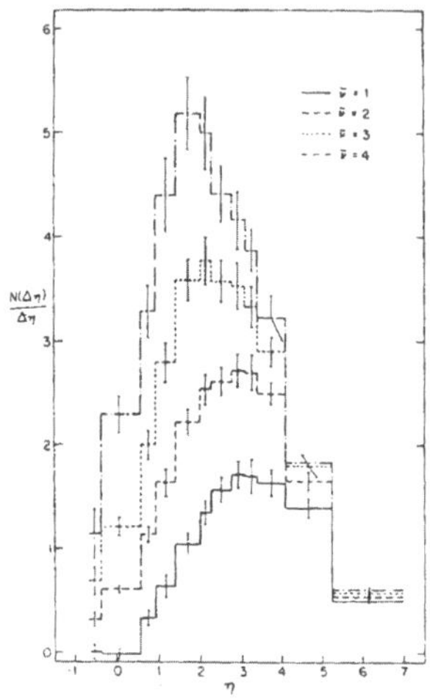

b)

Figure 10: a) $R_A = \langle N_{\rm ch}\rangle_{hA} \,/\, \langle N_{\rm ch}\rangle_{hp}$ as a function of the average thickness of each nucleus given in terms of the mean free path, $\overline{\nu} = \langle N_{\rm coll}\rangle$ [11] for 50 and 100 GeV/c h+A collisions; b) Charged particle multiplicity density, $dN_{\rm ch}/d\eta$, as a function of A (represented by $\overline{\nu}$) for 200 GeV/c p+A collisions [12].

Since the different projectiles, $h = \pi^+, K^+, p$ in Fig. 10a have different mean free paths, the fit to the same straight line in terms of $\overline{\nu}$ is convincing.

The other striking observation (Fig. 10b) [12] was that a relativistic incident proton could pass through e.g. $\nu = 4$ absorption-mean-free-paths of a target nucleus and emerge from the other side; and furthermore there was no intra-nuclear cascade of produced particles (a stark difference from what would happen to the same proton in a macroscopic 4 mean-free-path hadron calorimeter). In the forward fragmentation region of 200 GeV/c p+A collisions, within 1 unit of rapidity from the beam, $y^{\rm beam} = 6.06$, there was essentially no change in $dN_{\rm ch}/d\eta$ as a function of A, while at mid-rapidity ($\eta \approx y^{\rm cm}_{NN} = 3.03$), $dN_{\rm ch}/d\eta$ increased with A together with a small backward shift of the peak of the distribution resulting in a huge relative increase of multiplicity in the target fragmentation region, $\eta < 1$ in the laboratory system. These striking features of the ~ 200 GeV/c fixed target hadron-nucleus data ($\sqrt{s_{NN}} \sim 19.4$ GeV) showed the importance of taking into account the time and distance scales of the soft multi-particle production process including quantum mechanical effects.

5.1 The Wounded Nucleon Model

The observations in Fig. 10 had clearly shown that the target nucleus was rather transparent so that a relativistic incident nucleon could make many successive collisions while passing through the nucleus, and emerge intact. Immediately after a relativistic nucleon interacts inside a nucleus, the only thing that can happen consistent with relativity and quantum mechanics is for it to become an excited nucleon with roughly the same energy and reduced longitudinal momentum and rapidity. It remains in that state inside the nucleus because the uncertainty principle and time dilation prevent it from fragmenting into particles until it is well outside the nucleus. This feature immediately eliminates the possibility of a cascade in the nucleus from the rescattering of the secondary products. If one makes the further assumptions that an excited nucleon interacts with the same cross section as an unexcited nucleon and that the successive collisions of the excited nucleon do not affect the excited state or its eventual fragmentation products [13], this leads to the conclusion (c. 1977) that the elementary process for particle production in nuclear collisions is the excited nucleon, and to the prediction that the multiplicity in nuclear interactions should be proportional to the total number of projectile and target participants, rather than to the total number of collisions, or $R_A = \langle N_{\rm ch}\rangle_{hA} / \langle N_{\rm ch}\rangle_{hp} = \langle N_{\rm part}\rangle_{hA} / \langle N_{\rm part}\rangle_{hp} = (1+\overline{\nu})/2$, as observed. This is called the Wounded Nucleon Model (WNM) [14] and, in the common usage, Wounded Nucleons (WN) are called participants. In a later model from the early 1980's, the Additive Quark Model, AQM [15], constituent-quark participants were introduced; but the AQM is actually a model of particle production by color-strings in which only one color-string can be attached to a constituent-quark participant, effectively a projectile quark participant model.

5.2 Extreme Independent Models

The models mentioned above are examples of Extreme Independent Models in which the effect of the nuclear geometry of the interaction can be calculated independently of the dynamics of particle production which can be taken directly from experimental measurements. The nuclear geometry is represented by the relative probability, w_n per A+B interaction for a given number n of fundamental elements, in the present case, number of collisions ($N_{\rm coll}$), number of nucleon participants (wounded nucleon model-WNM [14]), number of constituent-quark participants (N_{qp}), number of color strings (AQM). The dynamics of particle production, the $N_{\rm ch}$ or E_T distribution of the fundamental element, is taken from the measured p+p data in the same detector: e.g. the measured $N_{\rm ch}$ distribution for a p+p collision represents: 1 collision; 2 participants (WNM); a predictable convolution of constituent-quark-participants (NQP), or projectile-quark-participants (AQM). Glauber calculations of the nuclear geometry (w_n) together with the p+p measurement provide a prediction for the A+B measurement in the same detector as the result of particle production by multiple independent fundamental elements.

I became acquainted with these models in my first talk at a Quark Matter conference (QM1984) where I presented measurements of transverse energy distributions from p+p and α+α interactions at $\sqrt{s_{NN}}$ =31 GeV at the CERN-ISR (Fig. 11a [18, 19]). The transverse energy, E_T, is a multiparticle variable defined as the sum

$$E_T = \sum_i E_i \sin\theta_i \qquad dE_T(\eta)/d\eta = \sin\theta(\eta)\, dE(\eta)/d\eta, \tag{2}$$

where θ is the polar angle, $\eta = -\ln\tan\theta/2$ is the pseudorapidity, E_i is by convention taken as the kinetic energy for baryons, the kinetic energy + 2 m_N for antibaryons, and the total energy for all other particles, and the sum is taken over all particles emitted into a fixed solid angle for each event.

The transverse energy was introduced by high energy physicists [16, 17] as an improved method to detect and study the jets from hard-scattering compared to high p_T single particle spectra by which hard-scattering was discovered in p+p collisions and used as a hard-probe in Au+Au collisions at RHIC. However, it didn't work as expected: E_T distributions, like $N_{\rm ch}$ distributions, are dominated by soft particles near $\langle p_T\rangle$ (e.g. see Ref. [8] for details).

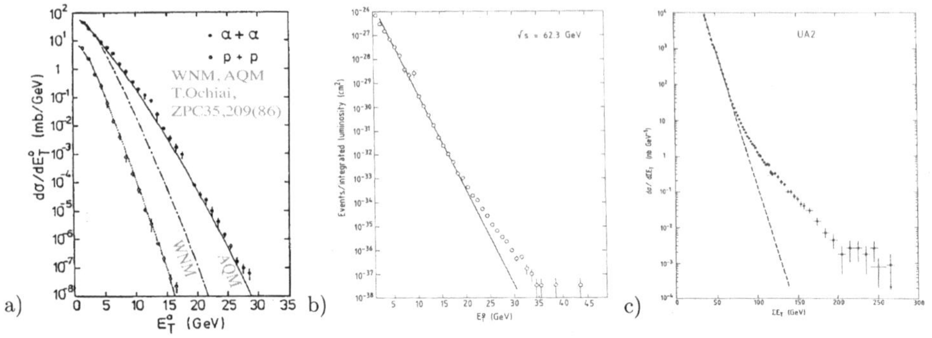

Figure 11: (a) E_T distributions in p+p, α+α [18] at $\sqrt{s_{NN}}$ =31 GeV, with AQM and WNM calculations [21]. (b),(c) E_T distributions with breaks indicating jets: (b) p+p $\sqrt{s}$ =62.3 GeV [22]; (c) $d\sigma/d\mathrm{E}_T$ (nb/GeV) vs. E_T for $\bar{p}$+p $\sqrt{s}$ =540 GeV [23].

Nevertheless, it was claimed at the conference [20], in comments to my talk, that the deviation from the WNM in Fig. 11a was due to jets, but in both proceedings [19, 20] it was demonstrated that [20] "there is no ... sign of jets. This indicates that soft processes are still dominant, and that we are still legimately testing the WNM at these high values of E_T." As shown in Fig. 11a, the the AQM [21], rather than the WNM, followed the data. Jets do appear in E_T distributions as a break $\lesssim 10^{-5}$ down in cross section (Figs. 11b,c).

5.3 E_T and $N_{\rm ch}$ distributions cut on centrality

At RHIC, following the style of the CERN SpS rather than the BNL-AGS fixed target heavy ion program, E_T and $N_{\rm ch}$ distributions were not generally shown. The measurements were presented cut in centrality in the form $\langle dN_{\rm ch}{}^{\rm AA}/d\eta\rangle/(\langle N_{\rm part}\rangle/2)$ vs. $\langle N_{\rm part}\rangle$ (Fig. 12), which would be a constant equal to $\langle dN_{\rm ch}{}^{\rm pp}/d\eta\rangle$ if the WNM worked. The measurements

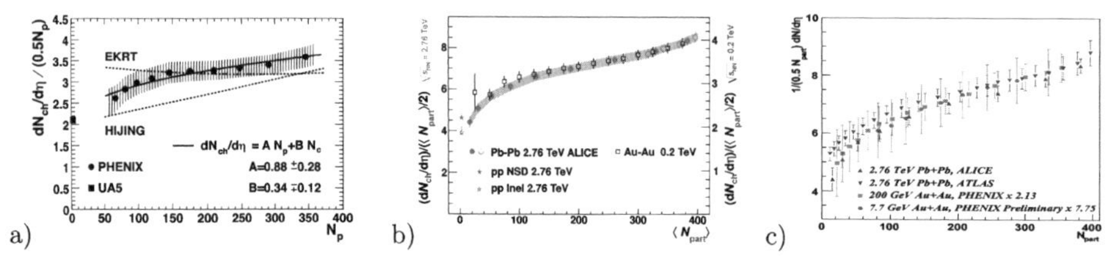

Figure 12: (a) PHENIX, Au+Au, $\sqrt{s_{NN}}$ =130 GeV [24]; (b) ALICE, Pb+Pb, $\sqrt{s_{NN}}$ =2.76 TeV [26]; (c) PHENIX preliminary Au+Au, $\sqrt{s_{NN}}$ =7.7 GeV compared to the data at larger $\sqrt{s_{NN}}$ [27].

clearly deviate from the WNM (Fig. 12a) [24]; so the PHENIX collaboration, inspired by the preceding article in the journal [25], fit their data to the two-component model:

$$dN_{\rm ch}{}^{AA}/d\eta = (dN_{\rm ch}{}^{pp}/d\eta)\,[(1-x)\,\langle N_{\rm part}\rangle/2 + x\,\langle N_{\rm coll}\rangle] \qquad (3)$$

where the $N_{\rm coll}$ term implied a hard-scattering component for E_T and $N_{\rm ch}$, known to be absent in p+p collisions [2] (recall Fig. 11). A decade later, the first measurement from Pb+Pb

[2]It was noted in Ref. [24] that hard-scattering was not a unique interpretation. The shape of the centrality dependences of $d\mathrm{E}_T{}^{AA}/d\eta$ parameterized as $N_{\rm part}{}^{\alpha}$ were the same for Pb+Pb at $\sqrt{s_{NN}}$ =17.6 GeV at the

collisions with $\sqrt{s_{NN}}$ =2.76 TeV at the LHC Fig. 12b [26], showed exactly the same shape vs. $N_{\rm part}$ as the RHIC Au+Au data at $\sqrt{s_{NN}}$ =200 GeV, although $\langle N_{\rm coll}\rangle$ is a factor of 1.6 larger and the hard-scattering cross section is more than a factor of 20 larger. This strongly argued against a hard-scattering component and for a nuclear geometrical effect, which was reinforced by a PHENIX preliminary measurement in Au+Au at $\sqrt{s_{NN}}$ =7.7 GeV (Fig. 12c) [27] which also showed the same shape for the evolution of $\left\langle dN_{\rm ch}{}^{\rm AA}/d\eta\right\rangle/(\langle N_{\rm part}\rangle/2)$ with $N_{\rm part}$ as the $\sqrt{s_{NN}}$ =200 and 2760 GeV measurements. It had previously been proposed that the number of constituent-quark participants provided the nuclear geometry that could explain the RHIC Au+Au data without the need to introduce a hard-scattering component [28]. However an asymmetric system is necessary in order to distinguish the NQP model from the AQM so the two models were applied to the RHIC d+Au data.

5.4 The number of constituent-quark participants model (NQP)

The massive constituent-quarks [29], which form mesons and nucleons (e.g. a proton=uud), are relevant for static properties and soft physics with $p_T\lesssim 1.4$ GeV/c. They are complex objects or quasiparticles [30] made of the massless partons (valence quarks, gluons and sea quarks) of DIS [31] such that the valence quarks acquire masses $\approx 1/3$ the nucleon mass with radii ≈ 0.3 fm when bound in the nucleon. With smaller resolution one can see inside the bag to resolve the massless partons which can scatter at large angles according to QCD. At RHIC, hard-scattering starts to be visible as a power law above soft (exponential) particle production only for $p_T > 1.4$ GeV/c at mid-rapidity (Fig 13a), where $Q^2 = 2p_T^2 = 4$ (GeV/c)^2 which corresponds to a distance scale (resolution) < 0.1 fm.[3]

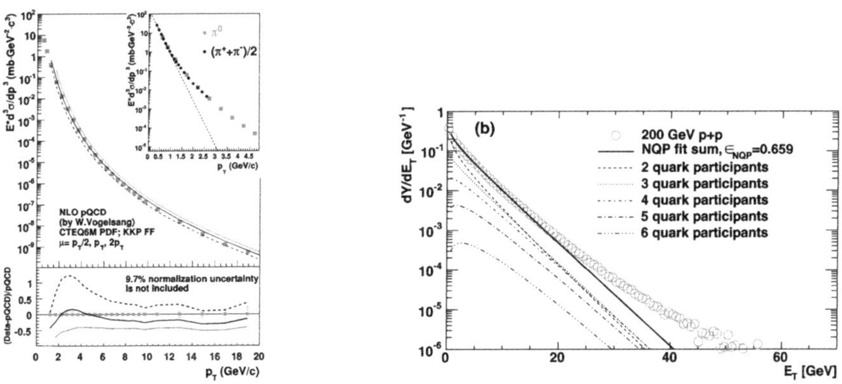

a)

Figure 13: (a) Invariant cross section of π^0 vs. p_T at mid-rapidity in p+p collisions at $\sqrt{s}$ =200 GeV [34]. The inset shows the transition from an exponental to a power-law in the range $1 < p_T < 2$ GeV/c (b) PHENIX deconvolution of p+p E_T distribution at $\sqrt{s}$ =200 GeV [4]

CERN SpS and Au+Au at $\sqrt{s_{NN}}$ =130 GeV, with $\alpha = 1.1$ and $\alpha = 1.16 \pm 0.04$, respectively. The LHC data 10 years later [26] gave $\alpha = 1.19 \pm 0.02$ for Pb+Pb at $\sqrt{s_{NN}}$ =2760 GeV, again the same shape.

[3]Shuryak and collaborators recently made similar arguments about resolution in separating hard from soft processes although their mechanism for soft particle and QGP production was 2 color strings per wounded nucleon from Pomeron exchange [32].

A standard Monte Carlo Glauber calculation is used to assemble the initial positions of all the nucleons. Then three quarks are distributed around the center of each nucleon according to the proton charge distribution $\rho(\vec{r}) \propto e^{-\sqrt{12}r/r_{\rm rms}}$, where $r_{\rm rms} = 0.81$ fm is the rms charge radius of the proton [33]. The q–q inelastic scattering cross section is adjusted to 9.36 mb at $\sqrt{s}$ =200 GeV to give the correct p+p inelastic cross section (42 mb) and then used in the A+B calculations.

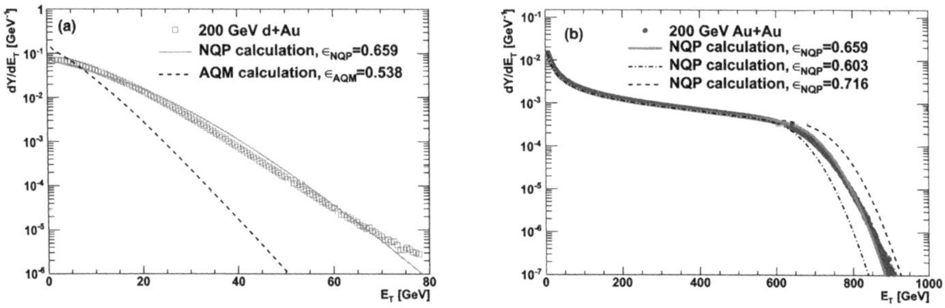

Figure 14: PHENIX NQP calculations [4] based on the E$_T$ distribution of a constituent-quark participant from Fig. 13b for: (a) d+Au (also AQM), (b) Au+Au E$_T$ distributions at $\sqrt{s_{NN}}$ =200 GeV.

Fig. 13b shows the deconvolution of the p+p E$_T$ distribution to the sum of 2–6 constituent-quark participants from which the E$_T$ distribution of a constituent-quark is determined and applied to d+Au (Fig. 14a) and Au+Au (Fig. 14b) reactions in the same detector.

The NQP calculations closely follow the measured d+Au and Au+Au E$_T$ distributions in shape and magnitude over a range of more than 1000 in cross section. A complete calculation was also done for the AQM which fails to describe the d+Au data (Fig. 14a). The conclusion is that the number of constituent-quark participants determines the $N_{\rm ch}$ and E$_T$ distributions and that the AQM calculation which describes the α–α data at $\sqrt{s_{NN}}$ =31 GeV (Fig. 11a) is equivalent to the NQP in the symmetric system.

The NQP model was also applied to the centrality-cut PHENIX data by making a plot of dE$_T/d\eta$ for a given centrality bin as a function of the number of constituent-quark participants N_{qp} in the bin for Au+Au collisions at $\sqrt{s_{NN}}$ = 62.4, 130 and 200 GeV (Fig. 15a).

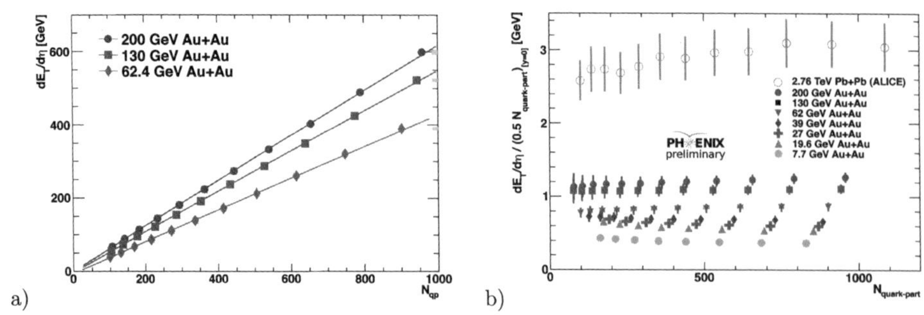

Figure 15: PHENIX [4]: (a) dE$_T/d\eta$ vs. N_{qp} ; (b) dE$_T/d\eta/(N_{qp}/2)$ vs. N_{qp}

The data for each $\sqrt{s_{NN}}$ are well described by a straight line and all are consistent with a zero intercept. This means that $d\text{E}_T/d\eta$ is strictly proportional to N_{qp} (Fig 15a) so that $d\text{E}_T/d\eta/(N_{qp}/2)$ vs. N_{qp} is a constant for $\sqrt{s_{NN}} > 39$ GeV (Fig 15b) even up to the LHC $\sqrt{s_{NN}} = 2.76$ TeV. This brought up a very interesting question, with a very important and fundamental answer.

Most experiments at RHIC, starting with PHENIX (Fig. 12) [24] had successfully fit their measurements of $d\text{E}_T/d\eta$ or $dN_{\text{ch}}/d\eta$ as a function of centrality (represented by N_{part}) to the two-component model (Eq. 3). Also, both the ATLAS [35] and ALICE [36] experiments at the LHC computed the ansatz, $[(1-x)\langle N_{\text{part}}\rangle/2 + x\langle N_{\text{coll}}\rangle]$, in event-by-event Glauber Monte Carlo calculations which fit their forward E_T measurements used to define centrality in Pb+Pb collisions. ALICE realized that the combination of the two components N_{part} and N_{coll} in the ansatz represented the number of emitting sources of particles, which they named "ancestors". Since the number of constituent-quarks N_{qp} also represents the number of emitting sources in a simple linear relationship, Bill Zajc of PHENIX suggested that "the success of the two component model is not because there are some contributions proportional to N_{part} and some going as N_{coll}, but because a particular linear combination of N_{part} and N_{coll} turns out to be an empirical proxy for the number of contsituent-quarks." We had a nice table of $\langle N_{\text{part}}\rangle$, $\langle N_{qp}\rangle$, $\langle N_{\text{coll}}\rangle$ as a function of centrality in Au+Au collisions at $\sqrt{s_{NN}}$ =200 GeV, so it did not take very long for me to verify the striking result that indeed it was true: with $x = 0.08$, the ratio $\langle N_{qp}\rangle/[(1-x)\langle N_{\text{part}}\rangle/2 + x\langle N_{\text{coll}}\rangle] = 3.88$ on the average and varies by less than 1% over the entire centrality range in 5% bins except for the most peripheral bin where it is 5% low (Table 1). This result clearly demonstrates that the ansatz works because the particular linear combination of N_{part} and N_{coll} turns out to be an empirical proxy for N_{qp} and not because the N_{coll} term implies hard-scattering.

Centrality	$\langle N_{part}\rangle$	$\langle N_{qp}\rangle$	$\langle N_{\text{coll}}\rangle$	ansatz	$\langle N_{\text{qp}}\rangle$/ansatz
0-5%	350.9 ± 4.7	956.6 ± 16.2	1064.1 ± 110.0	246.5	3.88
5-10%	297.0 ± 6.6	789.8 ± 15.3	838.0 ± 87.2	203.7	3.88
10-15%	251.0 ± 7.3	654.2 ± 14.5	661.1 ± 68.5	168.3	3.89
15-20%	211.0 ± 7.3	540.2 ± 12.3	519.1 ± 53.7	138.6	3.90
20-25%	176.3 ± 7.0	443.3 ± 10.4	402.6 ± 39.5	113.3	3.91
25-30%	146.8 ± 7.1	362.8 ± 12.2	311.9 ± 31.8	92.5	3.92
30-35%	120.9 ± 7.0	293.3 ± 11.0	237.8 ± 24.2	74.6	3.93
35-40%	98.3 ± 6.8	233.5 ± 9.2	177.3 ± 18.3	59.4	3.93
40-45%	78.7 ± 6.1	182.7 ± 6.8	129.6 ± 12.6	46.6	3.92
45-50%	61.9 ± 5.2	140.5 ± 5.3	92.7 ± 9.0	35.9	3.91
50-55%	47.6 ± 4.9	105.7 ± 5.5	64.4 ± 8.1	27.0	3.91
55-60%	35.6 ± 5.1	77.3 ± 6.8	43.7 ± 7.6	19.9	3.89
60-65%	26.1 ± 4.7	55.5 ± 7.1	29.0 ± 6.5	14.3	3.87
65-70%	18.7 ± 4.0	39.0 ± 6.7	18.8 ± 5.3	10.1	3.86
70-75%	13.1 ± 3.2	27.0 ± 4.9	12.0 ± 3.6	7.0	3.86
75-80%	9.4 ± 2.1	19.0 ± 3.2	7.9 ± 2.2	5.0	3.83
80-92%	5.4 ± 1.2	10.3 ± 1.5	4.0 ± 1.0	2.8	3.67
p+p	2	2.99 ± 0.05	1	1	2.99

Table 1: Verification that the ansatz, $[(1-x)\langle N_{\text{part}}\rangle/2 + x\langle N_{\text{coll}}\rangle]$, from Eq. 3, with $x = 0.08$, is a proxy for N_{qp}. The errors quoted on $\langle N_{\text{part}}\rangle$, $\langle N_{qp}\rangle$, $\langle N_{\text{coll}}\rangle$ are correlated and largely cancel in the $\langle N_{qp}\rangle$/ansatz ratio. For $x = 0.09$ the average $\langle N_{qp}\rangle$/ansatz=3.81, the maximum variation is less than 1.6%, but 4% low in the most peripheral bin.

The fact that the $\langle N_{qp}\rangle$/ansatz ratio drops from an average of 3.88 for Au+Au collisions to 2.99 for p+p collisions is also interesting. This is consistent with the PHOBOS [37] result that a fit of Eq. 3 to $\langle dN_{\rm ch}{}^{AA}/d\eta\rangle$ with $x = 0.09$ leaving $\langle dN_{\rm ch}{}^{pp}/d\eta\rangle$ as a free parameter gives the result $\langle dN_{\rm ch}{}^{pp}/d\eta\rangle = 2.70$ which is above the measured inelastic value of 2.29. The lower value of $\langle N_{qp}\rangle$/ansatz for p+p would then give a value of $2.70 \times 2.99/3.88 = 2.08$ (2.12 for $x = 0.09$) for $\langle dN_{\rm ch}{}^{pp}/d\eta\rangle$, much closer to the measured value. In that same paper, PHOBOS also noted that their data were consistent with a constant value of x from $\sqrt{s_{NN}}$ =19.6 to 200 GeV (more recently extended to $\sqrt{s_{NN}}$ =2.76 TeV [38]) which indicated that the fraction of hard-processes contributing to multiplicity did not increase over a huge range of $\sqrt{s}$ even though the hard-scattering cross section greatly increased over this same range.

5.5 Constituent-quark participants resolve several outstanding puzzles

PHOBOS also made some very nice measurements of the charged particle multiplicity over the full rapidity range, not just mid-rapidity. The total charged multiplicity $N_{\rm ch}|_{|\eta|<5.4}$ was measured and $N_{\rm ch}|_{|\eta|<5.4}/\langle N_{\rm part}/2\rangle$ plotted as a function of centrality for $\sqrt{s_{NN}}$ =19.6, 62.4,130, and 200 GeV (Fig. 16a) [39]. At first glance the data appear to follow the WNM because the multiplicity/per nucleon pair $N_{\rm ch}|_{|\eta|<5.4}/\langle N_{\rm part}/2\rangle$ appears to be constant in Au+Au collisions. However, the true believers, e.g. Ref. [40], claim that the WNM does not work because the value in Au+Au collisions is much larger than the p+p value shown; but "still the proportionality of these multiplicities to the number of participants holds" [40].

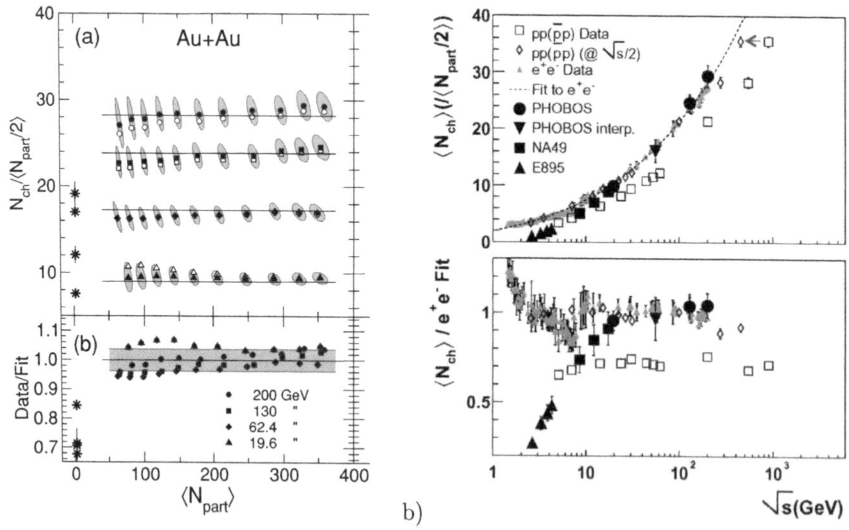

Figure 16: a) Total charged multiplicity per nucleon pair $N_{\rm ch}/\langle N_{\rm part}/2\rangle$ vs. centrality, $N_{\rm part}$, for the $\sqrt{s_{NN}}$ indicated. Open points are the measured $N_{\rm ch}|_{|\eta|<5.4}/(\langle N_{\rm part}/2\rangle$; solid points are extrapolated to $|\eta| \leq y_{\rm beam}$. b) Total charged multiplicity per nucleon pair in p+p and A+A collisions as a function of c.m. energy $\sqrt{s}$ compared to $e^+ + e^-$ collisions [41].

In fact, the difference between the $p+p$ and Au+Au values may be related to another interesting observation by PHOBOS [41] that the "leading particle effect" in $p+p$ collisions, as discovered by Zichichi and collaborators [42]—in which the total multiplicity at c.m. energy $\sqrt{s_{\rm pp}}$ is equal to that in e^+e^- collisions at $\sqrt{s_{\rm ee}} = \sqrt{s_{\rm pp}}/2$ (the "effective energy") (Fig. 16b) because the leading protons carry away half the $p+p$ c.m. energy—is absent in A+A collisions where the leading protons can reinteract. This observation seems to contradict the WNM, in which the key assumption is that what counts is whether or not a nucleon was struck, not how many times it was struck.

Both these effects can be reconciled by constituent-quark participants.

In the NQP model (Table 1), the $\langle N_{qp}/N_{\rm part}\rangle$ is 1.5 for a $p+p$ collision but rises to 2.27–2.73 (a factor of 1.51–1.82) for the more central (0-50%, $\langle N_{\rm part}\rangle > 60$) Au+Au collisions plotted in Fig. 16a. This would correspond to an increase in $N_{\rm ch}/\langle N_{\rm part}/2\rangle$ by a factor of ~ 1.5 as observed. It also might explain the slight rise of the open points with increasing $N_{\rm part}$. Similarly, the increase in "effective energy" for particle production shown by the increase in $N_{\rm ch}/\langle N_{\rm part}/2\rangle$ from $p+p$ to Au+Au collisions is due to an increase in the number of (constituent-quark) participants, not because of additional collisions of a given nucleon-participant. Furthermore, the factor 1.5 decrease in $N_{\rm ch}/\langle N_{\rm part}/2\rangle$ from Au+Au to $p+p$ corresponds to a reduction in $\sqrt{s}$ for the observed $N_{\rm ch}/\langle N_{\rm part}/2\rangle$ from 200 to 100 GeV $p+p$ collisions on Fig. 16b, the same factor of 2 discussed in the original measurement [41]. Thus, the NQP model rather than the WNM preserves the assumption in these "extreme-independent" participant models that successive collisions of a participant do not increase its particle emission while explaining these two interesting observations.

Another argument against the $N_{\rm part}$, $N_{\rm coll}$ ansatz representing actual hard-collisions rather than simply being a proxy for constituent-quark participants concerns the measurement of elliptic flow in central U+U collisions [43].

6 Collective Flow

For many years, since the days of the Bevalac [44], collective flow [45] has been observed in A+A collisions over the full range of energies studied, from incident kinetic energy of 100A MeV to c.m. energy of $\sqrt{s_{NN}} = 2.76$ TeV, and thought to be a distinguishing feature of A+A collisions compared to either $p+p$ or $p+$A collisions. Collective flow, or simply flow, is a collective effect which can not be obtained from a superposition of independent N+N collisions. I first present a short review (details can be found in previous ISSP proceedings [7, 9]) and then move on to the newer results.

Immediately after an A+A collision, the overlap region defined by the nuclear geometry is almond shaped (see Fig 17a) with the shortest axis along the impact parameter vector. The different pressure gradients along the short and long axes of the ellipse break the ϕ symmetry of the problem and create an azimuthal angular dependence of the semi-inclusive single particle spectrum with respect to the reaction plane, $\phi - \Phi_R$, which is represented by an expansion in harmonics [48], where the angle of the reaction plane Φ_R is defined to be along the impact parameter vector, the x axis in Fig. 17a:

$$\frac{Ed^3N}{dp^3} = \frac{d^3N}{p_T dp_T dy d\phi} = \frac{d^3N}{2\pi\, p_T dp_T dy}\left[1 + \sum_n 2v_n \cos n(\phi - \Phi_R)\right]. \tag{4}$$

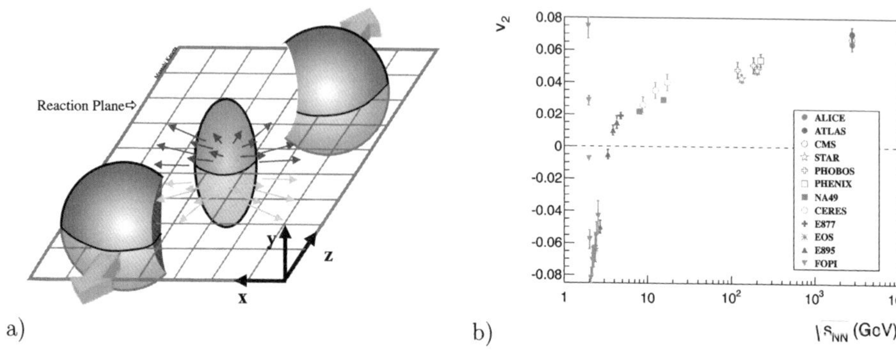

Figure 17: (a) Almond shaped overlap zone generated just after an A+A collision where the incident nuclei are moving along the $\pm z$ axis. The reaction plane by definition contains the impact parameter vector (along the x axis) [46]. (b) v_2 for charged particles integrated over p_T at $\sqrt{s_{NN}}$ =2.76 TeV for 20–30% centrality compared to the measurements at lower $\sqrt{s_{NN}}$ at the same centrality [47].

The Fourier coefficient v_2, called elliptic flow, is predominant at mid-rapidity. The evolution of v_2 with $\sqrt{s_{NN}}$ (Fig. 17b) [47] is the result of competing processes. At very low $\sqrt{s_{NN}}$ corresponding to values of $\sim 100A$ MeV [49] the main effect among many others is from nuclei bouncing off each other and breaking to fragments, which is sensitive to the equation of state of the nuclei—soft, like sponges, hard like billiard balls? The negative v_2 at larger $\sqrt{s_{NN}}$ is produced by the effective "squeeze-out" (in the y direction) of the produced particles by slow moving minimally Lorentz-contracted spectators (as in Fig. 17a) which block the particles emitted in the reaction plane. With increasing $\sqrt{s_{NN}}$, the spectators move faster and become more contracted so the blocking stops. The increase of v_2 with $\sqrt{s_{NN}}$ is generally described by hydrodynamics in the QGP region, but is also described by hadron transport theories for $\sqrt{s_{NN}} \lesssim 10$ GeV [50].

Flow measurements contributed two of the most important results about the properties of the QGP: i) the scaling of v_2 of identified particles at mid-rapidity with the number of constituent-quarks n_q in the particle—v_2/n_q scales with the transverse kinetic energy per constituent-quark, KE_T/n_q, because particles have not formed at the time flow develops; ii) the persistence of flow for $p_T > 1$ GeV/c which implied that the viscosity is small [51], perhaps as small as a quantum viscosity bound from string theory [52], $\eta/s = 1/(4\pi)$ where η is the shear viscosity and s the entropy density per unit volume. This led to the description of the "sQGP" produced at RHIC as "the perfect fluid".

New insight came in 2013, when measurements in p+Pb at LHC and d+Au at RHIC observed what looked very much like collective flow in these systems that were believed to be too small to support collective effects. This was the reason for the He^3+Au run at RHIC in 2014, to see whether triangular flow, v_3, would be more prominent with a 3 nucleon projectile. The improvement of the p+Au and d+Au measurements this year to identified pions and protons strengthened the case that the observed v_2 in these small systems is really hydrodynamic collective flow.

Figure 18a [53] shows the two-particle correlation function in d+Au (Eq. 5) fit with terms from c_1 to c_4 where the solid line is the fit and only the c_1 (dashes) and c_2 (dots) make significant contribution.

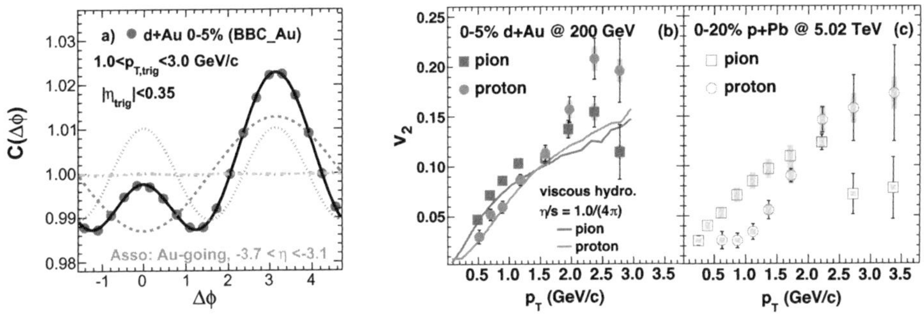

Figure 18: (a) PHENIX [53] two-particle azimuthal correlation function (Eq. 5) for $1.0 < p_{T,\mathrm{trig}} < 3.0$ GeV/c, $|\eta_{\mathrm{trig}}| < 0.35$, in central (0-5%) d+Au collisions at RHIC. $v_2(p_T)$ for identified $\pi^\pm$ $p^\pm$ by the standard reaction plane method (Eq. 4) for (b) central d+Au collisions at RHIC ($\sqrt{s_{NN}}$ =200 GeV) and (c) central p+Pb collisions at LHC ($\sqrt{s_{NN}}$ =5.02 TeV).

$$\frac{dN_{ta}}{d\phi_t d\phi_a} = C(\phi_t - \phi_a) \propto \left[1 + \sum_n 2\, c_n \cos n(\phi_t - \phi_a)\right], \qquad c_n \equiv (v_n)_t (v_n)_a \qquad (5)$$

The c_1 comes into play because the trigger particle t is a charged track at mid-rapidity while the associated particle a is a count in an MPC tower ($\delta\eta \times \delta\phi \approx 0.12 \times 0.18$) from π^0 or η meson decay photons at $-3.7 \leq \eta_{\mathrm{tower}} \leq -3.1$. Also, there is no evidence of a di-jet contribution because the large pseudorapidity gap between t and a is beyond that of a di-jet. Thus, the long-range correlation in Fig. 18a which is not seen in p+p comparison data but has the same properties as collective flow in Au+Au collisions is consistent with hydrodynamic collective flow in d+Au. Perhaps more convincing evidence for hydrodynamic flow is given in Fig. 18b,c where both at RHIC in d+Au (b) and LHC in p+Pb (c), the characteristic π, p mass splitting for $v_2(p_T)$ seen in Au+Au is observed [53]. The splitting occurs because, for a given transverse collective expansion velocity β, protons have a larger $p_T = \gamma\beta m$ than pions.

6.1 v_2 in U+U collisions and constituent-quark participants

Because Uranium nuclei are prolate spheroids, there is the interesting possibility of large v_2 in body-to-body central collisions which have a significant eccentricity and almond shape (Fig. 19a). Based on the assumption that the N_{part}, N_{coll} ansatz (Eq. 3) would describe the $dN_{\mathrm{ch}}/d\eta$ distribution in U+U collisions, it was predicted that for the highest $dN_{\mathrm{ch}}/d\eta$ (the most central collisions) the tip-to-tip configuration with much larger N_{coll} and small eccentricity (small v_2) would overtake the body-to-body configuration with large eccentricity corresponding to large v_2.

This led to two predictions: i) the tip-to-tip configuration would be selected by the most central collisions [54]; ii) these most central collisons would see a sharp decrease in v_2 with increasing $dN_{\mathrm{ch}}/d\eta$ [55, 56] called a cusp. This sharp decrease—represented by the bent line on the topmost U+U data (filled circles) in Fig. 19b (called a knee in Ref. [43])—is not observed. As discussed previously, this is because the N_{coll} term is not relevant for $dN_{\mathrm{ch}}/d\eta$ distributions, which also argues against the method proposed in Ref. [54] to select the tip-to-tip configuration.

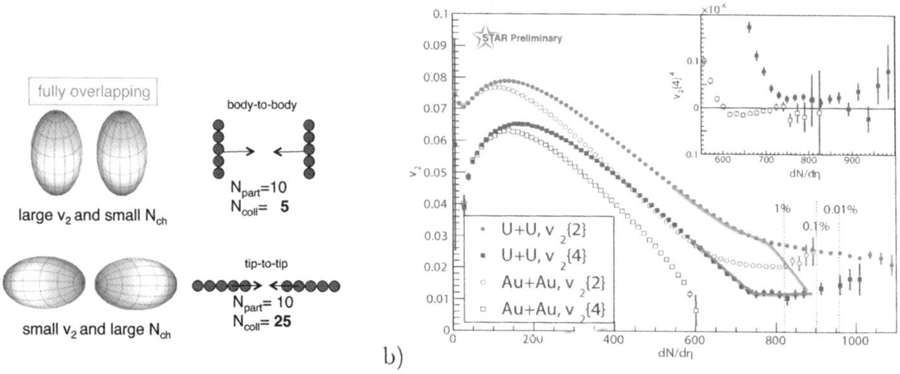

Figure 19: (a) Body-to-body and tip-to-tip configurations in U+U collisions with zero impact parameter. The different relation of N_{part} to N_{coll} is sketched next to each configuration. (Modified drawing from Ref. [43]). (b) STAR measurements of v_2 in Au+Au and U+U at $\sqrt{s_{NN}} \approx 200$ GeV as a function of $dN_{\text{ch}}/d\eta$ with upper percentiles of centrality for U+U indicated by vertical dashed lines [43].

7 RHIC Beam Energy Scan (BES)—in search of the critical point

In addition to discovering the QGP and measuring its properties, another objective of the RHIC physics program is to measure the phase diagram of nuclear matter and to determine the equation of state in the various phases and the characteristics of the phase transitions. Two of the many proposed phase diagrams of nuclear matter (e.g. see Ref. [57]) are shown in Fig. 20 together with the idealized trajectories of the evolution of the medium for Au+Au

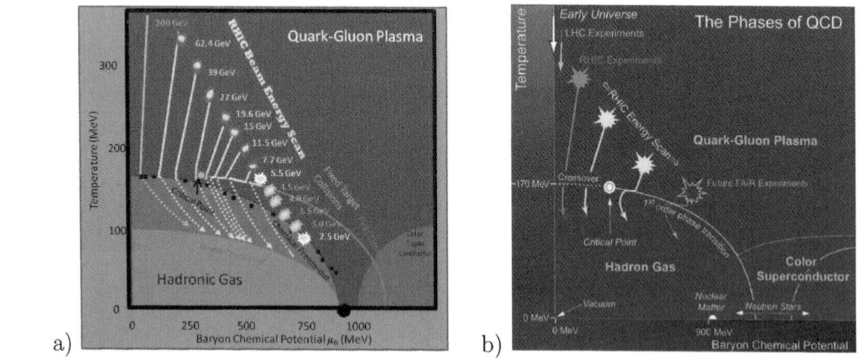

Figure 20: Proposed phase diagrams for nuclear matter: Temperature, T, vs Baryon Chemical Potential, μ_B. a)STAR's idea in 2013 [58]; b) STAR's more cautious idea in 2014 [59].

collisions at the $\sqrt{s_{NN}}$ proposed for the Beam Energy Scan at RHIC to search for a QCD critical point. The bursts represent the hottest and densest stage of the medium when thermal equilibrium is reached shortly after the collision. The axes are the temperature T

vs. the baryon chemical potential μ_B. The temperature for the transition from the Quark Gluon Plasma (QGP) to a hadron gas is taken as 170 MeV for $\mu_B = 0$ and the phase boundary is predicted to be a smooth crossover down to a critical point below which the phase boundary becomes a first order phase transition.

In an equilibrated thermal medium, particles should follow a Boltzmann distribution in the local rest frame [60]

$$\frac{d^2\sigma}{dp_L p_T dp_T} = \frac{d^2\sigma}{dp_L m_T dm_T} \propto \frac{1}{e^{(E-\mu)/T} \pm 1} \sim e^{-(E-\mu)/T} \quad , \tag{6}$$

where $m_T = \sqrt{p_T^2 + m^2}$ and μ is a chemical potential. In fact, the ratios of particle abundances (which are dominated by low p_T particles) for central Au+Au collisions at RHIC, even for strange and multi-strange particles, are well described [61] by fits to a thermal distribution,

$$\frac{d^2\sigma}{dp_L p_T dp_T} \sim e^{-(E-\mu)/T} \rightarrow \frac{\bar{p}}{p} = \frac{e^{-(E+\mu_B)/T}}{e^{-(E-\mu_B)/T}} = e^{-(2\mu_B)/T} \quad , \tag{7}$$

with similar expressions for strange particles. μ_B (and μ_S) are chemical potentials associated with each conserved quantity: baryon number, μ_B, (and strangeness, μ_S). Thus it is simple and instructive to estimate the $\bar{p}/p$ ratio from Fig. 20a, near the arrow, where I read $T = 160$ MeV, $\mu_B = 300$ MeV, $\sqrt{s_{NN}} \approx 30$ GeV, which gives $\bar{p}/p \approx 0.02$. Since the $\bar{p}/p$ ratio vs $\sqrt{s_{NN}}$ will be an important issue later, it is not a good idea to get this important information from a sketch in a proposal but from measurements and the best analysis (Fig. 21). The results

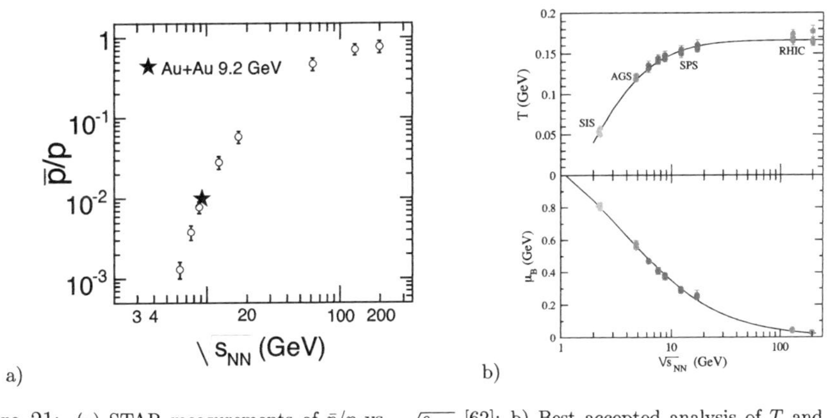

Figure 21: (a) STAR measurements of $\bar{p}/p$ vs. $\sqrt{s_{NN}}$ [62]; b) Best accepted analysis of T and μ_B vs $\sqrt{s_{NN}}$ [63].

are: i) the correct $\bar{p}/p$ ratio at $\sqrt{s_{NN}} = 30$ GeV is ~ 0.12 from Fig. 21a; ii) this ratio also corresponds to the correct $\mu_b \approx 170$ MeV at $\sqrt{s_{NN}} = 30$ GeV from Fig. 21b. The lesson is: if it looks more like art than like science, be skeptical and look in refereed journals for the correct numbers.

7.1 A press release during ISSP 2011

On June 23, 2011, shortly before I was to present my 2011 lectures, a press release from LBL arrived claiming that "By comparing theory with data from STAR, Berkeley Lab scientists and their colleagues map phase changes in the QGP" [64]. Since I was going to criticize in my lectures what I considered to be a particularly egregious case of "physics by press release" in the year 2000 by CERN (see Ref. [9]), I felt that I was obliged to quickly absorb and present in my talk the physics behind this latest press release, hopefully a "Highlight from RHIC".

The subject is "Fluctuations of conserved quantities", in this case the net baryon distribution taken as $p - \bar{p}$. Since there can be no fluctuations of conserved quantities such as net charge or net baryon number in the full phase space, one has go to "locally conserved quantities" [65] in small rapidity intervals to detect a small fraction of the protons and anti-protons which then fluctuates, i.e. varies from event to event. The argument is that, e.g. the fluctuation of one charged particle in or out of the considered interval produces a larger mean square fluctuation of the net electric charge if the system is in the hadron gas phase with integral charges than for the QGP phase with fractional charges.

However, while there are excellent statistical mechanical arguments about the utility of fluctuations of conserved quantities such as net baryon number as a probe of a critical point [66], there were, in 2011, no adequate treatments of the mathematical statistics of the experimental measurements. There are also additional problems such as short-range rapidity correlations in A+A collisions between like-particles induced by Fermi or Bose quantum statistics that must be reckoned with (e.g. see Refs. [67, 68]).

Theoretical analyses tend to be made by a Taylor expansion of the free energy $F = -T \ln Z$ around the critical temperature T_c where Z is the partition function, or sum over states, which is of the form

$$Z \propto e^{-(E-\sum_i \mu_i Q_i)/kT} \tag{8}$$

and μ_i are chemical potentials associated with conserved charges Q_i [66]. The terms of the Taylor expansion, which are obtained by differentiation, are called susceptibilities, denoted χ. The only connection of this method to mathematical statistics is that the Cumulant generating function in mathematical statistics for a random variable x is also a Taylor expansion of the ln of an exponential:

$$g_x(t) = \ln \langle e^{tx} \rangle = \sum_{n=1}^{\infty} \kappa_n \frac{t^n}{n!} \qquad \kappa_m = \left. \frac{d^m g_x(t)}{dt^m} \right|_{t=0} \quad . \tag{9}$$

Thus, the susceptibilities are Cumulants in mathematical statistics terms, where, in general, the Cumulant κ_m represents the $m^{\rm th}$ central moment, $\mu_m \equiv \langle (x-\mu)^m \rangle$, with all m-fold combinations of the lower order moments subtracted, where $\mu \equiv \langle x \rangle$. For instance, $\kappa_2 = \langle (x-\mu)^2 \rangle \equiv \sigma^2$, $\kappa_3 = \langle (x-\mu)^3 \rangle$, $\kappa_4 = \langle (x-\mu)^4 \rangle - 3\kappa_2^2$, $\kappa_5 = \langle (x-\mu)^5 \rangle - 10\kappa_3\kappa_2$. Two so-called normalized or standardized Cumulants are common in this field, the skewness, $S \equiv \kappa_3/\sigma^3$ and the kurtosis, $\kappa \equiv \kappa_4/\sigma^4 = \langle (x-\mu)^4 \rangle / \sigma^4 - 3$.

A sample [69] of STAR measurements of the distribution of net-protons in Au+Au collisions in the small interval $0.4 \leq p_T \leq 0.8$ GeV/c, $|y| < 0.5$ for different $\sqrt{s_{NN}}$ is shown in Fig. 22a. The moments in the form $\kappa\sigma^2 = \kappa_4/\kappa_2$ are shown from a previous STAR publication [70] in Fig. 22b while a plot, alleged to be of this same data, presented in the Lattice

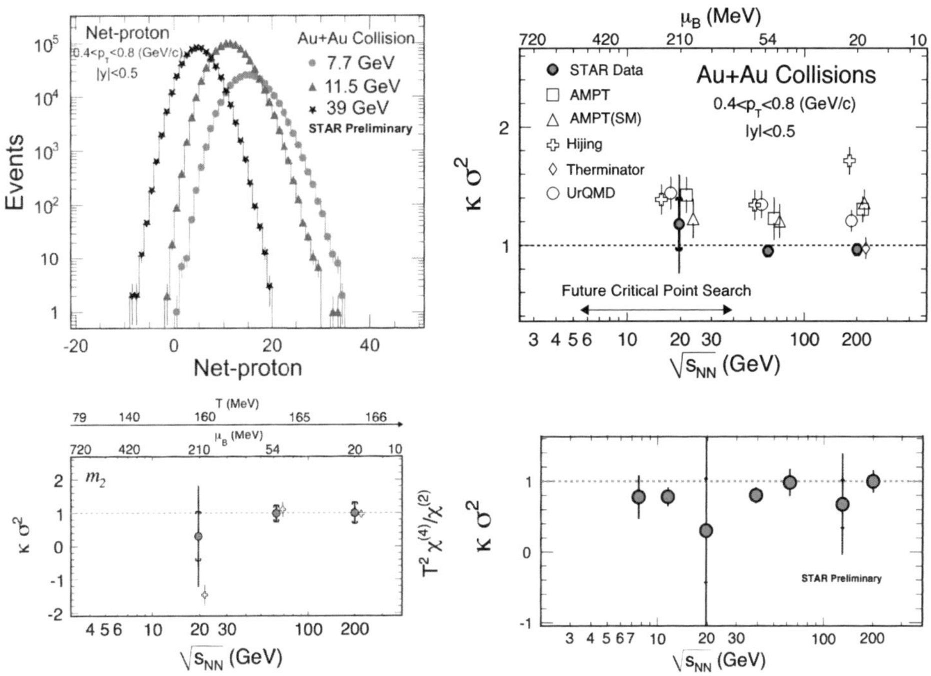

Figure 22: a) (top-left) STAR [69] distribution of event-by-event $p-\bar{p}$ at 3 values of $\sqrt{s_{NN}}$; b) (top-right) STAR published [70] measurements of $\kappa\sigma^2$; c) (bottom-left) Measurements from (b) as shown in Ref. [71] compared to the predicted ratio of susceptibilities (open crosses); d) (bottom-right) compilation [69] of STAR measurements of $\kappa\sigma^2$ for $p-\bar{p}$.

QCD theory publication that generated the press-release, is shown in Fig. 22c [71]; and a plot of the $\kappa\sigma^2$ from the data of Fig. 22c, combined with the results from Fig. 22b, is shown in Fig. 22d [69]. There are many interesting issues to be gleaned from Fig. 22.

The data point at 20 GeV in Fig. 22c is not the published one from (b), as stated in the caption [71], but the one from (d), which is different and with a much larger error. This, in my opinion, makes the data point look better compared to the predicted discontinuous value of $\kappa\sigma^2 = -1.5$ for the critical point at 20 GeV (open crosses) in contrast to the predictions of 1.0 for both 62.4 and 200 GeV. The published measurements in (b) together with the newer measurements in (d) are all consistent with $\kappa\sigma^2 = 1$; but clearly indicate the need for a better measurement at $\sqrt{s_{NN}} = 20$ GeV. Apart from these issues, the main problem of comparing Lattice QCD "data" to experimental measurements is that it is like comparing peaches to a fish, since the prediction is the result of derivatives of the log of the calculated partition function of an idealized system, which may have little bearing on what is measured using finite sized nuclei in an experiment with severe kinematic cuts. Maybe this is too harsh a judgement; but since this is the first such comparison (hence the press release), perhaps the situation will improve in the future. If a future measurement would show a significant huge discontinuity of $\kappa\sigma^2$ similar to the theoretical prediction at $\sqrt{s_{NN}} = 20$ GeV, then even I would admit that such a discovery would deserve a press release, maybe more!

7.1.1 If you know the distribution, you know all the moments and cumulants

When I first saw the measured distributions in Fig. 22a in 2011, my immediate reaction was that STAR should fit them to Negative Binomial distributions (NBD) so that they would know all the Cumulants. However, I subsequently realized that my favorite 3 distributions for integer random variables, namely, Poisson, Binomial, and Negative Binomial, are all defined only for positive integers (e.g. see Ref. [8] for details), while the number of net-protons on an event can be negative as well as positive, especially at higher c.m. energies. Thanks to Gary Westfall of STAR, in a paper presented at the Erice School of Nuclear Physics in 2012 [72], who found out that these three distributions fall into the class of "integer valued Lévy processes [73]" for which the Cumulants κ_j for the distribution $P(n-m)$ of the difference of samples from two such distributions, $P^+(n)$ and $P^-(m)$, with Cumulants κ_j^+ and κ_j^-, respectively, are [74, 73]:

$$\kappa_j = \kappa_j^+ + (-1)^j \kappa_j^- \quad , \tag{10}$$

so long as the distributions are not 100% correlated. This result is the same as if the distributions $P^+(n)$ and $P^-(m)$ were statistically independent. The first four Cumulants of the Poisson, Binomial and Negative Binomial distributions are given in Table 2.

Table 2: Cumulants for Poisson, Binomial and Negative Binomial Distributions

Cumulant	Poisson	Binomial	Negative Binomial
$\kappa_1 = \mu$	μ	np	μ
$\kappa_2 = \mu_2 = \sigma^2$	μ	$\mu(1-p)$	$\mu(1+\mu/k)$
$\kappa_3 = \mu_3$	μ	$\sigma^2(1-2p)$	$\sigma^2(1+2\mu/k)$
$\kappa_4 = \mu_4 - 3\kappa_2^2$	μ	$\sigma^2(1-6p+6p^2)$	$\sigma^2(1+6\mu/k+6\mu^2/k^2)$
$S \equiv \kappa_3/\sigma^3$	$1/\sqrt{\mu}$	$(1-2p)/\sigma$	$(1+2\mu/k)/\sigma$
$\kappa \equiv \kappa_4/\kappa_2^2$	$1/\mu$	$(1-6p+6p^2)/\sigma^2$	$(1+6\mu/k+6\mu^2/k^2)/\sigma^2$
$S\sigma = \kappa_3/\kappa_2$	1	$(1-2p)$	$(1+2\mu/k)$
$\kappa\sigma^2 = \kappa_4/\kappa_2$	1	$(1-6p+6p^2)$	$(1+6\mu/k+6\mu^2/k^2)$

7.2 The latest measurements have appeared without a press release

In the intervening period since 2011, the STAR collaboration has improved the preliminary measurements to publications and has improved the analysis by comparing to both Poisson and Negative Binomial distributions. Figure 23a [75] shows the STAR measurements of Cumulants of the net charge ($N^+ - N^-$) distributions from the "number of positive (N^+) and negative (N^-) charged particles within $|\eta| < 0.5$ and $0.2 < p_T < 2.0$ GeV/c on each event (after removing protons and antiprotons with $p_T < 400$ MeV/c) [75]". The corresponding Poisson and NBD Cumulants were calculated from the measured mean, μ, and variance, σ^2, of the N^+ and N^- distributions, respectively, and then calculated using Eq. 10. In contrast to Fig. 22, no non-monotonic behavior with $\sqrt{s_{NN}}$ is observed (or claimed) and the measurements of $S\sigma$ and $\kappa\sigma^2$ are all above the Poisson baseline. The $S\sigma$ measurements clearly favor the NBD.

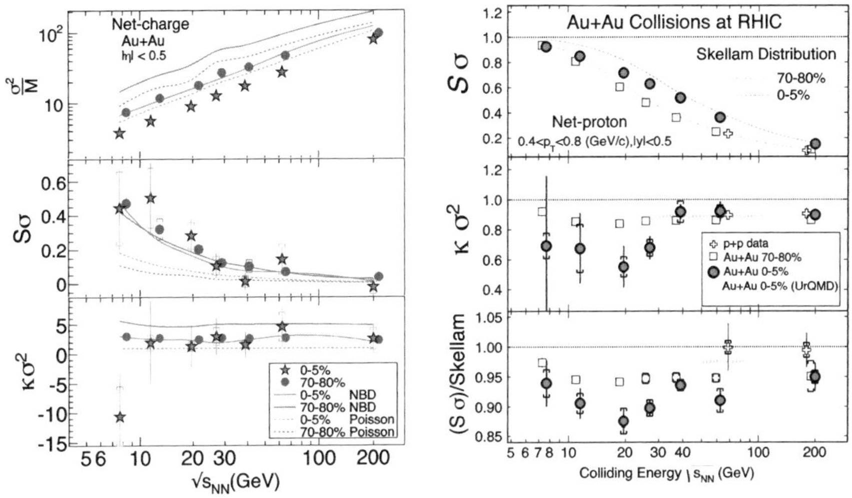

Figure 23: a) $\sqrt{s_{NN}}$ dependence of combinations of Cumulants in Au+Au (and p+p) from STAR: a) (left) net-charge Cumulants [75], where M is used to represent the mean, μ. b) (right) Cumulants of the $N_p - N_{\bar{p}}$ distributions [76], **where the error bars are statistical and the caps systematic errors**.

The situation is quite different for the net-proton ($N_p - N_{\bar{p}}$) Cumulants (Fig. 23b) [76] measured within $|y| < 0.5$ over the range $0.4 < p_T < 0.8$ GeV/c which covers roughly half the p_T spectrum. Here the measurements of $S\sigma$ and $\kappa\sigma^2$ are all below the Poisson baseline, denoted Skellam, which is the distribution of the difference between two Poissons and reflects "a system of totally uncorrelated, statistically random particle production" [76]. From Eq. 10 and Table 2 for a Poisson one can see that, for a Skellam, $S\sigma = \kappa_3/\kappa_2 = (\mu_p - \mu_{\bar{p}})/(\mu_p + \mu_{\bar{p}})$ which increases with decreasing $\sqrt{s_{NN}}$ because the $\bar{p}$ vanish ($\mu_{\bar{p}} = \langle N_{\bar{p}} \rangle \ll \mu_p$) so that the shape of the net distribution becomes dominated by the protons. This is easier to see in a plot of $\kappa\sigma^2$ vs μ_B with $\sqrt{s_{NN}}$ indicated (Fig. 24a) [59] which shows clearly that $\kappa\sigma^2$ starts dropping for $\sqrt{s_{NN}} < 39$ GeV where the $\bar{p}/p$ ratio drops below ~ 0.3 (recall Fig. 21).

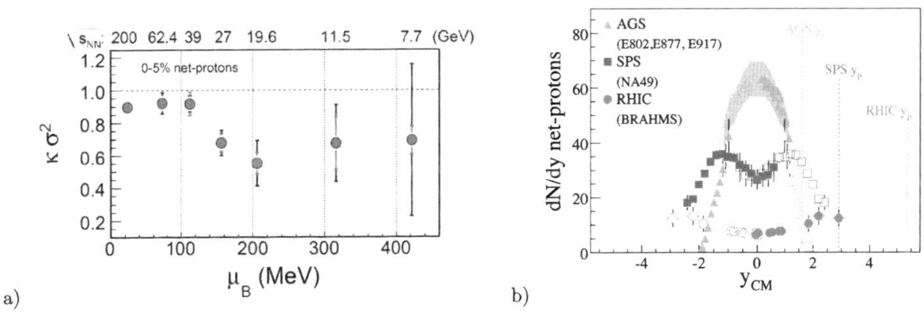

Figure 24: (a)(left) $\kappa\sigma^2$ vs. $\sqrt{s_{NN}}$ and μ_B [59]; b) (right) $(dN_p/dy - dN_{\bar{p}}/dy)$ for top 5% centrality at AGS (Au+Au, $\sqrt{s_{NN}}$ =4.9 GeV), SPS (Pb+Pb, $\sqrt{s_{NN}}$ =17.2 GeV) and RHIC(Au+Au, $\sqrt{s_{NN}}$ =200 GeV), beam rapidity in c.m. system $y_{\rm p} = 1.6, 2.9, 5.4$ [77].

The errors are still too large to determine whether or not $\kappa\sigma^2$ stays constant below $\sqrt{s_{NN}}$ =19.6 GeV but are sufficient to clearly rule out the value of $\kappa\sigma^2 = -1.5$ at $\sqrt{s_{NN}} \approx 20$ GeV predicted in Fig. 22 [71] which created the fuss in 2011. It is also important to point out that in addition to the vanishing of the anti-protons, the physics of the protons at mid-rapidity changes dramatically—the protons are no longer produced particles, which would conserve $N_p - N_{\bar{p}}$, but are the participants and fragments from the colliding nuclei which move to mid-rapidity and eventially stop as $\sqrt{s_{NN}}$ is reduced from 200, to 17.2 to 4.9 GeV (Fig. 24b) [77]. All these results indicate that the search for a QCD critical point at RHIC in the Beam Energy Scan (BES) in 2018-19 may not be as straightforward as originally assumed.

8 Jet quenching, RHIC's main claim to fame

The gold-plated signature for the QGP since 1986 [78] has been the suppression of J/Ψ because the color potential between $c, \bar{c}$ quarks would be screened (Debye screening) by all the free color charges in the medium so that the $c, \bar{c}$ would not be able to bind to form the J/Ψ. In fact the PHENIX experiment at RHIC was specifically designed to detect the J/Ψ at mid-rapidity at rest or with very low p_T (where the screening effect would be the largest) via the decay $J/\Psi \rightarrow e^+ + e^-$. J/Ψ suppression was reportedly observed several times at the CERN SpS fixed target heavy ion program starting with NA38 in O+U collisions in 1989 [79] but was plagued with many problems. The principal physics problem is that the J/Ψ does not follow the standard hard-scattering pointlike scaling in A+B collisions, $\sigma_{AB}^{J/\Psi} = (A \cdot B)^{\alpha} \cdot \sigma_{NN}^{J/\Psi}$ with $\alpha \equiv 1$, but is suppressed in cold nuclear matter (CNM) in p+A and A+B collisions, with $\alpha \approx 0.91$. Thus, the ultimate discovery by NA50 in Pb+Pb collisions at $\sqrt{s_{NN}}$ =17.2 GeV [80] in 1996–1998 was called "anomalous suppression" because it was below the CNM cross section dependence which was itself well below the hard-scattering pointlike scaling.

In 1998 at the QCD workshop in Paris [81], I found what I thought was a cleaner signal of the QGP when Rolf Baier asked me whether jets could be measured in Au+Au collisions because he had made studies in pQCD [82] of the energy loss of partons, produced by hard-scattering "with their color charge fully exposed", in traversing a medium "with a large density of similarly exposed color charges". The conclusion was that "Numerical estimates of the loss suggest that it may be significantly greater in hot matter than in cold. *This makes the magnitude of the radiative energy loss a remarkable signal for QGP formation*" [82]. In addition to being a probe of the QGP the fully exposed color charges allow the study of parton-scattering with $Q^2 \ll 1 - 5$ (GeV/c)^2 in the medium where new collective QCD effects may possibly be observed.

Because the expected energy in a typical jet cone $R = \sqrt{(\Delta\eta)^2 + (\Delta\phi)^2}$ in central Au+Au collisions at $\sqrt{s_{NN}}$ =200 GeV would be $\pi R^2 \times 1/2\pi \times d\text{E}_T/d\eta = R^2/2 \times d\text{E}_T/d\eta \sim 350$ GeV for $R = 1$, where the kinematic limit is 100 GeV, I said (and wrote [81]) that jets can not be reconstructed in Au+Au central collisions at RHIC—still correct after 16 years. On the other hand, hard-scattering was discovered in $p + p$ collisions at the CERN-ISR in 1972 with single particle and two-particle correlations, while jets had a long learning curve from 1977–1982 with a notorious false claim (e.g. see Refs. [8, 9]), so I said (and wrote [81]) that we should use single and two-particle measurements—which we did and it WORKED! The present solution for jets in A+A collisions (LHC 2010 and RHIC c.2014) is to take smaller cones, with 56 GeV in $R = 0.4$, 32 GeV in $R = 0.3$, 14 GeV in $R = 0.2$ at RHIC.

8.1 Jet quenching at RHIC — Suppression of high p_T particles

The discovery at RHIC [83] that π^0's produced at large transverse momenta are suppressed in central Au+Au collisions by a factor of ~ 5 compared to pointlike scaling from p+p collisions is arguably *the* major discovery in Relativistic Heavy Ion Physics. For π^0 (Fig. 25a) [84] the hard-scattering in p+p collisions is indicated by the power law behavior p_T^{-n} for the invariant cross section, $Ed^3\sigma/dp^3$, with $n = 8.1 \pm 0.1$ for $p_T \geq 3$ GeV/c. The Au+Au data at a given p_T can be characterized either as shifted lower in p_T by $\delta p'_T$ from the pointlike scaled p+p data at $p'_T = p_T + \delta p'_T$, or shifted down in magnitude, i.e. suppressed. In Fig. 25b, the suppression of the many identified particles measured by PHENIX at RHIC is presented as the Nuclear Modification Factor, $R_{AA}(p_T)$, the ratio of the yield of e.g. π per central Au+Au

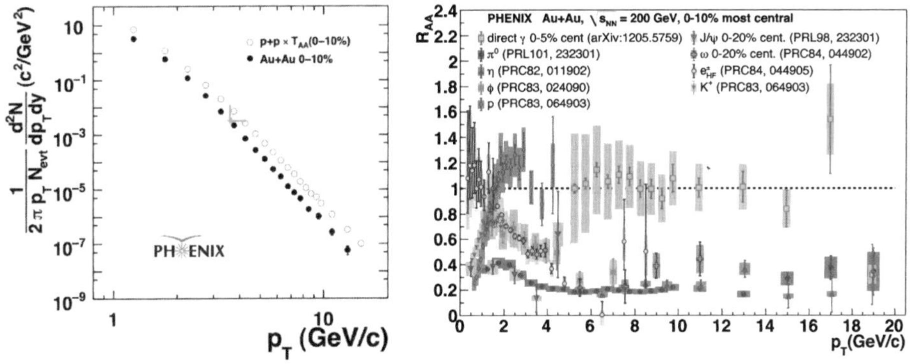

Figure 25: a) (left) Log-log plot of invariant yield of π^0 at $\sqrt{s_{NN}} = 200$ GeV as a function of transverse momentum p_T in p+p collisions multiplied by $\langle T_{AA}\rangle$ for Au+Au central (0–10%) collisions compared to the Au+Au measurement [84]. Vertical arrow is for $R_{AA}(p_T)$, horizontal arrow for $\delta p'_T$. b) (right) $R_{AA}(p_T)$ for all identified particles so far measured by PHENIX in Au+Au central collisions at $\sqrt{s_{NN}} = 200$ GeV.

collision (upper 10%-ile of observed multiplicity) to the pointlike-scaled p+p cross section at the same p_T, where $\langle T_{AA}\rangle$ is the average overlap integral of the nuclear thickness functions:

$$R_{AA}(p_T) = \frac{(1/N_{AA})\ d^2N^{\pi}_{AA}/dp_T dy}{\langle T_{AA}\rangle\ \ d^2\sigma^{\pi}_{pp}/dp_T dy} \quad . \tag{11}$$

The striking differences of $R_{AA}(p_T)$ in central Au+Au collisions for the many particles measured by PHENIX (Fig. 25b) illustrates the importance of particle identification for understanding the physics of the medium produced at RHIC. Most notable are: the equal suppression of π^0 and η mesons by a constant factor of 5 ($R_{AA} = 0.2$) for $4 \leq p_T \leq 15$ GeV/c, with suggestion of an increase in R_{AA} for $p_T > 15$ GeV/c; the equality of suppression of direct-single $e^{\pm}$ (from heavy quark (c, b) decay) and π^0 at $p_T \gtrsim 5$ GeV/c; the non-suppression of direct-γ for $p_T \geq 4$ GeV/c; the exponential rise of R_{AA} of direct-γ for $p_T < 2$ GeV/c [85], which is totally and dramatically different from all other particles and attributed to thermal photon production by many authors (e.g. see citations in Ref. [85]). For $p_T \gtrsim 4$ GeV/c, the hard-scattering region, the fact that all hadrons are suppressed, but direct-γ are not suppressed, indicates that suppression is a medium effect on outgoing color-charged partons likely due to energy loss by coherent Landau-Pomeranchuk-Migdal radiation of gluons, predicted in pQCD [82], which is sensitive to properties of the medium.

One nice advantage that hard-scattering and high p_T suppression have as a QGP probe compared to J/Ψ suppression is that although there is a CNM effect, it is an enhancment rather than a suppression; and as far as is known, the enhancement, historically called the Cronin effect [86], only occurs for baryons and not mesons at RHIC energies. Figure 26a

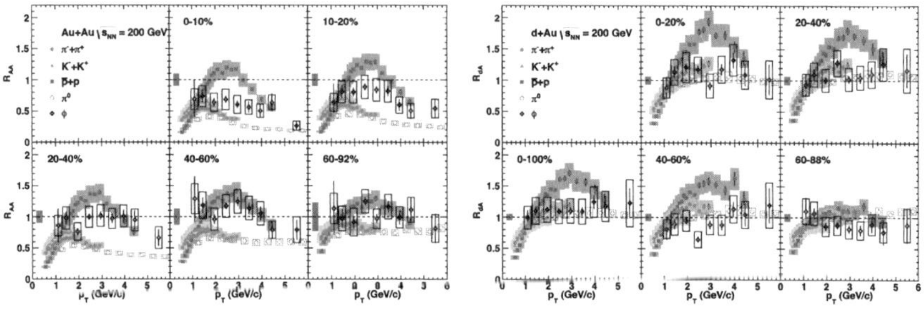

Figure 26: Measurements of R_{AA} of identified particles $\pi^{\pm}, K^{\pm}, p^{\pm}, \pi^0, \phi$ as a function p_T and centrality at $\sqrt{s_{NN}} = 200$ GeV [87]: a) (left) Au+Au; b) (right) d+Au.

shows R_{AA} in Au+Au for protons and mesons in the range $0.5 < p_T < 6.0$ GeV/c, where, in central collisions (0-10%), all the mesons are suppressed for $p_T > 2$ GeV/c while the protons are enhanced for $2 < p_T < 4$ GeV/c and then become suppressed at larger p_T. The d+Au results in Fig. 26b show no CNM effect for the mesons, $R_{AA} \approx 1$ out to $p_T = 6$ GeV/c; while the protons show a huge enhancement (Cronin effect) in all centralities except for the most peripheral (60-88%). At present, there is no explanation of the proton enhancement in either Au+Au or d+Au collisions, so π^0 and η are the favored hard-probes.

8.2 $\delta p'_T/p'_T$, the fractional shift in the p'_T spectrum

After more than a decade of using the ratio R_{AA}, we are now paying more attention to $\delta p'_T/p'_T$, the fractional shift of the p'_T spectrum, as an indicator of energy loss in the QGP Fig. 27 [88]. For a constant fractional energy loss, which is true at RHIC in the range $6 < p_T < 12$ GeV/c (as shown in Fig. 25a where the p+p reference and Au+Au measurement are parallel on a log-log plot) there is a simple relationship between R_{AA}, $\delta p'_T/p'_T$ and n, the power in the invariant p_T spectra:

$$R_{AA}(p'_T) = R_{AA}(p_T) = (1 - \delta p_T/p_T)^{n-2} \quad . \tag{12}$$

Using $\delta p_T/p_T$ is important for comparison to the LHC measurements where the power is $n \approx 6$ compared to $n = 8.1$ at RHIC, so that the same R_{AA} does not mean the same $\delta p_T/p_T$. Strictly $\delta p_T/p_T$ is not a measure of the parton energy loss in the QGP but is used as a proxy. Figure 27a shows that $\delta p_T/p_T$ at $p_T = 7$ GeV/c for RHIC and LHC both increase monotonically with centrality ($N_{\rm part}$) but is a factor of 2 to 1.4 larger at LHC, depending on centrality, a likely indication of a hotter and/or denser medium. Figure 27b attempts to determine whether $\delta p_T/p_T$ is a universal function of the charged particle density, $dN_{\rm ch}/d\eta$ at both RHIC and LHC. The dependence is not quite universal. A fit of $\delta p_T/p_T \propto (dN_{\rm ch}/d\eta)^{\alpha}$ gives $\alpha \approx 0.35$ at LHC and 0.55 at RHIC, although the data at $\sqrt{s_{NN}} = 200$ GeV and 2.76 TeV do appear to merge for $(dN_{\rm ch}/d\eta) \geq 300$. Hopefully, measurements of $\delta p_T/p_T$ will eventually lead to the determination of dE/dx of partons in the QGP.

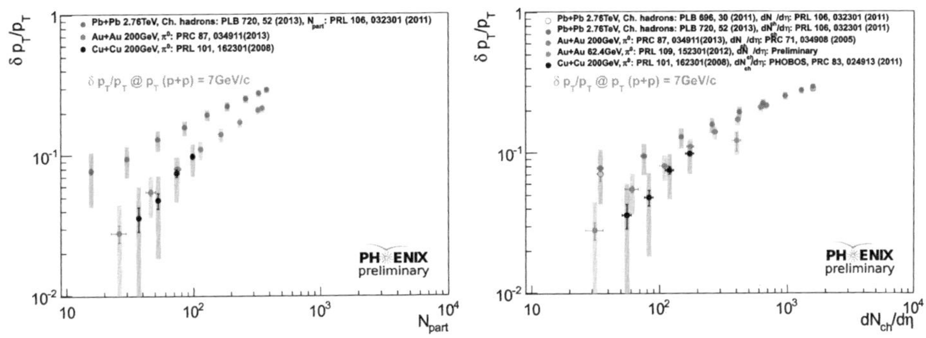

Figure 27: Plots from PHENIX [88] of $\delta p'_T/p'_T$ at $p'_T \equiv p_T(p+p) = 7$ GeV for π^0 (RHIC) and charged hadrons (LHC): a) as a function of centrality ($N_{\rm part}$), b) as a function of $dN_{\rm ch}/d\eta$.

8.3 At last: jet measurements in Au+Au at RHIC in 2014?

Some interesting new jet measurements in A+A collisions at RHIC were presented at Quark Matter 2014 in a plenary review talk on jets by Yen-Jie Lee who works on CMS [89]. Figure 28 shows that the STAR charged jets in a cone with $R = 0.2$ have much less suppression ($R_{AA} \gg 0.3$) than π^0 ($0.2 \leq R_{AA} \leq 0.3$) in the range $10 < p_T < 20$ GeV.

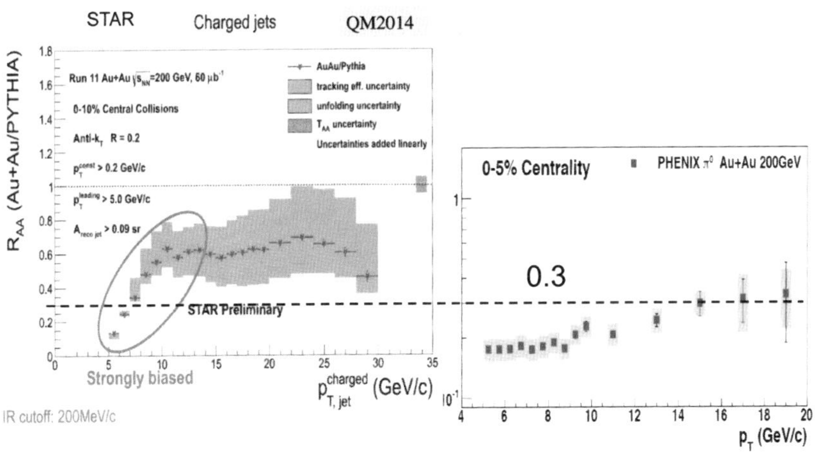

Figure 28: a) (left) STAR R_{AA} for charged jets at $\sqrt{s_{NN}}$ =200 GeV in central Au+Au collisions (see details in legend) compared to b) R_{AA} for PHENIX π^0. The dashed line at 0.3 is the maximum R_{AA} for π^0 in this p_T range.

This is quite different from jets at the LHC (Fig. 29) which have comparable or smaller R_{AA} than charged particles from jet fragmentation in the range $30 < p_T < 100$ GeV. Note that the γ, W and Z^0 bosons in Fig. 29b which are not coupled to color are not suppressed.

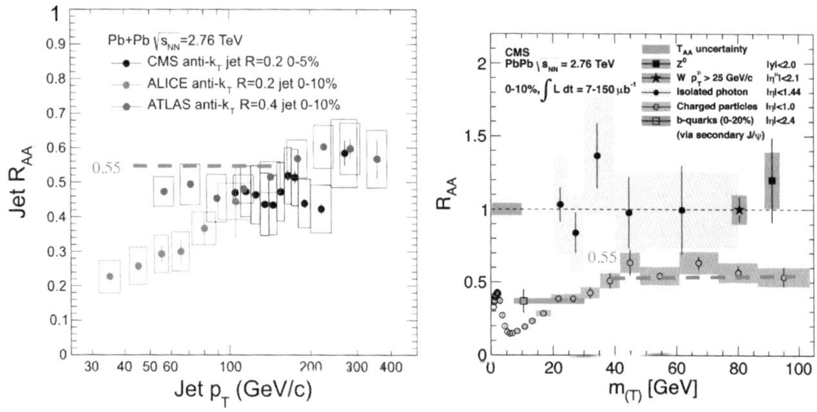

Figure 29: a) (left) R_{AA} for jets at $\sqrt{s_{NN}}$ =2.76 GeV by CMS and ALICE compared to b) CMS R_{AA} for charged hadrons ($R_{AA} \approx 0.55$), b-quarks and 3 favorite Electro-Weak Bosons [89].

For STAR, the disagreement of the jet and single particle R_{AA} gets worse as the jet cone is increased from R=0.2 to 0.3 to 0.4 (Fig. 30). Some people would say that this is great because all the jet fragments and/or any energy lost in the QGP by the originating parton have been captured in the R=0.4 cone. Skeptics like myself can hardly wait to see what happens when the jet cone is further increased. After 14 runs at RHIC, the jet learning curve in Au+Au central collisions still has a way to go.

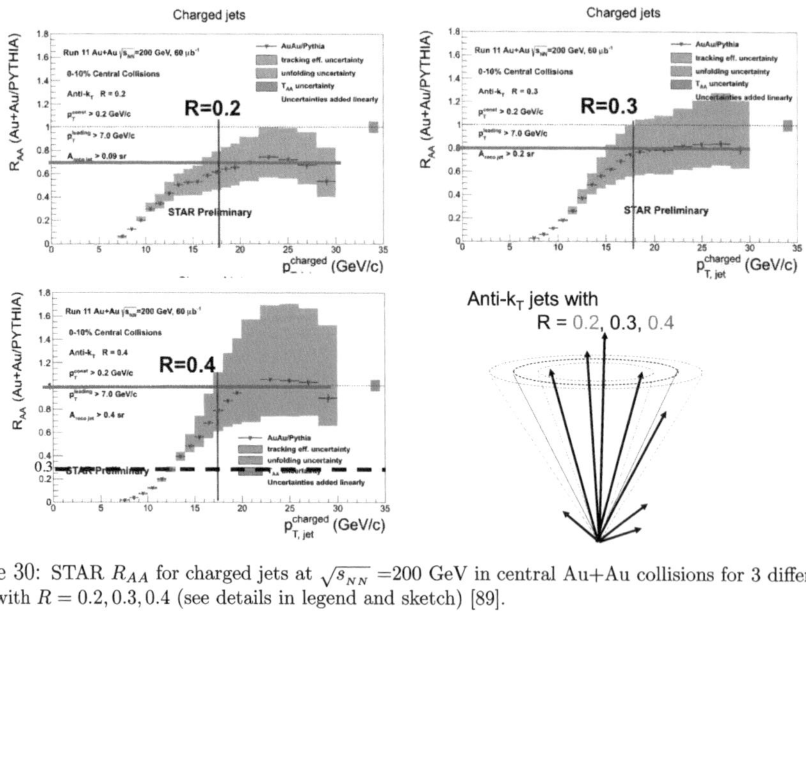

Figure 30: STAR R_{AA} for charged jets at $\sqrt{s_{NN}}$ =200 GeV in central Au+Au collisions for 3 different jet cones with $R = 0.2, 0.3, 0.4$ (see details in legend and sketch) [89].

The good news for the future is that a new detector, now called sPHENIX, to find jets by the more traditional method using hadron calorimetry has been proposed, is moving along on the approval process and is on the schedule at RHIC (Fig. 5b) for comissioning and full

Figure 31: a)(left) new detector in place of PHENIX; b) (right) beam's eye view with details [90].

installation in the PHENIX IR in 2019–20. It is based on the (made in Italy) BABAR superconducting solenoid from SLAC which became available when the B-factory in Italy was unfortunately cancelled. Design work is moving along quickly (Fig. 31) so now is the optimum time for new collaborators to join.

References

[1] See `http://bnlcompetition.science.energy.gov` for details.

[2] *Building for Discovery, Strategic Plan for U.S. Particle Physics in the Global Context*, `http://science.energy.gov/~/media/hep/hepap/pdf/May%202014/FINAL_P5_Report_053014.pdf`

[3] Stanley Wojcicki, Rev. Accl. Sci. Tech. *The Supercollider: The Pre-Texas Days*, **01** (2008) 259; *The Texas Days*, **02** (2009) 265.

[4] PHENIX Collaboration (S. S. Adler *et al.*), Phys. Rev. C**89** (2014) 044905.

[5] PHENIX Collaboration (K. Adcox *et al.*), Nucl. Phys. A**757** (2005) 184–283.

[6] M. Harrison, T. Ludlam and S. Qzaki (eds.), *The Relativistic Heavy Ion Collider Project: RHIC and its Detectors*, Nucl. Instrum. Methods A**499** (2003) 235–880.

[7] M. J. Tannenbaum, Int. J. Mod. Phys. A**26** (2011) 5299-5335, arXiv:1406.0830.

[8] Jan Rak and Michael J. Tannenbaum, *High p_T Physics in the Heavy Ion Era*, (Cambridge University Press, 2013).

[9] M. J. Tannenbaum, Int. J. Mod. Phys. A**29** (2014) 1430017, arXiv:1406.1100.

[10] W. Busza, *et al.*, Phys. Rev. Lett. **34** (1975) 836–839.

[11] J. E. Elias, *et al.*, Phys. Rev. D**22** (1980) 13–35.

[12] C. Halliwell, *et al.*, Phys. Rev. Lett. **39** (1977) 1499–1502.

[13] S. Frankel, Phys. Rev. C**48** (1993) R2170.

[14] A. Białas, M. Bleszyński, W. Czyż, Nucl. Phys. B**111** (1976) 461–476.

[15] A. Bialas, W. Czyz, and L. Lesniak, Phys. Rev. D**25** (1982) 2328.

[16] W. J. Willis, Report CRISP-72-15, BNL-16841, Upton, NY, 1972.

[17] J. D. Bjorken, Phys. Rev. D**8** (1973) 4098.

[18] BCMOR Collaboration, (A.L.S. Angelis *et al.*), Phys. Lett. B**141** (1984) 140.

[19] BCMOR Collaboration, (Michael J. Tannenbaum *et al.*), Lect. Notes Phys. **221** (1985) 174.

[20] AFS Collaboration, (Bruce Callen *et al.*), Lect. Notes Phys. **221** (1985) 133.

[21] T. Ochiai, Z. Phys. C**35** (1987) 209.

[22] CMOR Collaboration, (A.L.S. Angelis *et al.*), Nucl. Phys. B**244** (1984) 1.

[23] L. DiLella, Ann. Rev. Nucl. Part. Sci. **35** (1985) 107.

[24] PHENIX Collaboration, (K. Adcox *et al.*), Phys. Rev. Lett. **86** (2001) 3500.

[25] X.-N. Wang and M. Gyulassy, Phys. Rev. Lett. **86** (2001) 3496.

[26] ALICE Collaboration, (K. Aamodt *et al.*), Phys. Rev. Lett. **106** (2011) 032301.

[27] PHENIX Collaboration, (J. T. Mitchell *et al.*), PoS CPOD2013 (2013) 003.

[28] S. Eremin and S. Voloshin, Phys. Rev. C**67** (2003) 064905.

[29] M. Gell-Mann, Phys. Lett. **8** (1964) 214.

[30] E. V. Shuryak, Nucl. Phys. B**203** (1982) 116.

[31] M. Breidenbach *et al.*, Phys. Rev. Lett. **23** (1969) 935.

[32] T. Kalaydzhyan and E. Shuryak, Phys. Rev. C**90** (2014) 014901.

[33] R. Hofstadter, F. Bumiller, and M. R. Yerian, Rev. Mod. Phys. **30** (1958) 482.

[34] PHENIX Collaboration, (S. S. Adler *et al.*), Phys. Rev. D**76** (2007) 051006(R).

[35] ATLAS Collaboration, (G. Aad *et al.*), Phys. Lett. B**707** (2012) 330.

[36] ALICE Collaboration, (B. Abelev *et al.*) Phys. Rev. C**88** (2013) 044909.

[37] PHOBOS Collaboration, (B. B. Back *et al.*), Phys. Rev. C**70** (2004) 021902(R).

[38] L. Zhou and G. S. F. Stephans, Phys. Rev. C**90** (2014) 014902.

[39] PHOBOS Collaboration, (B. Alver *et al.*), Phys. Rev. C**83** (2011) 024913.

[40] H. Bialkowska, Acta Phys. Polon. **B37** (2006) 3415.

[41] PHOBOS Collaboration (B. B. Back, *et al.*), Phys. Rev. C**74** (2006) 021901(R).

[42] M. Basile, ... A. Zichichi, *et al.*, Phys. Lett. B**95** (1980) 311–312.

[43] STAR Collaboration (Hui Wang *et al.*), Flow measurements and selection of body-body and tip-tip enhanced samples in U+U collisions, presented at *Quark Matter 2014*, Darmstadt, Germany, May 19, 2014. Also see `http://dx.doi.org/10.1016/j.nuclphysa.2014.09.111`.

[44] H. A. Gustafsson, *et al.*, Phys. Rev. Lett. **52** (1984) 1590.

[45] J.-Y. Ollitrault, Phys. Rev. D**46** (1992) 229–245, Nucl. Phys. A**638** (1998) 195c–206c.

[46] PHENIX Collaboration (M. Kaneta *et al.*), J. Phys. G**30** (2004) S1217–S1220.

[47] U. Heinz and R. Snellings, Ann. Rev. Nucl. Part. Sci. **63** (2013) 123.

[48] S. A. Voloshin, A. M. Poskanzer and R. Snellings, arXiv:0809.2949v2.

[49] FOPI Collaboration (A. Andronic *et al.*), Nucl. Phys. A**679** (2001) 765–792.

[50] M. Nasim, L. Kumar, P. K. Netraknti and B. Mohanty, Phys. Rev. C**82** (2010) 054908.

[51] D. Teaney, Phys. Rev. C**68** (2003) 034913.

[52] P. K. Kovtun, D. T. Son, and A. O. Starinets, Phys. Rev. Lett. **94** (2005) 111601.

[53] PHENIX Collaboration (A. Adare *et al.*), arXiv:1404.7461.

[54] A. Kuhlman and U. Heinz, Phys. Rev. C**72** (2005) 037901.

[55] P. Filip, R. Lecnicky, H. Masui and N. Xu, Phys. Rev. C**80** (2009) 054903.

[56] S. A. Voloshin, Phys. Rev. Lett. **105** (2010) 172301.

[57] J. M. Pawlowski, Equation of state and phase diagram of strongly interacting matter, presented at *Quark Matter 2014*, Darmstadt, Germany, May 22, 2014.

[58] STAR Collaboration, *RHIC Beam Use Request For Runs 14 and 15*, May 28, 2013. `https://indico.bnl.gov/getFile.py/access?resId=0&materialId=2&confId=632`

[59] STAR Collaboration, *Studying the Phase Diagram of QCD Matter at RHIC*, June 1, 2014. `https://indico.bnl.gov/getFile.py/access?resId=0&materialId=2&confId=764`

[60] F. Cooper and G. Frye, Phys. Rev. D**10** (1974) 186–189.

[61] STAR Collaboration, (J. Adams *et al.*), Nucl. Phys. A**757** (2005) 102183.

[62] STAR Collaboration, (B. I. Abelev *et al.*), Phys. Rev. C**81** (2010) 024911.

[63] J. Cleymans, H. Oeschler, K. Redlich and S. Wheaton, Acta Phys. Polon. Supp. **3** (2010) 533–538, arXiv:0911.0526.

[64] `http://newscenter.lbl.gov/news-releases/2011/06/23/when-matter-melts/`.

[65] M. Asakawa, U. Heinz, and B. Müller, Phys. Rev. Lett. **85** (2000) 2072.

[66] e.g. see V. Koch, PoS(CFRNC2006)008

[67] W. A. Zajc *et al.*, Phys. Rev. C**29** (1984) 2173.

[68] M. J. Tannenbaum, Phys. Lett. B**347** (1995) 431.

[69] Presentation by T. J. Tarnowsky at *Quark Matter 2011*, Annecy, France, May 23, 2011; STAR Collaboration (T. J. Tarnowsky *et al.*), J. Phys. G**38** (2011) 124054.

[70] STAR Collaboration (M. M. Aggarwal *et al.*), Phys. Rev. Lett. **105** (2010) 022302.

[71] S. Gupta, X. Luo, B. Mohanty, H. G. Ritter and N. Xu, Science **332** (2011) 1525.

[72] STAR Collaboration (G. Westfall *et al.*), Search for the QCD Critical Point, *International School of Nuclear Physics, 33rd Course* Erice, Sicily, Italy 16–24 September 2011.

[73] O. E. Barndorff-Nielsen, D. G. Pollard and N. Shephard, Quant. Finance **12** (2012) 587–605.

[74] T. J. Tarnowsky and G. D. Westfall, Phys. Lett. B**724** (2013) 51.

[75] STAR Collaboration (L. Adamczyk *et al.*), Phys. Rev. Lett. **113** (2014) 092301.

[76] STAR Collaboration (L. Adamczyk *et al.*), Phys. Rev. Lett. **112** (2014) 032302.

[77] BRAHMS Collaboration (I. G. Bearden *et al.*), Phys. Rev. Lett. **93** (2004) 102301.

[78] T. Matsui and H. Satz, Phys. Lett. B**178** (1986) 416.

[79] NA38 Collaboration (C. Baglin *et al.*), Phys. Lett. B**220** (1989) 471; L. Kluberg, Nucl. Phys. A**488** (1988) 613c.

[80] NA50 Collaboration (M. Gonin *et al.*), Nucl. Phys. A**610** (1996) 404c.

[81] R. Baier, M. J. Tannenbaum in *Proc. IV Workshop on Quantum Chromodynamics*, Paris, France, 1–6 June 1998, eds. H. M. Fried and B. Müller (World Scientific,1999).

[82] See R. Baier, D. Schiff and B. G. Zakharov, Ann. Rev. Nucl. Part. Sci. **50** (2000) 37–69, and references therein.

[83] PHENIX Collaboration (K. Adcox *et al.*), Phys. Rev. Lett. **88** (2002) 022301.

[84] PHENIX Collaboration (S. S. Adler *et al.*), Phys. Rev. C**76** (2007) 034904.

[85] PHENIX Collaboration (A. Adare, *et al.*), Phys. Rev. Lett. **104** (2010) 132301.

[86] J. W. Cronin *et al.*, Phys. Rev. D**11** (1975) 3105.

[87] PHENIX Collaboration (A. Adare, *et al.*), Phys. Rev. C**88** (2013) 024906.

[88] Presentation by T. Sakaguchi at *10th International Workshop on High p_T Physics in the RHIC/LHC era*, Nantes, France, 9–12 September 2014.

[89] Yen-Jie Lee, Experimental results on jets in ultra-relativistic nuclear collisions, presented at *Quark Matter 2014*, Darmstadt, Germany, May 23, 2014.

[90] PHENIX Collaboration (J. Nagle, *et al.*) sPHENIX Science Overview, presented at *Department of Energy sPHENIX Science Review*, BNL, Upton, NY, USA, July 1, 2014.

LECTURER: M. J. TANNENBAUM

Highlights from BNL and RHIC 2014

Scientific Secretaries: M. Kofarago, A. Laszlo

DISCUSSION

- *Savitskyi:*

Bottomium production could be used for research of color screening as well as charmonium production. Although at B-meson factories statistics of Y-resonances is large, there is still no evidence of color deconfinement from these results. Why is the charmonium production special? Why don't we consider the bottomonium production and quarkonium state of the φ-meson?

- *Tannenbaum:*

It is a very good question and hard to answer. The φ did not have the new quantum number, charm, so it was never considered for the same analysis. The φ looks like the pions in many aspects, it flows like it is made out of two quarks and when it is suppressed it is suppressed like the kaons.

- *Korthals-Altes:*

To prove deconfinement you need an object which has a well defined color electric string between the quarks that constitute it and then the screening that you want to see in the plasma becomes clearly visible. So the screening radius of the φ could be too large or too small for the measurement.

- *Kovalenko:*

Antiproton to proton ratio is rapidity dependent. Would that mean that by measuring observables at different rapidity ranges we explore different parts of phase diagram?

- *Tannenbaum:*

In general, yes. Several experiments at energy ~5 GeV (at AGS) made such measurements and some interesting results were obtained. Indeed, this method was considered by experiments to extend exploration of the phase diagram to higher baryochemical potential by looking at forward and backward rapidities where the projectile and target nucleons are located.

- *Drozhzhova:*

How the parameters of the phase diagram (temperature and baryochemical potential) are assigned to observables in an experimental manner for a given nucleus-nucleus collision with given energy?

- *Tannenbaum:*

The temperature is defined in terms of inverse slope parameter of the low transverse momentum part of the transverse mass spectra. The rationale behind this is that it is experimentally seen to be exponential in the low transverse momentum regime. If one theoretically identifies it with a Boltzmann spectrum, then in that picture the inverse slope parameter indeed means the temperature. The picture is complicated by the fact that these inverse slopes are slightly particle type dependent, but that is most commonly explained by radial flow in the expanding medium. Given the temperature, the baryochemical potential is determined via the abundances of identified hadrons (p,pi,K and their anti-particles). For example the antiproton to proton yield ratio corresponds to exp(-(E+mu_B)/T)/exp(-(E-mu_B)/T)=exp(-2mu_B/T) in a Boltzmann picture. For the details, see the experimental papers cited in my talks in the previous years of Erice School.

- *Drozhzhova:*

What is the resolution of the Zero Degree Calorimeter in the PHENIX experiment?

- *Tannenbaum:*

It was designed and constructed in 1990's, so I don't remember its resolution. Typical hadron calorimeters have resolution $\Delta E/E \sim 100\%/\mathrm{sqrt}(E) + 10\%$, so 20% for 100 GeV neutrons and no problem in separating individual neutrons. See arXiv:nucl-ex/0008051v1

- *Ivanytskyi:*

The number of produced J/Ψ particles is rather small, so it could be analyzed by a microcanonical ensemble instead of the grand canonical one. There are arguments that the J/Ψ suppression can be explained without QGP formation. What do you think about these arguments?

- *Tannenbaum:*

I don't know about such ensembles. That is the business of theorists. The J/Ψ data have been there for eight years, but no-one actually looked into that aspect.

- *Ivanytskyi:*

Hadronisation is in some sense similar to the recombination of electrons and protons in the early universe. So can we expect some effect for prompt photons which is similar to the formation of the microwave background?

- *Tannenbaum:*

I am not sure that we could detect these effects in the collisions with our current detectors. The QGP is completely transparent to gamma-rays, the lowest energy we have measured is 400 MeV. These photons don't interact with the medium at all; although it might be different for ~1 volt photons (red photons). We know the density, the charges in the medium and the dimensions so maybe a cosmologist could calculate something but it hasn't been studied yet. Maybe you can do it.

- *Zennamo:*

You mentioned that during the sPHENIX upgrade you wanted to put a Cerenkov detector in the front. I was wondering about the performance of this Cerenkov detector. Especially in the light of the detector used in the MIPP experiment, that didn't work.

- *Tannenbaum:*

We have been using Cerenkov counters ever since I've been in this field, which I can also show you there, called the DIRC. This detector which you refer to has been already used by BaBar over ten years in similar conditions that we will have. If the principle is correct, the engineering is correct, the testing is correct, then it works. What can I tell you about MIPP? Get better guys there. Our RICH detector in PHENIX works beautifully, that is why we can detect J/Psi-s at rest that nobody else can..

- *Zennamo:*

Is there a dedicated method to determine the jet energy scale in A-A collisions or you use the one in p-p?

- *Tannenbaum:*

I'm glad that you asked me that. I will come back to a slide that I skipped over in my talk. Here is what I said in 1998 to Rolf Baier at a QCD workshop in Paris when he asked me whether jets could be measured in Au+Au collisions at RHIC because he had a prediction of a QCD medium-effect on colored partons in a hot dense medium with lots of unscreened color charge. Since the expected energy from soft particles in a typical R=1 jet cone in central Au+Au collisions at sqrt{s}=200 GeV is 350 GeV, while the maximum energy Jet is only 100 GeV, I said that Jets can not be reconstructed in Au+Au central collision at RHIC---which is still true after 16 years. So the answer is make the cone smaller. However at that time in 1998, if you had lived through what I call the Jet fiasco with a long learning curve from 1972-1982, including false claims of jet observation in a 1 steradian (R=1) calorimeter (see my Book), you would be very leery about reducing the size of a jet cone. Since I had participated in the discovery of hard scattering in p-p collisions at the CERN-ISR in the early 1970s, where single and two-particle measurements were used to map out hard-scattering, I said that we would use the same techniques at RHIC, which we did and it worked. Of course the solution (as shown at the LHC in 2010 and perhaps soon at RHIC) is to take smaller cones: R=0.4 with 60 GeV, R=0.3 with 34 GeV, R=0.2 with 15 GeV; but as I showed you from the STAR data of R_[AA} versus jet cone size compared to single particle R_{AA}, it is still very confusing even now with 14 years of learning curves when you see these weird results. Now, we have some experts in the audience who know how to do it. Dennis is one of them. So the answer is yes, after 14 years, we learned how to do it; but the people at the heavy ion experiments at LHC were really lucky because when jets turned out to be the interesting result, which Brian Cole noticed on the event display as soon as they had the first heavy ion event: instead of seeing two jets--he saw one jet all the time. The LHC high energy physicists had spent years calibrating the detectors for p-p collisions. The difference between the heavy ion and the p-p collisions is the issue of the large background of soft particles in the jet cone, which fluctuates, so how do you measure it. This is the main issue between heavy-ion and p-p collisions but they do a good job of it at LHC. At RHIC it's going to be harder because we are closer to the limit: the multiplicity is only half that at the LHC, but the jet spectrum at RHIC is at much lower pT. The answer is that it has been shown to be done at the LHC.

- *Boren:*

How is the plot of emission duration explained by anti-protons becoming negligible?

- *Tannenbaum:*

There were two different plots, I think you are talking about the phase diagram. The chemical potential is measured by measuring the anti-proton to proton ratio as a function of the collision energy. At 200 GeV, at mid rapidity, they are almost equal. If you go to forward and backward rapidity where the 197 nucleons from each Au nucleus are as somebody suggested previously, you get a much higher chemical potential but what you see at mid-rapidity are produced protons and antiprotons. The points on the phase diagram are plotted at the measured mid-rapidity chemical potential If you know the temperature of the freeze-out which is nearly the same, 160 MeV, everywhere, the famous exp-{6pT} exponential, and the pion to proton ratio or the positive kaon to negative kaon ratio you can calculate the chemical potential.

- *Russo:*

You showed that the nuclear modification factor for heavy flavor electrons was enhanced at mid rapidity in d+Au in central collisions. Since, at the LHC, the definition of centrality in p+Pb collisions has turned out to be quite challenging, I would like you to comment on this enhancement and how is the approach to centrality in d+Au at RHIC.

- *Tannenbaum:*

That's an excellent question but I answered it last year in great detail, so see my articles referenced above; but I'll explain it to you. Last year's data show RdA for identified particles, π, K, η, φ, p out to 10 GeV/c, where RdA=1 for all the particles except for protons where there is a huge enhancement for pT between 1 and 3 GeV/c in d+Au In Au+Au we had previously seen the same thing, what we called the baryon anomaly, an enhancement of protons in this pT range in Au+Au while all identified mesons are suppressed There were many ideas, but nobody really knew what it was. Now that we see an even bigger enhancement (as well as v2) in d+Au, people begin to think that it is due to radial flow. From the meson results, we believe that the centrality in dAu works well out to 10 GeV/c measured particles. The heavier charm particles seem to be enhanced just like the protons and the hyperons, so it seems to fit this pattern. The issue on the centrality is that beyond 10 GeV/c particles, the central collision results are reasonable but the peripheral results look ridiculous, greatly enhanced. Previously, we always assumed that centrality measured in forward calorimeters was independent of what was detected at mid-rapidity. But now I believe that if you trigger on 20,30,40 GeV/c jets or particles, where by momentum conservation there is another roughly equal pT opposite in azimuth., so e.g. 80 GeV/c has been removed from the energy that could have gone in the forward calorimeter. We feel that this modifies the relation between centrality and energy deposited in the forward calorimeter that we calibrate in minimum bias collisions; but we think that the centrality in dAu is valid up to 10 GeV/c particles, so we also believe that the enhancement that you noted for charm particles in central d+Au collisions is correct.

«ETTORE MAJORANA» FOUNDATION AND CENTRE FOR SCIENTIFIC CULTURE
INTERNATIONAL SCHOOL OF SUBNUCLEAR PHYSICS
STATUS OF THEORETICAL UNDERSTANDING AND OF EXPERIMENTAL POWER FOR LHC PHYSICS AND BEYOND
50TH ANNIVERSARY CELEBRATION OF THE QUARK WITH MURRAY GELL-MANN AND GEORGE ZWEIG
ERICE: 24 JUNE – 3 JULY 2014

INTERNATIONAL SCHOOL OF SUBNUCLEAR PHYSICS
Talk by Horst WENNINGER
The QGCW Project
Quark-**G**luon-**C**olored-**W**orld

Some 5 years ago, when discussing RHIC heavy ion experiments during the ERICE ISSP, the idea was proposed by Prof. A. Zichichi to study with different particle probes the "Quark–Gluon–Colored–World" (QGCW), which is totally different from our world made of QCD colorless baryons and mesons.

How to study the New World

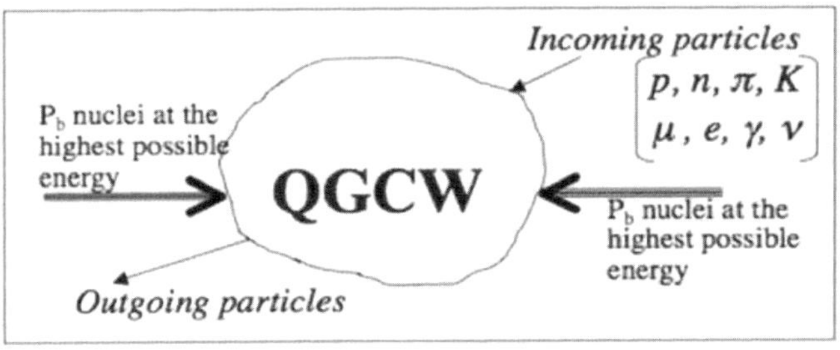

The idea was further outlined in his contribution paper presented during the 80th birthday celebration of Marry Gell-Mann in Singapore during spring 2010 and presented by me in a talk at the ISSP 2010

CONFERENCE IN HONOUR OF
MURRAY GELL-MANN'S 80TH BIRTHDAY CELEBRATIONS

MURRAY GELL-MANN
AND THE LAST FRONTIER
OF LHC PHYSICS:
THE QGCW PROJECT

A. Zichichi
INFN and University of Bologna, Italy
CERN, Geneva, Switzerland
Enrico Fermi Centre, Roma, Italy
Ettore Majorana Foundation and Centre for Scientific Culture, Erice, Italy

24–26 February 2010
Nanyang Executive Centre, Nanyang Technological University, Singapore

INTERNATIONAL SCHOOL OF SUBNUCLEAR PHYSICS
WHAT IS KNOWN AND UNEXPECTED AT LHC
48th Course, ERICE-SICILY: Directors: G. 'T HOOFT - A. ZICHICHI
29 AUGUST - 7 SEPTEMBER 2010

THE QGCW PROJECT:
Technological Challenges to Study the New World
« status report »

Horst Wenninger

At that occasion I used a slide from **Paul Sorensen** (RHIC) to illustrate different bases and the time scales of heavy ion collisions of order ***10^{-24} seconds***.

We then concluded that one might start using probes produced together with the system they probe: i.e.: QCD matter probes such as *jets* and *heavy quarkonia* but in parallel we start a development work at 2 levels on detectors and on beam timing / synchronisation to prepare the future.

Since 2010 the results from RHIC and LHC heavy ion experiments have shown many interesting properties of the QGCW. The important contribution to the results obtained with instruments such as the LAA - MRPC (base of Bologna TOF) – which allowed to measure precisely the properties of sub-nuclear particles coming out from the QGCW in ALICE.

While the time-resolution achievable by particle detectors is continuously being improved new development in MRPC technology are looking very promising: now 15 ps already achieved, more work has to go into the topic of timing and synchronisation.

Advanced R&D on the topic of beam synchronisation has started also for FAIR (Facility for Antiproton and Ion Research) at GSI. The "Bunch-phase Timing System (BuTiS)" for FAIR concentrates on the research of thermal stability properties of optical fibres. The work is done by the communications institute of TU Darmstadt under contract with GSI. The term "bunch phase" shall emphasize the kind of timing precision that is dealt with in the system.

The FAIR - Facility for Antiproton and Ion Research overview of R&D for the new FAIR detectors and accelerator systems provide also interesting opportunities for young talents of this ISSP School, as it is an international project, co-financed by its member-states. At present a merger of the FAIR project into the GSI laboratory is prepared.

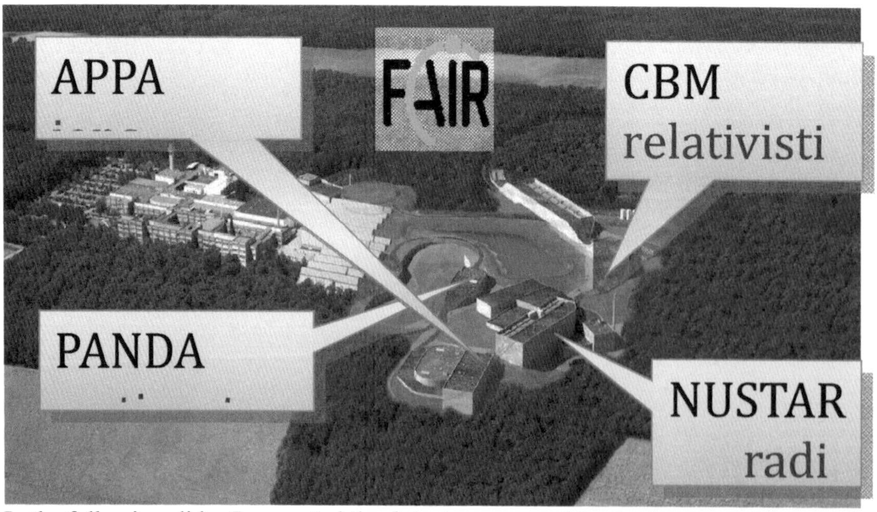

In the following slides I presented the detectors prepared by the 4 communities which will share the beams of the new FAIR laboratory.

I recall the statement by Professor A. Zichichi in his OPENING LECTURE:

We should be prepared with powerful experimental instruments,
technologically at the frontier of our knowledge, to discover
Totally Unexpected Events

«ETTORE MAJORANA» FOUNDATION AND CENTRE FOR SCIENTIFIC CULTURE
TO PAY A PERMANENT TRIBUTE TO GALILEO GALILEI, FOUNDER OF MODERN SCIENCE AND TO ENRICO FERMI, THE "ITALIAN NAVIGATOR", FATHER OF THE WEAK FORCES

INTERNATIONAL SCHOOL OF SUBNUCLEAR PHYSICS
52nd Course: *STATUS OF THEORETICAL UNDERSTANDING AND OF EXPERIMENTAL POWER FOR LHC PHYSICS AND BEYOND*
Erice, 24 June – 3 July 2014

COMPLEXITY AND NEW PHYSICS

ANTONINO ZICHICHI
Pontifical Academy of Sciences, Vatican City
INFN and University of Bologna, Italy
CERN, Geneva, Switzerland
Enrico Fermi Centre, Rome, Italy
World Federation of Scientists, Beijing, Geneva, Moscow, New York

Paul A.M. Dirac Lecture Hall – Patrick M.S. Blackett Institute
Erice, 1 July 2014

COMPLEXITY AND NEW PHYSICS

ANTONINO ZICHICHI

ABSTRACT

Complexity is ill-defined: more than 60 definitions of Complexity exist. The simplest one is not followed by the Logic of Nature: two examples are GUT and Supersymmetry. The fundamental property of Complexity is the number of experimental evidences for its existence: these are only two, **AFB** & **UEEC**. Both of them exist at the two extreme levels of Complexity: the highest one (History) and the lowest one (Science). Complexity exist, no matter the energy and other properties into play. It is therefore crucial to know what Complexity can tell us on the existence of New Physics.

COMPLEXITY AND NEW PHYSICS

How it happens that Complexity brings us to think about New Physics. Let us first review what is Complexity.

Then we will see what Complexity can tell us about **New Physics**.

There are many definitions of Complexity $\gtrsim 60$.

The one I like most is the following. The author is Gerardus 't Hooft.

DEFINITION NUMBER 2 (Gerardus 't Hooft)

Bad ones:

1) Chaos.
2) The need for lengthy calculations.
3) The need for many distinct variables.

Better ones:

4) Unexpected difficulty when attempting to describe something in a precisely formulated theory.
5) What is left over after all systematic approaches failed.

But it could also be that: Complexity is an excuse for sloppy thinking.

The simplest one: **Complexity is the opposite of Simplicity**.

1st Question: does the Logic of Nature prefer Simplicity?

Here is the answer coming from Supersymmetry. How many times Nature needs to violate Simplicity in order to have Supersymmetry in the real world? At least five times (Figure 1).

THE PLATONIC SUPERSYMMETRY

The **Gauge Principle** should generate a
Gauge Force ⟶ **Gauge Bosons**
If *NATURE* was platonically *SUPERSYMMETRIC*
Supersymmetry Transformation should generate **Gauginos**

1st **DEVIATION FROM PLATONIC SIMPLICITY**
OUR FERMIONS ARE NOT THE GAUGINOS

2nd **DEVIATION FROM PLATONIC SIMPLICITY**
THE FUNDAMENTAL FERMIONS ARE OF TWO DIFFERENT CLASSES: LEPTONS AND QUARKS

3rd **DEVIATION FROM PLATONIC SIMPLICITY**
THERE IS NOT ONLY ONE BUT THREE FAMILIES OF FUNDAMENTAL FERMIONS

4th **DEVIATION FROM PLATONIC SIMPLICITY**
THE FUNDAMENTAL FERMIONS BECOME MIXED WHEN THE WEAK FORCES ARE SWITCHED ON: MIXINGS EXIST

5th **DEVIATION FROM PLATONIC SIMPLICITY**
THERE ARE DIFFERENT MIXINGS

Figure 1

The crucial point – not to be forgotten – is that the **number** of **experimental evidences** for the **existence of Complexity is two** (see later).

What about the Deviations from Simplicity needed in order to have GUT ≡ the Unification of the Gauge couplings $\alpha_1, \alpha_2, \alpha_3$.

Here is the answer. The simplest way to get GUT (the point where all fundamental forces are together: Grand Unification Theory) would be the straight line. But the real world does not follow this "platonic" straight line (blue points in Figure 2).

PLATONIC VERSUS REAL GUT

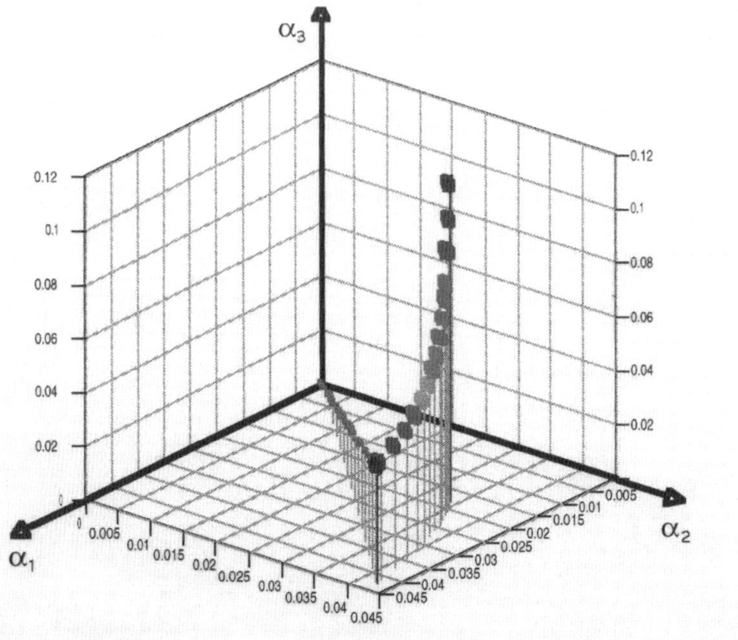

Figure 2

The "big" red points **represents the real GUT**. They have a sequence of 100 GeV in energy. The last point where the "ideal" platonic straight line intercepts the theoretical prediction is at the energy of the Grand Unification. This corresponds to $E_{GU} = 10^{16.2}$ GeV. Other detailed information on the theoretical inputs: the number of fermionic families, N_F, is 3; the number of Higgs particles, N_H, is 2. The input values of the gauge couplings at the Z^0-mass is α_3 $(M_Z) = 0.118 \pm 0.008$; the other input is the ratio of weak and electromagnetic couplings also measured at the Z^0-mass value: $\sin^2 \theta_W (M_Z) = 0.2334 \pm 0.0008$.

The Platonic GUT is the straight line of the "dotted" blue points.

The sequence of points (the big red points), in steps of 100 GeV, is very different from the Platonic line (dotted blue points).

The way nature goes is reported by the sequence of the big red points which are the result of the mathematics reported in Figure 3.

THE UNIFICATION OF ALL FUNDAMENTAL FORCES AND THE EGM EFFECT

A more extensive way to illustrate GUT is reported in Figure 4, where the three lines (α_1^{-1}, α_2^{-1}, α_3^{-1}) are the result from calculations executed with a supercomputer using the following system of equations:

$$\mu \frac{d\alpha_i}{d\mu} = \frac{b_i}{2\pi} \alpha_i^2 + \sum_j \frac{b_{ij}}{8\pi^2} \alpha_i \alpha_j$$

α_i, α_j *(with $i = 1, 2, 3$; and $J = 1, 2, 3$ but $i \neq j$).*

This is a system of three coupled non-linear differential equations where **the existence of the Superworld is included**. This system describes how the gauge couplings (α_1, α_2, α_3) vary with "μ", the basic parameter which depends on the energy of the elementary process, from the maximum level of Energy (Planck Scale) to the energy level of our world. **During more than ten years (from 1979 to 1991), no one had realized that the energy threshold for the existence of the Superworld was strongly dependent on the "running" of the masses.**

This is now called: **the EGM effect** (from the initials of Evolution of Gaugino Masses). This effect produces a factor 700 in the threshold for the lightest supersymmetry particle. No one knows the value of the threshold Energy for the production of the lightest supersymmetry particle. The EGM effect lowers this value by a factor 700. Suppose that somebody would say that the threshold is at 700 TeV, thanks to the EGM effect this value is going to be 1 TeV.

Figure 3

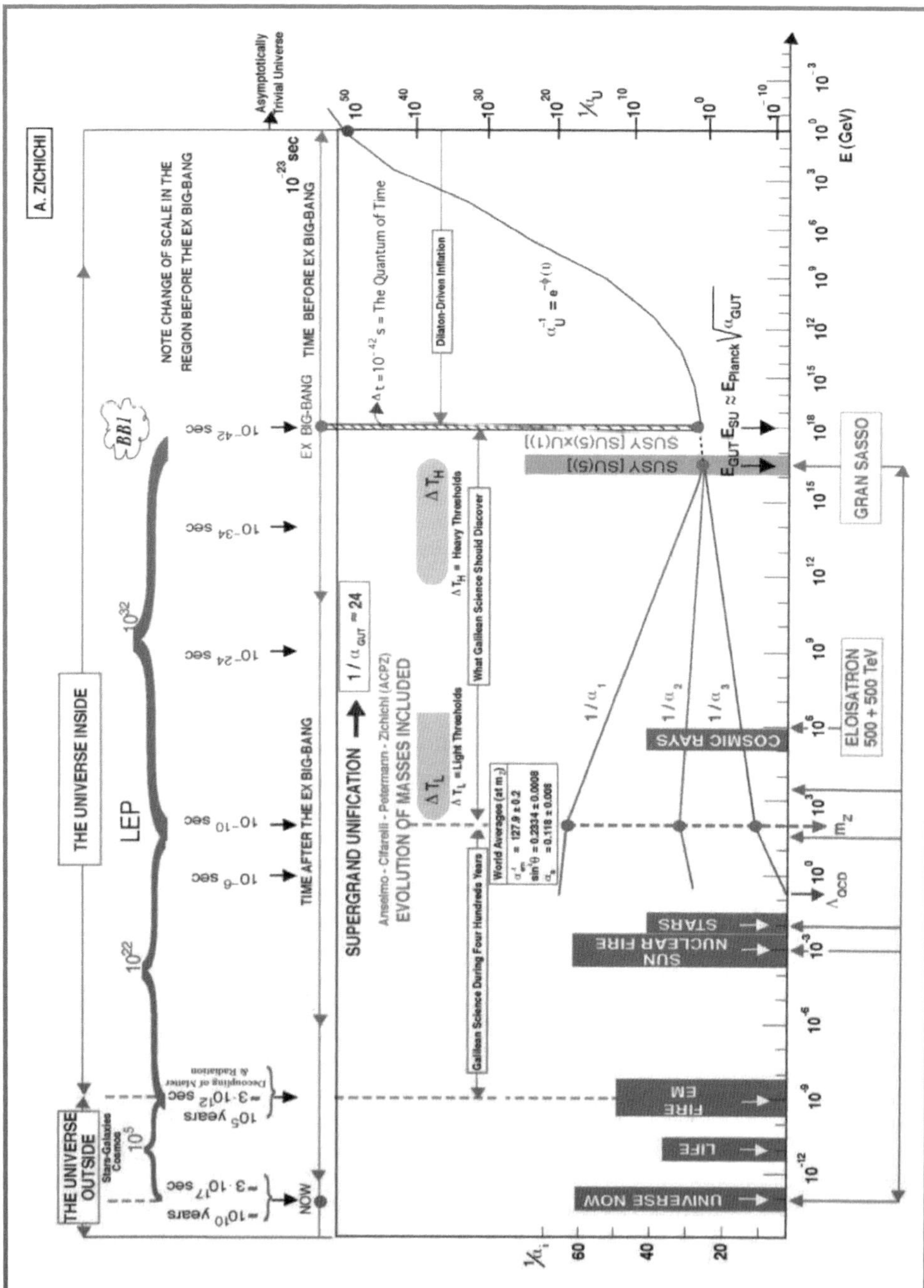

Figure 4: The present status of the unification of the three gauge couplings $\alpha_1\ \alpha_2\ \alpha_3$ is reported. This is the most accurate description of the unification whose mathematics is in Figure 3.

The first time where it was pointed out that Supersymmetry is needed for the convergence of the gauge couplings,

$$\alpha_1^{-1};\ \alpha_2^{-1};\ \alpha_3^{-1}\ ,$$

is at the EPS Conference, Geneva 1979, Proceedings; and Nuovo Cimento Rivista 2 (1979) 1 (A. Zichichi) [1].

We have been engaged to look for the **lightest Supersymmetric particle (LEP)**. Unfortunately no evidence from LHC (so far).

People claim that the novelty is **Holism.** If this was true the **Reductionism should be over**: we should introduce **Holism** in our research work.

How the Logic of Nature goes from the Fundamental Constituents to History is illustrated in Figure 5.

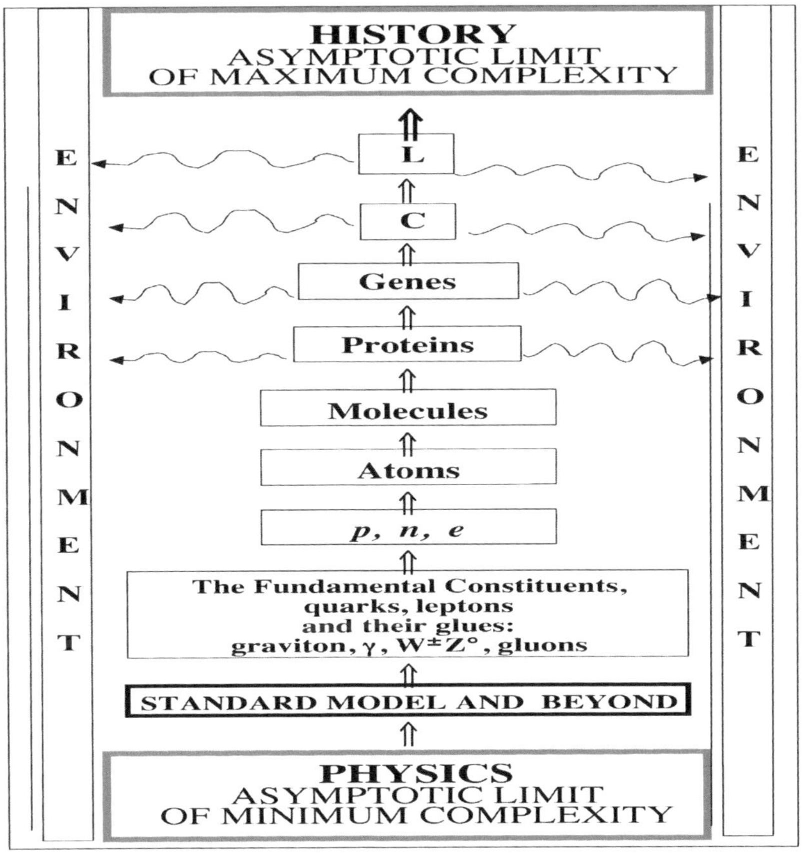

Figure 5

The reason for introducing "Holism" is the interaction with the environment. But the interaction with the environment starts at the level of Proteins. Nevertheless the new guiding line should be, according to many fellows, Complexity.

The Logic of Nature does not follow the theoretical asymptotic lowest degree of Complexity: Simplicity. The proof is given by the non existence of Platonic GUT (Unification of the Fundamental Forces (Figure 2)) and the non existence of the Platonic Supersymmetry (Figure 1).

Basic Question: despite the more than 60 definitions of Complexity how many classes of experimental evidences exist for Complexity? The answer is only **two**.

As reported in Figure 6 there are very many fields where Complexity exist. Despite the enormous number of fields **the only certainty about Complexity** is the existence of the **two experimentally observable effects**: **AFB** and **UEEC**.

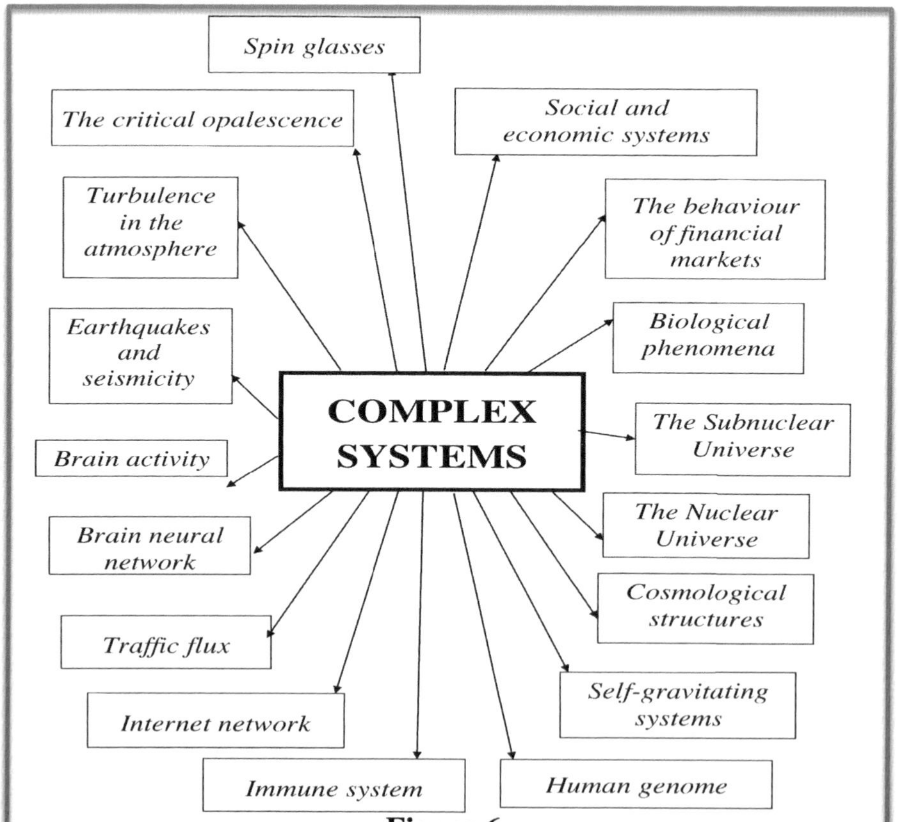

Figure 6

Let us describe the so-called **AFB** phenomenology (Anderson-Feynman-Beethoven).

Anderson because he published a one page long paper in 1972.

Feynman because in 1972 he told to me at Fermi-Lab, during the World Physics Conference, that we should discuss this problem at the Subnuclear Physics School in Erice. Feynman did not know the Anderson paper, but he realized that a problem had to be solved. Exactly the problem discussed by Anderson in his short paper.

Why Beethoven? Because I wanted to let everybody understand what the problem is.

Beethoven and the laws of acoustics.

Beethoven could compose superb masterpieces of music without any knowledge of the laws governing acoustic phenomena. But these masterpieces could not exist if the laws of acoustics were not there.

The Problem of Minimal Life and QED.

To study the mechanisms governing the smallest possible living cell, people working in the field do not need to know the laws of electromagnetic phenomena whose advanced formulation is QED (Quantum ElectroDynamics). All mechanisms needed for life are, to a great extent, examples of electromagnetic processes. If QED was not there, Life could not exist.

Nuclear physics and QCD.

Proton and neutron interactions appear as if a fundamental force of nature is at work: the nuclear force, with its rules and its regularities. These interactions ignore that protons and neutrons are made with quarks and gluons.

Nuclear physics does not appear to care about the existence of Quantum ChromoDynamics (QCD), the fundamental force acting between quarks and gluons at the heart of the subnuclear world.

Nuclear physics ignores QCD but all phenomena occurring in nuclear physics have their roots in the interactions of quarks and gluons.

In other words, protons and neutrons behave like Beethoven: they interact and build up nuclear physics without any clear link with the laws governing QCD.

The world where we are and the Superworld.

The most recent example of AFB type phenomenon comes from the frontier of our scientific knowledge: the Superworld. What we call the **World could *apparently* not care less about the existence of the Superworld**, whose foundation is not in the four dimensional Space–Time but in a superspace with 43 dimensions. And now let us see what are the **UEEC** events.

The UEEC (Unexpected Events which produce Enormous Consequences) has been invented by Historians who realized that History is dominated by UEEC events.

UEEC events and Nuclear Physics – The Yukawa goldmine.

Let me dedicate some attention to discuss UEEC events in nuclear physics. Why Nuclear Physics? Because it is considered standard wisdom the fact that nuclear physics is based on perfectly sound theoretical predictions. People forget the impressive series of UEEC events discovered in what I have decided to call the "Yukawa goldmine" [2] on the occasion of the centennial celebration of the birth of Hideki Yukawa in Japan.

The opening Ceremony had the great honour of the presence of the Emperor and the Empress.

Let me quote just three UEEC in the field of Nuclear Physics.

1 The first experimental evidence for a cosmic ray particle believed to be the Yukawa meson was a lepton: the muon.

2 The decay-chain, $\pi \to \mu \to e$, of the 1st fundamental object of the Nuclear Forces, the π–meson was found to break the symmetry laws of Parity and Charge Conjugation.

3 The intrinsic structure of the Yukawa particle was found to be governed by a new Fundamental Force of Nature, Quantum ChromoDynamics: QCD.

None of these three UEEC events was predicted.

And now History and UEEC.

The most spectacular UEEC in History is **Sarajevo**. As you know on 28 June 1914 in Sarajevo a man has been assassinated. Many Celebrations in Sarajevo have been organized to celebrate the 100th Anniversary of this event from which the 1st World War was generated. Historians call Sarajevo-type events all UEEC events which have occurred in History.

What is History? The highest level, the Asymptotic limit, of Complexity.

Question: does the lowest limit of Complexity exist? The answer is yes. This limit is Science.

Following the definition of Enrico Fermi, the **queen of all Sciences** is **Physics**. And here comes a very interesting point: Do these two extreme limits of Complexity share common properties? The answer is yes. **History** and **Science** share the same experimental evidence: UEEC events and AFB phenomenology. Figure 7 illustrates the presence of UEEC and AFB in History (the highest degree of Complexity) and in Science (the lowest degree of Complexity).

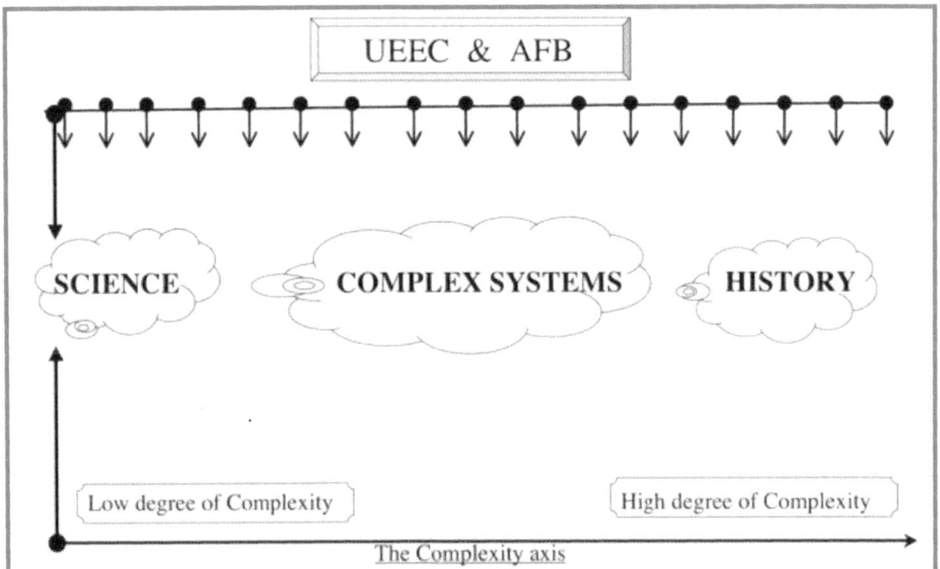

Figure 7

The presence of AFB phenomenology has been reported for cases much simpler than History, like Beethoven and the problem of minimal life.

What happens in History does not appear as being related with the Logic of Nature discovered by Science. But without the Laws discovered by Science, History could not exist.

For Science, Nuclear Physics with the Yukawa goldmine is the best proof for the existence of Complexity.

And now let us discuss Virtual History:

Historians know that in 1932 the first example of "Virtual" effect has been discovered by P.M.S. Blackett. This effect was not the simplest one. The simplest example being the Lamb-shift discovered in 1947.

What is a **Virtual effects**? No instrument – even the most powerful we can imagine – can ever observe "**directly**" a Virtual effect.

Nevertheless this "Virtual effect" does generate rigorously measurable phenomena, the most spectacular one being the anomalous magnetic moment of the electron:

$$\text{the } (g-2)_e.$$

History and **Science** have a property which is common to both of them: **Evolution. In fact**:

- **History** is defined to be EWRL ≡ ***E****volution of the* ***W****orld in its* ***R****eal* ***L****ife*.
- **Science** is defined to be EBUS ≡ ***E****volution of our* ***B****asic* ***U****nderstanding of the laws governing the world in its* ***S****tructure*.

Historians say that if any UEEC **would not have** taken place, our **History** would be different. This different History is what they call **Virtual History**.

With Science we will see that this is not true.

Virtual Physics is exactly the Physics of our world including what will be discovered in the next millennia. But **Virtual History** is totally different from **Real History**.

This basic difference exists despite the fact that History and Science are both dominated by (AFB) phenomenology and **UEEC events**. The UEEC are shown in Figure 8.

UEEC EVENTS IN HISTORY AND IN SCIENCE

	In History ≡ EWRL		In Science ≡ EBUS
I	What if Julius Caesar had been assassinated many years before?	*I*	What if Galileo Galilei had not discovered that $F = mg$?
II	What if Charles VII had not been able to win the 100 years war?	*II*	What if Newton had not discovered that $F = G\dfrac{m_1 \cdot m_2}{R_{12}^2}$?
III	What if America had been discovered a few centuries later?	*III*	What if Maxwell had not discovered the unification of electricity, magnetism and optical phenomena, which allowed him to conclude that light is a vibration of the EM field?
IV	What if Napoleon had not been born?	*IV*	What if Becquerel had not discovered radioactivity?
V	What if Louis XVI had been able to win against the 'Storming of the Bastille'?	*V*	What if Planck had not discovered that $h \neq 0$?
VI	What if the 1908 Tunguska Comet had fallen somewhere in Europe instead of Tunguska in Siberia?	*VI*	What if Lorentz had not discovered that space and time cannot both be real?
VII	What if the killer of the Austrian Archduke Franz Ferdinand had been arrested the day before the Sarajevo event?	*VII*	What if Einstein had not discovered the existence of time-like and space-like real worlds? Only in the time-like world, simultaneity does not change, with changing observer.
VIII	What if Lenin had been killed during his travelling through Germany?	*VIII*	What if Rutherford had not discovered the nucleus?
IX	What if Hitler had not been appointed Chancellor by the President of the Republic of Weimar Paul von Hindenburg?	*IX*	What if Hess had not discovered cosmic rays?
X	What if Pyotr Kapitza accepted to be the leader of the USSR H-bomb Project as wanted by Stalin?	*X*	What if Dirac had not discovered his equation, which opens new horizons, including the existence of the antiworld?
XI	What if Nazi Germany had defeated the Soviet Union?	*XI*	What if Fermi had not discovered weak forces?
XII	What if Karol Wojtyla had not been elected Pope, thus becoming John Paul II?	*XII*	What if Fermi and Dirac had not discovered the Fermi–Dirac statistics?
XIII	What if Gorbachev had not been defeated by Yeltsin?	*XIII*	What if Yukawa had not proposed the existence of a "meson" in order to have the nuclear glue?
XIV	What if the USSR had not collapsed?	*XIV*	What if the 'strange particles' had not been discovered in the Blackett Lab?

Figure 8

Nevertheless in Physics we can make theoretical predictions. In History no.

Theoretical predictions can be formulated always **after** but never **before** the unexpected discovery, UEEC event, has occurred. Recall the 200 years of discoveries in Electricity, Magnetism and Optics: a series of UEEC events.

It is after all these UEEC events that Maxwell was able to find his famous 4 equations. Now we have QED (Quantum ElectroDynamics) whose predictions are there after all UEEC allowed the latest version of QED to be formulated.

Can Fundamental New Physics be theoretically predicted from the knowledge of Virtual Physics? The answer is **no**. Exactly like in History. But in History no one is able to translate in a rigorous mathematical formalism the status of our world in its real life.

Let us be clear. If a **UEEC** event had not occurred the History in our world would be different. In Physics the absence of a UEEC event would not change the real world.

If Planck had not discovered $h \neq 0$ somebody else would have discovered it. No matter what UEEC event we can pick up from Figure 8; sooner or later this **UEEC event** would have been discovered.

Virtual Physics is the most rigorous mathematical formulation of all we have so far experimentally been able to find.

What Complexity tells us is that New Physics goes far beyond what we believe is the frontier of our knowledge.

The most advanced frontier of our knowledge is what we call **SM&B** as illustrated in the following Figure 9, where the five basic steps are examples of UEEC and AFB.

Let us recall few old examples.

No Physicist had the idea of proposing that (Space–Time) had to be a complex quantity. This was discovered by H. Lorentz on the properties of the 4 Maxwell equations.

Dirac started to think about his equation because he wanted to provide the reason why the "electron" discovered by J.J. Thomson in 1897 was not a scalar particle but a particle with spin. Why the spin was there? From trying to find the solution for this problem came the **antielectron**.

SM&B

THE STANDARD MODEL AND BEYOND

① RGEs (α_i ($i \equiv 1, 2, 3$); m_j ($j \equiv q, l, G, H$)) : $f(k^2)$.

- GUT ($\alpha_{GUT} \cong 1/24$) & GAP ($10^{16} - 10^{18}$) GeV.
- SUSY (to stabilize $m_F/m_P \cong 10^{-17}$).
- RQST (to quantize Gravity).

② Gauge Principle (hidden and expanded dimensions).

— How a Fundamental Force is generated: SU(3); SU(2); U(1) and Gravity.

③ The Physics of Imaginary Masses: SSB.

— The Imaginary Mass in SU(2)×U(1) produces masses ($m_{W^\pm}$; m_{Z^0} ; m_q; m_l), including $m_\gamma = 0$.

— The Imaginary Mass in SU(5)⇒SU(3)×SU(2)×U(1) or in any higher (not containing U(1)) Symmetry Group ⇒ SU(3)×SU(2)×U(1) produces Monopoles.

— The Imaginary Mass in $SU(3)_c$ generates Confinement.

④ Flavour Mixings & CP ≠ , T ≠ .

— No need for it but it is there.

⑤ Anomalies & Instantons.

— Basic Features of all Non-Abelian Forces.

Note:

q	≡	quark and squark;	m_F	≡	Fermi mass scale;
l	≡	lepton and slepton;	m_P	≡	Planck mass scale;
G	≡	Gauge boson and Gaugino;	k	≡	quadrimomentum;
H	≡	Higgs and Shiggs;	C	≡	Charge Conjugation;
RGEs	≡	Renormalization Group Equations;	P	≡	Parity;
GUT	≡	Grand Unified Theory;	T	≡	Time Reversal;
SUSY	≡	Supersymmetry;	≠	≡	Breakdown of Symmetry Operators.
RQST	≡	Relativistic Quantum String Theory;			
SSB	≡	Spontaneous Symmetry Breaking.			

The five basic steps in our understanding of nature. ① The renormalization group equations (RGEs) imply that the gauge couplings (α_i) and the masses (m_j) all run with k^2. It is this running which allows GUT, suggests SUSY and produces the need for a non point-like description (RQST) of physics processes, thus opening the way to quantize gravity. ② All forces originate in the same way: the gauge principle. ③ Imaginary masses play a central role in describing nature. ④ The mass-eigenstates are mixed when the Fermi forces come in. ⑤ The Abelian force QED has lost its role of being the guide for all fundamental forces. The non-Abelian gauge forces dominate and have features which are not present in QED.

Figure 9

Why in 1964 (Englert, Brout, Higgs, Guralnik, Hagen, Kibble) proposed the existence of an imaginary mass in the Lagrangian? Because they wanted to explain why all known particles (with $m = 0$ and $m \neq 0$) were never scalar on Lorentz Space–Time and in all other possible intrinsic Space–dimensions you can imagine.

All we can be sure of is that we **must be technologically powerful** in order **to find something** which **no one is at present able to predict**. Recall all UEEC events from Galilei to SM&B.

UEEC EVENTS, FROM GALILEI UP TO SM&B

In Figure 10 there is a sequence of **UEEC** events from Galilei to Fermi-Dirac and the "strange particles". In Figures 11, 12, 13 from Fermi-Dirac to the construction of the Standard Model. The Standard Model and Beyond (**SM&B**) has already been illustrated in Figure 9.

"UEEC"
TOTALLY UNEXPECTED DISCOVERIES
FROM GALILEI TO THE "STRANGE" PARTICLES

I	Galileo Galilei discovery of F = mg.
II	Newton discovery of $F = G\frac{m_1 \cdot m_2}{R_{12}^2}$.
III	Maxwell discovers the unification of electricity, magnetism and optical phenomena, which allows him to conclude that light is a vibration of the EM field.
IV	Planck discovery of $h \neq 0$.
V	Lorentz discovers that Space and Time cannot be both real.
VI	Einstein discovers the existence of Time-like and Space-like worlds. Only in the Time-like world, simultaneously does not change, with changing observer.
VII	Rutherford discovers the nucleus.
VIII	Hess discovers the cosmic rays.
IX	Dirac discovers his equation, which opens new horizons, including the existence of the antiworld.
X	Blackett discovers the "vacuum polarizations" (the first example of Virtual Phenomena).
XI	Fermi discovers the weak forces.
XII	Fermi–Dirac and Bose–Einstein discover two totally different statistical laws.
XIII	The "strange particles" are discovered in the Blackett Lab.
XIV	Heavy mesons with positive and negative strangeness are discovered in the Blackett Lab.

Figure 10

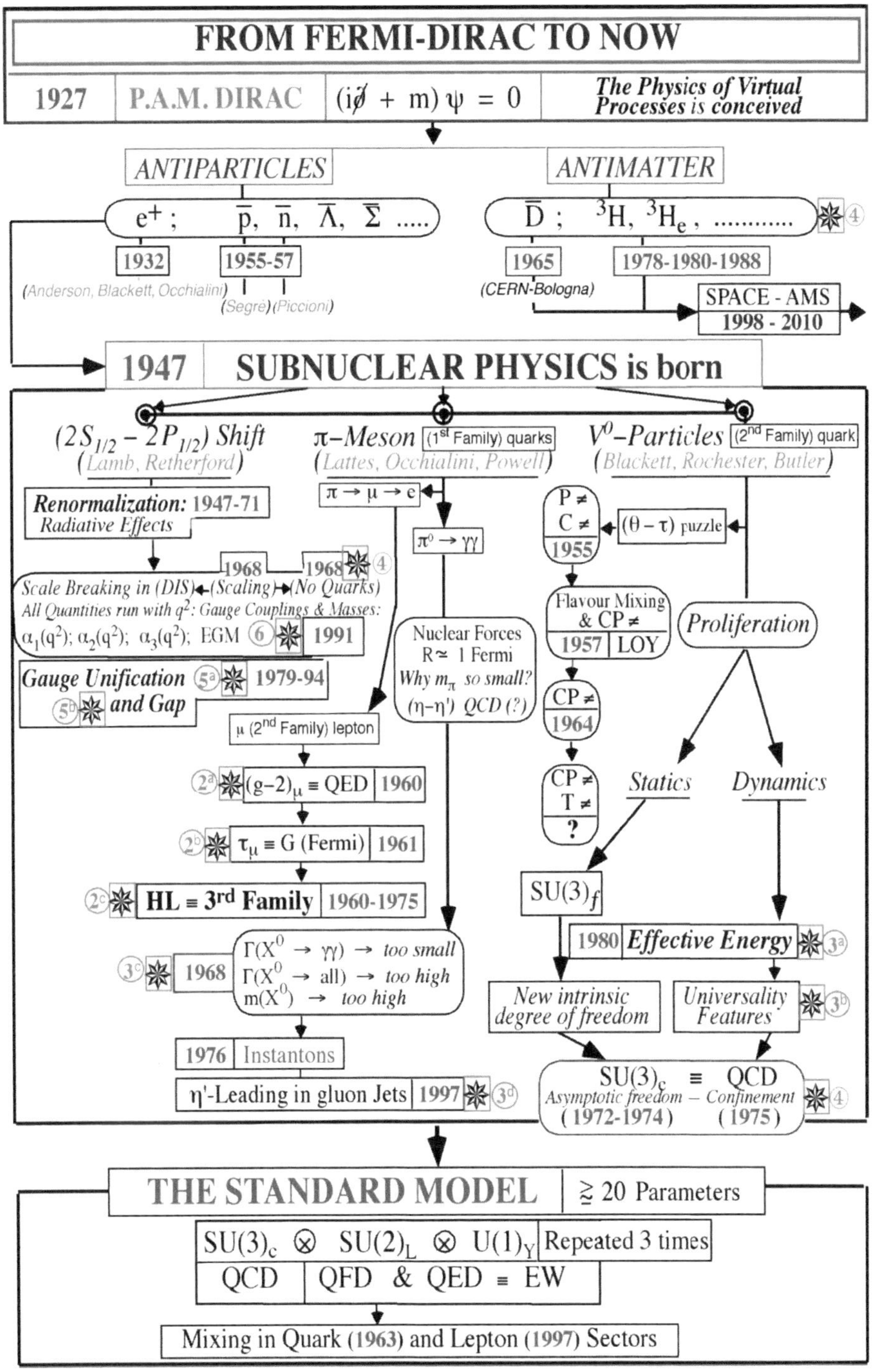
FROM FERMI-DIRAC TO NOW
1927
P.A.M. DIRAC
(i∂̸ + m) ψ = 0
The Physics of Virtual Processes is conceived
ANTIPARTICLES
ANTIMATTER
e+ ; p̄, n̄, Λ̄, Σ̄ .....
D̄ ; 3H, 3He , ...
4
1932
1955-57
1965
1978-1980-1988
(Anderson, Blackett, Occhialini)
(Segrè) (Piccioni)
(CERN-Bologna)
SPACE - AMS
1998 - 2010
1947
SUBNUCLEAR PHYSICS is born
(2S1/2 – 2P1/2) Shift
(Lamb, Retherford)
π–Meson
(1st Family) quarks
(Lattes, Occhialini, Powell)
V0–Particles
(2nd Family) quark
(Blackett, Rochester, Butler)
Renormalization: 1947-71
Radiative Effects
π → μ → e
π0 → γγ
P ≠
C ≠
1955
(θ – τ) puzzle
1968
1968
4
Scale Breaking in (DIS) (Scaling) (No Quarks)
All Quantities run with q2: Gauge Couplings & Masses:
α1(q2); α2(q2); α3(q2); EGM
6
1991
Flavour Mixing
& CP ≠
1957
LOY
Proliferation
Nuclear Forces
R ≃ 1 Fermi
Why mπ so small?
(η–η') QCD (?)
Gauge Unification
5a
1979-94
5b
and Gap
CP ≠
1964
μ (2nd Family) lepton
2a
(g–2)μ ≡ QED
1960
CP ≠
T ≠
?
Statics
Dynamics
2b
τμ ≡ G (Fermi)
1961
2c
HL ≡ 3rd Family
1960-1975
SU(3)f
1980
Effective Energy
3a
3c
1968
Γ(X0 → γγ) → too small
Γ(X0 → all) → too high
m(X0) → too high
New intrinsic degree of freedom
Universality Features
3b
1976
Instantons
SU(3)c ≡ QCD
Asymptotic freedom – Confinement
(1972-1974) (1975)
4
η'-Leading in gluon Jets
1997
3d
THE STANDARD MODEL
≳ 20 Parameters
SU(3)c ⊗ SU(2)L ⊗ U(1)Y
Repeated 3 times
QCD
QFD & QED ≡ EW
Mixing in Quark (1963) and Lepton (1997) Sectors

Figure 11

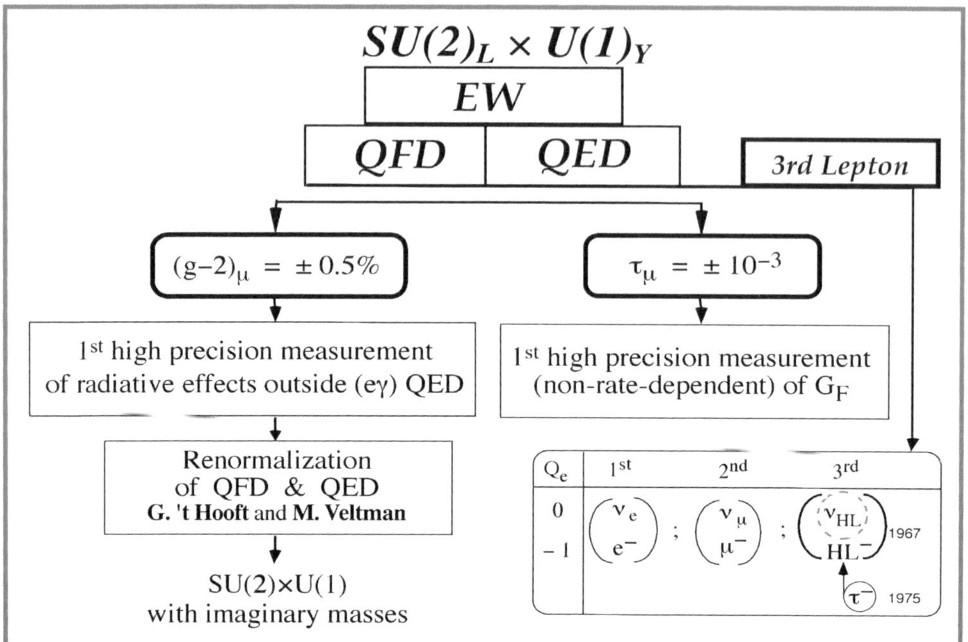

Figure 12: Details from Figure 11, concerning $SU(2)_L$ and $U(1)_Y$.

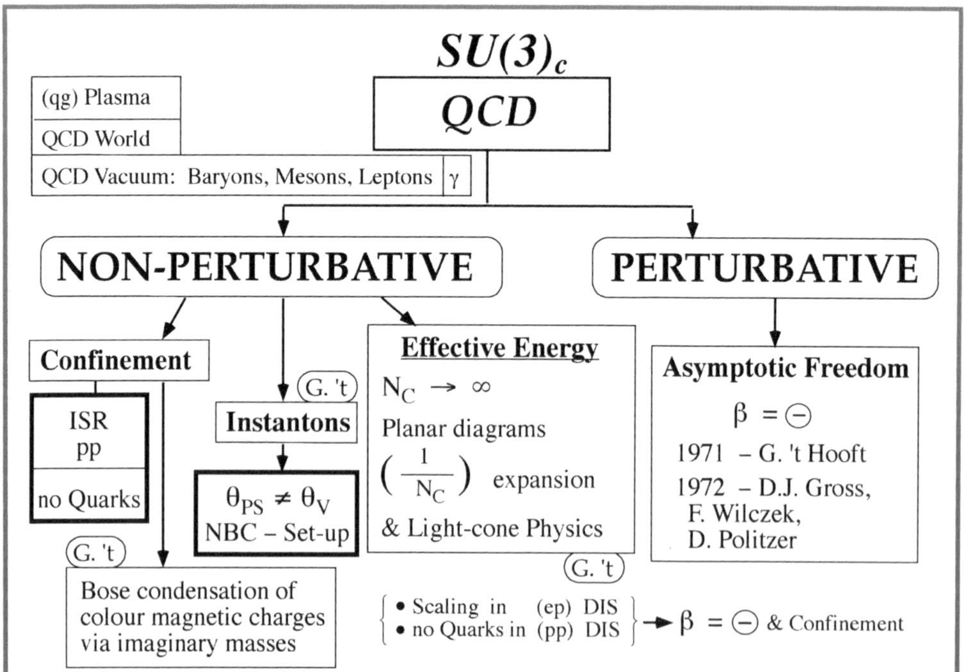

Figure 13: Details from Figure 11, concerning $SU(3)_C$.

CONCLUSION

*** *** *** ***

Reductionisms is the only way we have to investigate the **Logic of Nature.**

*** *** *** ***

Complexity in the real world exists, no matter the mass-energy and Space–Time scales considered [3].

Therefore the only possible prediction is that:

- **Totally Unexpected Effects** should **show up**.
- **Effects**, which are impossible to be predicted on the basis of **present knowledge**.

We should be prepared with **powerful experimental instruments**, **technologically at the frontier of our knowledge**, to discover Totally Unexpected Events in all laboratories, the world over (including CERN in Europe, Gran Sasso in Italy, and other facilities in Japan, USA, China and Russia). All the pieces of our Physics could not have been discovered if the experimental technology was not at the frontier of our knowledge.

Recall the technology from Blackett (**1932**) (e^+e^-) in Cosmic rays using Wilson Chambers up to MRPC (**2014**) with 20ps ≡ ΔT ≡ World Record [4].

REFERENCES

[1] A. Zichichi, EPS Conference, Geneva 1979, Proceedings; and Nuovo Cimento Rivista 2, 1 (1979).

[2] *From the Yukawa Particle to the QGCW*
A. Zichichi, in Proceedings of the *Symposium for the Centennial Celebration of Hideki Yukawa*, International Nuclear Physics Conference, Tokyo, Japan, June 3-8 (2007).

[3] *The Logic of Nature and Complexity*
A. Zichichi, presented at
International Conference on "Quantum [un]speakables" in Commemoration of John S. Bell, International Erwin Schrödinger Institut (ESI), Universität Wien (Austria), 10-14 November 2000
27th Sessions of the International Seminars on Planetary Emergencies, Erice (Italy), 24 August 2002

40th Course of the International School of Subnuclear Physics, Erice (Italy), 6 September 2002

Pontificia Academia Scientiarum, The Vatican, Rome (Italy), 8-11 November 2002

The joint Session: 6th Course of the International School of Biological Magnetic Resonance; 2nd Workshop on Science and Religion of the Advanced School of History of Physics; 10th Workshop of the International School of Liquid Crystals; Erice (Italy), July 2003

28th Sessions of the International Seminars on Planetary Emergencies, Erice (Italy), August 2003

31st, 32nd and 33th Course of the International School of Solid State Physics, Erice (Italy), July 2004

42nd International School of Subnuclear Physics, Erice (Italy), August - September 2004

International School on Complexity, 1st Workshop on Minimal Life, Erice (Italy), December 2004

INFN-Alice Meeting, University of Catania (Italy), January 2005

Trinity College, Dublin (Ireland), February 2005

Department of Physics, University of Padova (Italy), March 2005

43th Course of the International School of Subnuclear Physics, Erice (Italy), 6 September 2005

Italian Physics Society (SIF) XCI Annual National Congress, University of Catania (Italy), September 2005

Desy, Hamburg, 22 November 2005

INFN Eloisatron Project "The 1st Physics ALICE Week", Erice (Italy), 5 December 2005

50th Anniversary of INFN Bologna - ALICE Week, Bologna (Italy), 21 June 2006

44th International School of Subnuclear Physics, Erice (Italy), August-September 2006

45th International School of Subnuclear Physics, Erice (Italy), August-September 2007

47th International School of Subnuclear Physics, Erice (Italy), August-September 2009

50th International School of Subnuclear Physics, Erice (Italy), June-July 2012

51st International School of Subnuclear Physics, Erice (Italy), June-July 2013.

[4] *The problem of (CPT) Invariance in Experimental Physics and the Time of Flight (TOF) World Record*
A. Zichichi, in Proceedings of the 2013-Erice Subnuclear Physics School *Reflections on the Next Step for LHC*, World Scientific (2015).

DISCUSSIONS

LECTURER: I. ANTONIADIS

The Mass Hierarchy and Physics Beyond the Standard Theory

Scientific Secretaries: A.Katanaeva, A.Salakh

DISCUSSION

- *Savitskyi:*

You have mentioned that there is a project of a 100 km collider (VHE-LHC) with 16T magnets, why is such a project considered when there is a much cheaper project by Fermilab of a 200 km collider with 4T magnets? What is the estimated cost of the VHE-LHC project?

- *Antoniadis:*

The VHE-LHC collider is one of the possible projects of CERN, there are study groups which discuss various options and work on choosing the optimal one at the end of the second phase of LHC. The exact cost is not known at the moment, but the amount is constrained by the annual budget of CERN which is around one billion Swiss francs per year.

- *Ferrara:*

How much do the future plans depend on new discoveries on the second stage of LHC? Supposed, for instance, new particles and supersymmetry will be discovered at the second stage of LHC. What would be the best option for a future machine then? I think these plans should depend on the outcome of the second stage of LHC.

- *Antoniadis:*

Yes, obviously. And also they depend on what will happen with the ILC project. If, for instance, there is discovery of supersymmetry or something else, then there would be a lot of discussions on what is the best way to continue after LHC. I remind you that there is at CERN another proposal, the CLIC, which is a linear collider, but with different technologies of acceleration. It can go up in energy much higher than ILC, up to a few TeV. And then, in the context of FCC, that is Future Circular Colliders, it could be a lepton circular collider before the hadron collider. As you see, there are several options, which would be put on the table and be discussed in connection with the physics status at that moment.

- *'t Hooft:*

I have just a remark in this connection. There could be competition from non-western world, I mean Chinese physicists, who would like to see the new gigantic accelerator being built near Beijing. I don't know how they are going to push through this programme. But it would be interesting, I think, if there would be more competition in this field.

- *Antoniadis:*

Certainly. Actually FCC could also be Future Chinese Collider.

- *Inguglia:*

I think that when deciding on future investments in collider physics we shouldn't underestimate the importance of lower energy facilities.

- *Antoniadis:*

Yes, certainly, I agree. Actually at the LHC there are 4 experiments, and two of them are of the lower energies physics.

- *Tannenbaum:*

You said that CERN was supposed to do projects. What would happen if, by some miracle, when we run the LHC in the next two years, a totally new particle (for instance, one particular supersymmetric particle) is discovered, which solves all the outstanding questions of the Standard Model and there is nothing left to search? What do you think would be the next project then?

- *Antoniadis:*

I think it is too optimistic. We all know very well that science, especially physics, has no end. As we go on we discover new phenomena and every discovery brings new questions.

- *'t Hooft:*

Just one supersymmetric particle would be somewhat strange. What you really want is dozens of them.

LECTURER: Z. BERN

Recent Developments in Scattering Amplitudes

Scientific Secretaries: O. Fischer, M. Zoccali

DISCUSSION

- *Ponomarev:*

As I understand, in gravitational amplitudes, you replace the infinite series of vertices by only one cubic vertex. How do you do that and how do you control that they produce the same result?

- *Bern:*

When you look at the ordinary formalism, the role of the higher-order terms is to preserve the gauge invariance; in a sense the higher-order terms don't really carry new information. This is something Richard Feynman understood: if you start with the three-vertex of gravity you can figure out all the infinite-order contact terms that you need to add, in order to get current conservation and Ward identities to work. Even from Feynman's viewpoint, we understand that there really is no new information in the higher point vertices, but everything is derivable from the 3-vertex and basic consistency principles. The on-shell formalisms goes a step further: not only do the higher-order vertices not carry any new information, but we never have to actually look at the higher-point vertices. In on-shell recursion, the main reason why you don't have to worry about the infinite number of vertices has to do with the excellent large-z behavior of the gravity amplitudes under the BCFW shifts that allow you to apply Cauchy's theorem to derive the amplitudes starting only from the three vertices. It may look a bit like magic from the traditional Feynman viewpoint, however in the on-shell construction there is no need for any of those higher-order contact terms.

- *Ponomarev:*

What is the kinematic counterpart of the Jacobi identity?

- *Bern:*

I assume you are asking what the underlying infinite-dimensional Lie algebra behind the kinematic algebra might be. That is not well understood, except for one particular case: there is a paper by Monteiro and O'Connell (see arXiv:1203.0944) for the case of self-dual field theory. Looking at the self-dual case, they find exactly what the infinite dimensional algebra is in Yang-Mills theory: it is a hidden area-preserving diffeomorphism invariance, so in this case it is very explicit. The unfortunate part is nontrivial go to field configurations that are not self-dual. It is much more obscure, although people have made some progress on this.

- *Ponomarev:*

Also, why is BMS symmetry a symmetry of gravitational amplitudes? It is an approximate symmetry, am I right?

- *Bern:*

BMS symmetry can be understood as a piece of diffeomorphism invariance. I think it is fair to call it an exact symmetry. At the leading level it has no loop corrections, but at subleading level it does related to infrared singularities.

- *Ponomarev:*

What is the counterpart for YM?

- *Bern:*

I don't know. Your question is related to the fact that that there ought to be other ways of deriving soft theorems by different techniques, rather than from the BMS symmetry. In fact, we have just finished a paper, a few days ago, showing how one can use gauge invariance to derive subleading soft theorems, which perhaps is not too much of a surprise in relation to your question.

- *Cerchiai:*

What happens to the gauge group indices when you use the color kinematic duality?

- *Bern:*

The color factors of course depend on the gauge group; what we do is to go from the gauge theory to the gravity theory by replacing the color factor with a kinematic factor. The only property of the gauge group we make use of is the fact that it satisfies the Jacobi identity, or equivalently the fact that there is an underlying Lie algebra. The gauge group does not matter at all because it is only the algebraic properties that are important.

- *Cerchiai:*

There have been some approaches to the computation of Feynman diagrams using techniques from cohomology or algebraic geometry: as an example, Marcolli (cf. Feynman motives, World Scientific, 2010) and Lee, Pomeranski (cf. arXiv:1308.6676) wrote how to compute the number of master integrals by using Morse theory, classifying them as critical points of polynomials. Can you comment on this?

- *Bern:*

There certainly are many interesting approaches to loop integrals. However generally there is very little out there that can really help us at the high loop orders that we are most interested in. Those techniques, can be quite helpful at say two loops, it is unclear how they can really help, at least today, at higher loops.

- *Cerchiai:*

One last question: can you comment on the relation with the squaring of YM theories that M. Duff is doing?

- *Bern:*

What we do is for very simple supergravity theories. In particular, N=8 supergravity is N=4 times N=4 super-Yang-Mills theory, or N=4 supergravity is N=4 as N=4 times N=0 super-Yang-Mills theory. Duff and collaborators systematically go through more sophisticated supergravity theories. It is the same kind of idea, except that what they do is much more sophisticated in terms of the supergravity models that they look at. Another difference is that we are focused on the scattering amplitudes while they are focused on the categorization of the theories.

- *Cerchiai:*

Do you need supersymmetry to do these “squarings”, or could you think of doing them even without SUSY?

- *Bern:*

One can definitely do the same constructions without SUSY. We have a paper with my student J. Nohle and my postdoc S. Davies; we squared *pure* YM; by doing the state count the theory includes a graviton, an antisymmetric tensor and a dilaton. At least the 1-loop 4-point case seems to work, and it certainly works at tree level for an arbitrary number of gravitons. SUSY is not necessary for this, at least for the examples that we have been able to work out.

- *Hughes:*

Do you know why the color kinematic duality works? It is a bit different from saying “gravity is YM squared”?

- *Bern:*

The answer is very likely that there is a hidden symmetry in YM theory that we haven’t fully understood yet. The paper that shows this most clearly is that by Monteiro and O’Connell, where they tell you exactly what the hidden symmetry is, at least for the special case of self-dual field configurations. However they have a paper that goes beyond that, despite not being quite explicit enough; perhaps within a year or two there will be a better explanation. I think it has to be a hidden symmetry, which will not be visible in the conventional YM Lagrangian, but in a much more complicated version of the YM Lagrangian that differs from the standard one by a nontrivial field redefinition. The duality between color and kinematics seems to be more fundamental than the statement that “gravity is YM squared”. Even string theory, which is helpful for understanding “gravity is YM squared”, has so far not offered any clarity on the underlying reason behind the duality between color and kinematics.

- *Prochazka:*

How does the color-kinematics duality manifest on the Lagrangian level?

- *Bern:*

It manifests in a fairly obscure way. You can start with the ordinary YM Lagrangian and add a zero to it in a very special way, namely adding terms that vanish under the color Jacobi identity. Say at five points you have to add a special five-point interaction that happens to be zero, but looks to be non-zero if you do not use the color Jacobi identity. If you keep adding such terms in a particular way order by order, you can make the duality between color and kinematics manifest in amplitudes generated by Feynman diagrams of this Lagrangian. Unfortunately, this Lagrangian becomes very complicated. From this modified Yang-Mills Lagrangian one can obtain a gravity Lagrangian that manifests the double-copy structure. This gravity Lagrangian generates the same scattering amplitudes as the Einstein-Hilbert Lagrangian, but is nontrivially related to it under some set of obscure field redefinitions. Due to the complexity of the new Lagrangian, we have not understood this systematically to all orders yet but only through six points.

- *Minkowski:*

Assume that the method you developed allows you to prove that all non-trivial counter terms (or divergences) to be absent from the theory. What can one conclude from this?

- *Bern:*

The key thing that we would study is the underlying reason for this finiteness. Possibly we would then be able can transfer the finiteness property to much more realistic theories than *N*=8 SUGRA. The reason behind order by order finiteness is what we are after, not the finiteness itself. The underlying reason for the finiteness surely would be a crucial for understanding quantum gravity. Even though perturbation theory itself is of course not the entire story of quantum gravity, an order-by-order ultraviolet cancellation would surely have underlying reason of central importance to our understanding of quantum gravity.

LECTURER: M.J. DUFF

Gravity as the Square of a Gauge Theory

Scientific Secretaries: A. Leithes, M. Zoccali

DISCUSSION

- *Cerchiai:*

When you compute the magic pyramid with Yang Mills theories, the space-time symmetry is given by the Lorentz group, SL(2,A), which corresponds to the second row of the magic square appearing for the rank-2 Jordan algebra (see paper by Barton, Sudbery). In addition, when you construct the magic pyramid for the conformal case, the space-time symmetry corresponds to the third row of the rank-2 Jordan algebra. Did you consider the other rows of that magic square of the rank-2 Jordan algebra, although I don't know to what physical theories that would correspond?

- *Duff:*

No, I don't think we have.

- *Prochazka:*

So the theories that you are squaring had to be conformal and massless, in order to get a conformal theory from the squaring procedure?

- *Duff:*

We did two different things. Firstly, we squared Yang-Mills, which in general is not conformal (except in four dimensions), then we squared the conformal theories from which we get a different picture. Now, I should say these six dimensional theories are rather mysterious because they don't have a graviton; they are perfectly respectable supermultiplets and they have a graviton when you dimensionally reduce them but they don't have a graviton themselves. In other words, the representation that appears in six dimensions is bigger than the graviton, it has a graviton plus more, however when you dimensionally reduce you get the familiar graviton plus matter fields. So these are theories that my colleague, Chris Hull at Imperial College, has been looking at, like four comma zero, which has this E(6,6), and they are rather mysterious – we don't have Lagrangians for them – and this one if it exists would be even more mysterious. But I interrupted you; what was your follow up?

- *Prochazka:*

But the outcome that you get, the gravity part, is conformal, right?

- *Duff:*

No. Squaring the usual Yang-Mills gives you the usual supergravity, for example squaring N=4 Yang-Mills gives you N=8 supergravity, which is not conformal. These supergravities have the usual Einstein-Hilbert Lagrangian for gravity plus couplings to matter, so they are not conformally invariant.

- *Ponomarev:*

With your squaring procedure, you can reproduce on-shell and off-shell spectra of supergravities from SYM. Am I right in thinking that you cannot promote this to the level of the action, so you cannot construct the supergravity action if you didn't know it in advance?

- *Duff:*

Very good question. We would really love to be able to reconstruct the gravitational Lagrangian starting from the YM Lagrangian. We have not yet been able to do that; that is ongoing research. What I have shown you today is the group theory that underlies the construction, however we haven't shown why you get the dynamics of gravity from the dynamics of YM. That is a more difficult question, but it is a very important one of course.

LECTURER: F. FERRONI

The INFN vision for Neutrino Physics

Scientific Secretaries: A. Esposito, S. Ahmad

DISCUSSION

- *Prochazka:*

Is there any possibility of neutrino oscillation with massless neutrino? Because theoretically if we work with PT Symmetric Hamiltonian means, there are papers which show that you could possibly have neutrino oscillation phenomena with massless neutrino. Is there any experimental evidence that could rule out such models?

- *Ferroni:*

In neutrino oscillation the probability a neutrino changing its flavor depends on the difference between the square of the masses of the neutrino mass eigenstates. For example, the mass squared difference between the mass states ν_1 and ν_2 is denoted by Δm_{12}^2. Experimentally, mass squared difference have been measured and it is nonzero i.e., $\Delta m^2 \neq 0$ ($\Delta m_{12}^2 \neq 0$, $\Delta m_{13}^2 \neq 0$, $\Delta m_{23}^2 \neq 0$). So we know that definitely ν_1 and ν_2 are differently massive, and also ν_1 and ν_2 are differently massive relative to ν_3. Experimentally, it is solved that the neutrinos are massive but theoretically.if we put massless neutrinos then the oscillation frequency goes to zero. I am not aware of the possibility you mention and I need help from a theorist !

- *Answer given by G. 't Hooft:*

Theoretically, Lorentz invariance would not allow the massless neutrino oscillation. As the theory put in is the special theory of relativity, and Lorentz invariance would not allow it to say Because if the thing is massless then there is no time unit. So, we can't time the interference effect, and that would be a problem. So, the answer is, it would not be compatible with the Lorentz invariance. But, for the exact theoretical possibilities, it needs to be discussed further.

- *Belov:*

There were claims about detecting the neutrinoless double beta decay in the past. Why were they not accepted?

- *Ferroni:*

The claim came from the HEIDELBERG-MOSCOW experiment on the neutrinoless double beta decay led by Prof. H. V. Klapdor-Kleingrothaus. The experiment published the half-life number corresponding to the upper limit to the Majorana neutrino mass. But, there were subset of the people to that experiment who claimed in other paper that indeed the decay was observed. In my opinion the things are still not very clear because, of the forest of the peaks, some unexplained) that is seen around the Q-value. But, by now GERDA experiment is able to directly check the results of HEIDELBERG-MOSCOW experiment since the GERDA experiment is using the same diodes/ actually the same detectors plus some others they got from experiment IGEX. So, I think that the results that GERDA has produced, in the latest paper is now at the limit of saying that they can disprove the HdM experiment claim. One more year and the issue will be solved.

- *Belov:*

What is the advantage of your experiment on detecting the neutrinoless double beta decay over the SNO experiment in Sudbury (Canada)?

- *Ferroni:*

Your question has to do with the bolometers or even calorimeters made by solid materials vs. liquid scintillations detectors or whatever else come to your mind. The point is that the bolometers and Ge calorimeters have the best energy resolution you can think of (FWHM resolution is of few KeV over 3MeV). So, the region of interest is really very narrow. You can hope with such a detector to aim at zero background experiment. Now, as soon as you approximate the zero background experiment, you eliminate the square root dependence which is fundamental because, then you improve much faster your reach. On the contrary, with the big detectors made by Liquid Scintillators, you are bound to have an energy resolution which can be 3% or 4% or something like that. There is a second point though that even if you go to zero background from external sources you cannot reduce the background which is due to the normal two neutrino two electron decay which goes below the peak, an irreducible background you cannot eliminate.

- *Bettini:*

I have just one comment. The construction for the next programs will probably start in at least 10 years from now (or even more), so, by that time, we will probably know the sign of Δm^2 and also, if we are lucky, CP violation may be larger. The point is that only after the study of these quantities (if the data from T2K are confirmed and also waiting for NOVA) we will be able to choose the better technique for the baselines.

- *Ferroni:*

I agree. I give a fair chance of success to the combination of T2K with NOVA. The point is that this is the same line of thought that is behind the "shall we decide to do the linear collider now or should we wait for LHC results?" If you follow this way, you will never start the linear collider because, so far, with the results available, you could indeed decide now. If you want to study the Higgs coupling then you can already decide for a "yes". But then someone says "let's wait, maybe a new particle will appear telling us if it's better to build the factory with a 700 GeV machine rather than a 250 GeV machine and so forth". My humble opinion is that we wait up to the point where the so-called supersymmetric particle has gone beyond our horizon then the ILC will never be built because we lost the opportunity. At a certain point some decisions have to be taken. Politically, and again this is strictly my opinion, the U.S. have got a problem of what to do in their country as a contribution to the global physics. Neutrino, at this moment, looks like the best opportunity.

- *Massarotti:*

Some exotic decay in B and K sectors can happen thanks to the presence of Majorana neutrinos. Which are the complementary aspects of these studies with respect to the double-β decay in the neutron sector? [To clarify] Majorana decays can also be studied in B and K sectors. For example, $K^+ \text{-->} \pi^- \mu^+ \mu^+$. So, I want to know which are the complementary aspects between these kinds of decays and the one studied at Gran Sasso Laboratory.

- *Ferroni:*

In an ordinary particle decay, having a Majorana or a Dirac neutrino does not make any real difference. For example, in K --> π ν anti-ν, whether the neutrino is Dirac or Majorana does not play any role. Double-β decay is unique to my knowledge so far because for certain you might discover something else in the sense that it happens with such a low Q^2 (if you like) that you have a chance to flip the neutrino helicity. Since the neutrino is massive, if it is a Dirac particle, this would never happen because the lepton number is conserved globally while, if the neutrino does not carry that information, it can happen. The only thing is that you have to flip it and it can only happen realistically at low Q^2. Unfortunately, inside a nucleon you also have to pay the G_F^2 and also the fact that they have to happen closely in space and time. This brings you to 10^{26} years for the event to happen. To my knowledge there is no experiment done with normal decays or production that can tell you anything about the Dirac/Majorana nature of the neutrino, at least in practice.

LECTURER: J. KNOBLOCH

Probing Subnuclear Structure with Neutrinos

Scientific Secretaries: Shakeb Ahmad, T. Drozhzhova

DISCUSSION

- *Haidt:*

You have given a very nice talk this morning and I still remember most of the transparencies you have shown. I would like to add few remarks to this presentation this morning. It was in 1970, that the charged current programme started in Gargamelle and at the same time also in the US National Laboratory (later called Fermilab). Of course that was the beginning and everything was concentrated on charged currents and in 1971 Gerardus 't Hooft has demonstrated that the Glashow Salam Weinberg (GSW) model was renormalisable, and at that moment we got awake and theoreticians came to us and said that you have to show whether neutral currents exist or not. The interesting thing is that, already after 3 years the measurement of $\sin^2\Theta_w$ was precise enough to use it as input to the GSW model in order to predict – as you mentioned this morning - the masses of W and Z. This was the trigger to the pbar-p programme initiated by Rubbia, Cline and McIntyre. Within the short time of seven years they discovered the weak bosons and measured their masses at the predicted value. Another interesting aspect is that neutrino experiments, which usually had the reputation of experiments with small statistics, particularly for the bubble chamber, entered 1981 in a phase of precision measurements. It turned out that the measurement of $\sin^2\Theta_w$ interpreted at lowest order and at first order differed by 0.01. So, if the experiments were able to make a measurement with the precession of 0.005, then one could test the radiative corrections and, indeed, you CDHS and also CHARM have succeeded in that. This was a great achievement and gave the most precise measurement from neutrino of the order of about 1% from apparatus and about 1% from theory mainly from structure functions. Now, together with the radiative corrections, and also top mass and Higgs mass entered into the discussion and it is remarkable that with the data available it was possible to predict the top mass which was much later observed at that value. Today the Electroweak Way stands out. Its beginning with the weak neutral currents was, as I remember well, much criticized by Steinberger at that time, however, a few years ago he said that the greatest contribution to physics was the discovery of neutral current. Thank you.

- *Knobloch:*

Thank you, not much to add to that, you were looking back, but it is also interesting to see that the neutrino DIS scattering still interesting there is the NuTeV experiment which, a couple of years ago, made a big noise with their 3σ deviation on the comparison of $\sin^2\Theta_w$ with a prediction from theory including the masses of Z and W and to my knowledge - maybe someone else knows it better - there is no resolution to this, although there are many explanations. Once you make such a paper and pronounce something that is out of the standard model or out of the standard wisdom, immediately people jump on it is always extremely difficult from outside to reanalyze the data of an experiment.

- *Haidt:*

I criticized their fit procedure. It consisted in an iteration of a fit to the calibration of their apparatus and the physics parameters. This is a dangerous circle.

- *Knobloch:*

I said this morning that I did not discuss any newer experiment because I am not so much up-to date with it. On the other hand there are a number of arguments on the NuTeV experiment results. The one I find quite interesting is that the electron neutrino background in the beam and the electron in a charged current reaction can hide in the hadronic shower. They claim to be able to handle this, but it is a difficult business.

- *'t Hooft:*

Which measurement is the most accurate neutrino measurement?

- *Knobloch:*

The most accurate is always relative to what others do. So, at the time certainly the Weinberg angle was a very precise measurement. Other measurements had not been done before like the charm contribution to the mixing matrix and so on. So, at that time this was the most precise measurement but of course now modern experiments has overcome this.

- *Haidt:*

The question of 't Hooft has various aspects, one of which regards the measurement of the couplings of Z to d-dbar and Z to u-ubar. The combination of all neutrino and antineutrino experiments succeeded in measuring both the left and the right handed couplings of the u and d quarks.

- *Knobloch:*

Yes, not only a question of precision but also fundamental questions like this V+A versus V-A, which is shown here, the other question of the longitudinal structure function that was measured. I had to select certain number of things.

LECTURER: H. NICOLAI

Another look at Quadratic Divergencies

Scientific Secretaries: M. Csanád, Á. Lukács

DISCUSSION

- Savitskyi:

Are there any predictions in the low energy supersymmetric models regarding the range of masses? Where do you expect the observation of supersymmetric particles?

- Nicolai:

There are many supersymmetric models, and supersymmetric extensions of the Standard Model. These models are constrained by the fact that supersymmetric particles have not been observed so far. If you had asked this question 20 years ago, people would have answered that supersymmetric particles should pop up very close to the electroweak scale. This would have been a very natural expectation in view of the fact that low energy supersymmetry was introduced to stabilize the electroweak scale. However, as you have heard today, there are so far no indications of any superpartners, or in fact, any new physics at all. This means that in these models, the masses of the superpartners have to be pushed further up, in order not to be in conflict with experiment. This, of course, can be done, because there is quite some freedom in this SUSY model building, but there are no real predictions for the masses or couplings of the superpartners, beyond the prediction of their mere existence. For their mass spectra there is a wide range possible, at least up to a point. Namely, if the masses get pushed up too far, the idea loses credibility (at least for me) because if supersymmetry was really the reason for the stability of the observed hierarchy there should be something not too far from the electroweak scale. On the other hand, it is still possible to construct supersymmetric models which are compatible with present experimental data. In the standard model, we had a definite prediction for the W and Z mass, but we did not have a prediction for the Higgs mass, that had to be determined experimentally. There were nevertheless some constraints, as for instance a too large Higgs mass would have indicated the breakdown of the Standard Model at around 10 TeV (and also the measured value of 126 GeV is at the upper limit for SUSY models). The more embellishments you introduce in your Beyond the Standard Model, the more indirect constraints of this type you may get. And by making the model more baroque you will eventually clash with a very successful principle, the principle of Occam's razor, which says that if for some observations, there are several explanations, usually the one with the minimal number of assumptions it the best.

- Prochazka:

You had a figure, showing the running of the coupling constants, the Λ_1, showing that two of them were growing with scale, in the UV.

- Nicolai:

As you see, there are three couplings in this model; one of them, Λ_1 is the Standard Model Higgs self-coupling, which should not be too far from its Standard Model value, then there is Λ_3, which describes the mixing, and finally Λ_2, which describes the self-interaction of the new scalar. Now, one of our basic assumptions is that there should be no Landau poles, because a Landau pole would mean that the running couplings diverge at that scale. E.g., if the Higgs mass would have come out too large, then the Higgs self-coupling would have been large, and following the RG running, you would have run into a Landau pole far below the Planck scale. Now, when I say that we use this requirement as an input, I have to be more precise: we require that there should be no Landau poles below the Planck scale. This is because for theories with interacting scalar fields and Yukawa couplings, there will always be sectors of the theory, which are not asymptotically free; it is only the non-abelian gauge couplings which are asymptotically free, i.e., go to zero in the UV. By contrast, the Yukawa couplings and scalar self-coupling will explode eventually, and actually this is what you see here (slide 19). Our requirement is simply to postpone this beyond the Planck scale. So when we say that we want to salvage the Standard Model to the Planck scale, that is not the same as making it consistent up to infinite energies (as would be the case for instance in a scenario based on asymptotic safety). The idea is that when you reach the Planck scale, standard Quantum Field Theory is going to be replaced by some other theory, Quantum Gravity, which is presumably not a Quantum Field Theory. So we do not get rid of the Landau poles altogether, we just push them beyond the Planck scale, and then invoke Quantum Gravity to solve the problem there.

- *Prochazka:*

But even then, you would have some strongly interacting regime below that, wouldn't you?

- *Nicolai:*

No, not in the case of this theory. It has been put into it, as a requirement. We try to make sure that couplings remain reasonably small, and that nothing blows up below the Planck scale. Yes, beyond the Planck scale, things will start to blow up.

- *'t Hooft:*

What do you think of the possibility of conformal symmetry being spontaneously broken? In a sense, one may argue, that Einstein-Hilbert gravity breaks conformal symmetry spontaneously, if you write $g_{\mu\nu} = \Omega^2 \hat{g}_{\mu\nu}$, where Ω is a scalar field with a vacuum expectation value (VEV) of $1/G_N$.

- *Nicolai:*

The question, which you might ask anyway, is that if the Standard Model is conformally invariant, then is the theory, which it is embedded into, also conformally invariant? There have been several suggestions in this direction, for example, it has been suggested that it is not Einstein gravity, but conformal gravity coupled to a scalar, that gets a vacuum expectation value, and then, as a result, one gets effective Einstein gravity. There are ways to do this, for instance one where the conformal factor is considered a scalar field. There is also Weyl invariant gravity, where the ultimate gravity theory is based on an expression of the Weyl tensor squared, which would be conformally invariant. A third possibility is that one could embed this theory into non-conformal gravity, such that only when one takes the decoupling limit, when the gravitational coupling disappears, one is left with an effective conformal theory. We have checked this at the classical level for the example of maximally supersymmetric N=4 Yang Mills theory which is the conformally invariant theory par excellence. Namely, you can embed this theory into gauged N=4 supergravity, which is not conformally invariant, but when you take the limit kappa goes to zero, then you are left with a conformally invariant theory. For this to work, you must make sure, that the gravitational coupling kappa in this theory up there does not appear with inverse powers, because then you could not take this limit. In summary, I should say, that we do not have a definite answer to this question, but it seems perfectly possible for a non-conformal theory at the Planck scale to give rise to a classically conformally invariant theory at low energy, in the limit where gravity decouples.

- *Minkowski:*

In connection with the question of Gerard 't Hooft, let me mention the "Achilles heel" of supersymmetry including a non-Abelian, or even Abelian gauge sector, where the trace anomaly, i.e. the trace of Θ is proportional to $\beta/g(F)^2$, is indeed signaling SUSY and conformal symmetry breaking. In this context, this question is really pertinent, and in your presentation, you avoided this, because there is no supersymmetry below the Planck mass for you.

- *Nicolai:*

I think it is fair to say, that the problem of supersymmetry breaking is as unsolved as ever, because, if you consider established supersymmetry breaking mechanisms, like D-term breaking or F-term breaking in globally supersymmetric theories, they will not do. The other feature, which I already mentioned on slide 10, is that a distinct feature of supersymmetry is, that once you have it, it is very hard to get rid of. It is precisely the opposite with the conformal symmetry, which is generically broken by quantum anomalies. As for supersymmetry my expectation would be, that there is some, as of now, unknown, dynamical mechanism to break it. People have speculated about fermion condensates or other such ideas, but I think it has to be something more sophisticated. Maybe I should say this, because I grew up with Julius Wess, and at that time, the consensus was that although supersymmetry is very beautiful, it can only be part of the answer. At that time, we did not have a good way to break it, and this is still true. The way it is done now is to introduce explicit mass terms by hand, which I do not find very appealing. That is unfortunately where we are.

- *Inguglia:*

I wanted to comment on the first question regarding the mass range. Actually it is true that things have to agree with the experiment. There are current searches going on by the ATLAS collaboration and I suspect by CMS as well. They are looking for extra Higgs particles, for example, and these searches are done by the B physics group. They suspect that this object can have quite low masses. In this sense this low mass range is not necessarily excluded. Another thing I wanted to add was about the energy scale or mass scale, depending on the model. You should also look at rare processes in the Standard Model, such as B_s going to two muons. Unfortunately, as you all know probably there is nothing appearing there as excess in these branching rations and these decays. So it seems like that for the most generic SUSY models the mass scale or energy scale has to be very high.

- *Nicolai:*

When Pierre Fayet proposed the first supersymmetric extension of the Standard Model in the late 70ies, the expectation (at least among the believers) was that the photino would turn up at 3 GeV. So you see we have come a long way with these expectations and predictions! Of course in this case the particles are very weakly interacting, so you have some more leeway. But the structure of these models is such that not only do you have these weakly interacting particles, but there are always some other particles being forced on you by supersymmetry that may not be weakly interacting but do undergo strong interactions (like new Higgses coupling to quarks), and they would be much harder to hide. So if you don't see anything it will be difficult to argue that supersymmetry is there nevertheless.

- *Krog:*

Let's just go back to the model for a moment. The way I understand it is that when you have some assumptions for your heavy neutrinos, it basically fixes your Yukawa couplings.

- *Nicolai:*

The treatment of the Yukawa-couplings is fairly standard here. If you take those fields and including chiral neutrinos and this new scalar field and you simply write down the most general renormalizable and in this case classically conformal Lagrangian, and this is it.

- *Krog:*

My question is the following. You have the new Yukawa-couplings and the two new scalar couplings. And now you want to do an RG running of your couplings. So you have to make some assumptions on the values. Do you know what would happen if you would not fix the Yukawa-couplings to these specific neutrino-friendly values?

- *Nicolai:*

Actually, because we don't know what these values are we simply scan the parameter space and look for ranges of values compatible with the assumptions. And as I said the interactions with the new scalar are not new. This kind of interaction already appeared in the days of grand unified theories (GUTs) based on SO(10) where you naturally have a right handed neutrino. But the difference here is that you don't need to choose this scalar to have a VEV of the order of the GUT scale. You can assign this VEV to be not far from the electroweak scale, but nevertheless get the values for the light neutrino masses sufficiently small. The trick that does this is shown on slides 15 and 16. The scalar sector is shown on slide 16, i.e., how scalars obtain a VEV. On slide 15, you see that besides the scalar fields, the Yukawa couplings also enter the neutrino masses, Y^{ν} into the Dirac, and Y^{M} into the Majorana mass term. Now, if you appropriately rescale the theory, the VEV's are of unit order, and the Dirac mass term squared (m_D^2) becomes $\sim Y^{\nu}$ squared, and the Majorana masses become $M \sim Y^{M}$. Taking the former to be 10^{-6} you get the required factor on the order of 10^{-12}.

- *Krog:*

If you relax the condition that you want to have realistic neutrino masses, are then there any interesting areas in the phase-space that you investigated?

- *Nicolai:*

Well, if you relax that condition then you certainly must have very small values for the light neutrinos. We accept it as a fact that the light neutrinos must be under 1 eV, but what changes compared to other investigations in the context of GUT, is that the heavy neutrinos are not heavier than 10^{10} or 10^{15} GeV, they have masses in the 1 TeV range. Then the question is if you can see them. It would be something like missing energy. It also depends on the precise mass spectrum of the theory whether or not a decay with this missing energy signature is possible. But we just have this range of parameters, so in some ranges it is possible and in some other ranges it is not. There is no precise prediction, only order of magnitudes.

- *Krog:*

Let me have another question. In your case you summarize the Veltman conditions basically at the Planck-scale. I have seen other cases when they are taken at the electroweak scale. You said there is a way to extrapolate between the two?

- *Nicolai:*

This is the one new element in our construction, because what Veltman did back then in 1982 is that he simply took the physical (renormalized) couplings at the electroweak scale. He added them up and noticed that if the top quark had a certain value, then the quadratic divergencies would cancel. From this he derived the prediction for the top quark mass, which was not the one that was measured later. From that point of view the idea does not work. But what we say here is that the natural scale to impose this requirement is in the unified theory where there are (presumably) no divergencies and no singular coupling parameters. The coefficient of the quadratic divergence is then just a coefficient that you can set to zero. What is unusual is that we are dealing with non-singular bare couplings here, and by assuming finiteness, our requirement becomes a renormalization group invariant statement. But you can, as a sort of auxiliary construction, introduce renormalized and running parameters. Then you see that this condition would change as if you express it as a function of the running couplings. Thus running couplings are only an auxiliary device to facilitate calculations. If you had a theory that was finite, with finite parameters, there would be no need for running couplings. It's just a device to improve perturbation theory if you cannot do exact calculations. This is perhaps the main new idea here.

- *Miyazaki:*

I'm interested in the decay width of the standard Higgs boson. You said that it is narrower in this model. How does the doublet mixing make the widths narrower? What is the mechanism here?

- *Nicolai:*

It is just a standard calculation where you work out the Feynman diagrams.

- *Miyazaki:*

Is it a general thing happening in doublet models?

- *Nicolai:*

Well I think it is a pretty general feature. You would have it in other models as well. In fact as you heard this morning in Bertolucci's talk, that to measure Higgs boson decay width precisely is quite a challenge for LHC. If you had an electron-positron collider, you could really go to that point and measure this width extremely precisely. Just like LEP did with the Z^0 giving a prediction for the number of neutrino species.

CLOSING CEREMONY

DIPLOMAS

Twenty six Diplomas were open for competition among the participants. They have been awarded as follows:

- John S. BELL Diploma to:
  Valentin SLEPUKHIN
  ITEP, Moscow, Russian Federation

- Patrick M.S. BLACKETT Diploma to:
  Teresa LENZ
  Hamburg University, Germany

- Nicola CABIBBO Diploma to:
  Angelo ESPOSITO
  Columbia University in the City of New York, USA

- James CHADWICK Diploma to:
  Juri FIASCHI
  University of Southampton, UK

- Sidney COLEMAN Diploma to:
  Peter William MILLINGTON
  School of Physics and Astronomy, University of Manchester, UK

- Richard H. DALITZ Diploma to:
  Pavel A. BELOV
  St. Petersburg State University, Saint Petersburg, Russian Federation

- Paul A.M. DIRAC Diploma to:
  Vladimir PROCHAZKA
  The University of Edinburgh, UK

- Bruno FERRETTI Diploma to:
  Oliver FISCHER
  Dept. of Physics Particles and Cosmology group
  University of Basel, Switzerland

- Richard P. FEYNMAN Diploma to:
  Vladimir KOVALENKO
  St. Petersburg State University, Saint Petersburg, Russian Federation

- Vladimir N. GRIBOV Diploma to:
  Oksana IARYGINA
  Taras Shevchenko National University, Kyiv, Ukraine

- Robert HOFSTADTER Diploma to:
  Dennis V. PEREPELITSA
  Physics Dept., Brookhaven National Laboratory, Upton, NY, USA

- Gunnar KÄLLEN Diploma to:
  Oliver GOULD
  Imperial College London, UK

- Seymour J. LINDENBAUM Diploma to:
  Mykola SAVITSKYI
  National University of Kyiv, Kyiv, Ukraine

- Yuval NE'EMAN Diploma to:
  Alla SEMENOVA
  National Research Centre "Kurchatov Institute" B.P. Konstantinov, Petersburg Nuclear Physics Institute, Saint Petersburg, Russian Federation

- Giuseppe P.S. OCCHIALINI Diploma to:
  Akira MIYAZAKI
  CERN fellow, Geneva, Switzerland, and The University of Tokyo, Japan

- Oreste PICCIONI Diploma to:
  Samuel BOREN
  University of Kansas, Lawrence, KS, USA

- Bruno PONTECORVO Diploma to:
  Tatiana DROZHZHOVA
  St. Petersburg State University, Saint Petersburg, Russian Federation

- Giampietro PUPPI Diploma to:
  Shakeb AHMAD
  Aligarh Muslin University, Aligarh (U.P.), India

- Isidor I. RABI Diploma to:
  Riccardo RUSSO
  INFN and University of Turin, Italy

- Giulio RACAH Diploma to:
  Jens Frederik Colding KROG
  University of Southern Denmark, Odense, Denmark

- Bruno ROSSI Diploma to:
  Maddalena GIULINI
  Universität Heidelberg, DE

- Julian S. SCHWINGER Diploma to:
  Leo HUGHES
  Imperial College London, UK

- Victor F. WEISSKOPF Diploma to:
  Luca PARISI
  University of Salerno, Italy

- Eugene P. WIGNER Diploma to:
  Cohl FUREY
  Perimeter Institute for Theoretical Physics, Waterloo, ON, Canada

- Bjorn H. WIIK Diploma to:
  Anastasiia REBESH
  Bogolyubov Institute for Theoretical Physics of the NAS of Ukraine, Kyiv, Ukraine

- Chien Shiung WU Diploma to:
  Olga BONDARENKO
  KVI-CART, University of Groninger, The Netherlands

NEW TALENTS AWARDS

The Awards for the best 'New Talent – Speakers' have been attributed to:

- **Leo HUGHES**
  Imperial College London, UK
  MATHEMATICAL PHYSICS

- **Jens Frederik Colding KROG**
  University of Southern Denmark, Odense, Denmark
  THEORETICAL PHYSICS

- **Olga BONDARENKO**
  KVI-CART, University of Groninger, The Netherlands
  EXPERIMENTAL PHYSICS

- **Maddalena GIULINI**
  Universität Heidelberg, Germany
  DETECTOR PHYSICS

Award for the BEST SCIENTIFIC SECRETARY to:
Shakeb AHMAD
Aligarh Muslin University, Aligarh (U.P.), India

Award for the BEST STUDENT to:
Joseph ZENNAMO
University of Chicago, IL, USA

BRUNO ZUMINO special Scholarship for the BEST STUDENT to:
Joseph ZENNAMO
University of Chicago, IL, USA

PARTICIPANTS

Aidar ABZALOV
Russia
Faculty of Mathematics, HSE
Moscow, Russian Federation

Shakeb AHMAD
India
Aligarh Muslin University, Aligarh (U.P.),
India

Azar AHMADOV
Azerbaijan
Baku State University, Azerbaijan

Muhammad AJAZ
Pakistan
Abdul Wali Khan University Mardan
Khyber Pakhtoonkhwa, Pakistan

Evgeny ANDRONOV
Russia
St. Petersburg State University
Saint Petersburg, Russian Federation

Ignatios ANTONIADIS
France
Centre de Physique Théorique ,
Ecole Polytechnique, Palaiseau, France

Nikita ASTRAKHANTSEV
Russia
MIPT and ITEP, Moscow, Russian Federation

Aleksandr AVDOSHKIN
Russia
MIPT and ITEP, Moscow, Russian Federation

Pavel A. BELOV
Russia
St. Petersburg State University
Saint Petersburg, Russian Federation

Zvi BERN
USA
UCLA, Los Angeles, CA, USA

Sergio BERTOLUCCI
Italy
CERN, Geneva, Switzerland

Alessandro BETTINI
Italy
INFN & Padoa University, Italy and
Canfranc Underground Lab, Spain

Olga BONDARENKO
Ukraine
KVI-CART, University of Groninger,
The Netherlands

Samuel Steed BOREN
USA
University of Kansas, Lawrence, KS, USA

Alexander BYLINKIN
Russia
ITEP, Moscow, Russian Federation

Bianca Letizia CERCHIAI
Italy
University of Milano, Italy

Luisa CIFARELLI
Italy
INFN & University of Bologna, Italy

Yasser CORRALES MORALES Cuba	University of Turin, Italy
Màtè CSANAD Hungary	Eötvös Loránd University, Budapest, Hungary
Vitalii DROHAN Ukraine	Bogolyubov ITP and Physics Faculty Kyiv State University, Kyiv, Ukraine
Tatiana DROZHZHOVA Russia	St. Petersburg State University Saint Petersburg, Russian Federation
Michael J. DUFF UK	Imperial College London, UK
Angelo ESPOSITO Italy	Columbia University in the City of New York, USA
Sergio FERRARA Italy	CERN, Geneva, Switzerland and LNF–INFN, Frascati, Italy
Fernando FERRONI Italy	INFN, Rome, Italy
Juri FIASCHI Italy	University of Southampton, UK
Oliver FISCHER Germany	Dept. of Physics Particles and Cosmology group University of Basel, Switzerland
Harald FRITZSCH Germany	Ludwig Maximilians Universitaet, Muenchen, Germany
Cohl FUREY Canada	Perimeter Institute for Theoretical Physics Waterloo, ON, Canada
Pavlo GAVRYLENKO Ukraine	Kyiv State University, Bogolyubov ITP, Kyiv, Ukraine
Paolo GIUBELLINO Italy	University of Turin, Italy
Maddalena GIULINI Italy	Universität Heidelberg, DE
Oliver GOULD UK	Imperial College London, UK
Barbara GUERZONI Italy	University of Bologna, Italy

Nikolay GULITSKIY
Russia
St. Petersburg State University
Saint Petersburg, Russian Federation

Dieter HAIDT
Germany
EPJC–DESY,
Hamburg, Germany

Leo HUGHES
UK
Imperial College London, UK

Oksana IARYGINA
Ukraine
Taras Shevchenko National University,
Kyiv, Ukraine

Gianluca INGUGLIA
Italy
Queen Mary University of London, UK

Iryna IVANCHENKO
Ukraine
Bogoliyubov Institute for Theoretical Physics
Kyiv, Ukraine

Oleksii IVANYTSKYI
Ukraine
Bogolyubov Institute for Theoretical Physics
National Academy of Sciences, Kyiv, Ukraine

Juansher JEJELAVA
Georgia
Andronikashvili Institute of Physics
TSU, Tbilisi, Georgia

Sonia KABANA
Greece
Subatech, Nantes, France

Polina KAKIN
Russia
St. Petersburg State University
Saint Petersburg, Russian Federation

Daniil KALINOV
Russia
Faculty of Mathematics, HSE
Moscow, Russian Federation

Georgios KARANANAS
Greece
Ecole Polytechnique de Lausanne,
Switzerland

Nikolaos KARASTATHIS
Greece
National Technical University of Athens
Zografou Campus, Greece

Alisa KATANAEVA
Russia
St. Petersburg State University
Saint Petersburg, Russian Federation

Aian KHANAKHMEDOV
Georgia
Physics Faculty, Baku State University
Baku, Azerbaijan

Aleksandr KLEMENTEV
Russia
St. Petersburg State University
Saint Petersburg, Russian Federation

Juergen KNOBLOCH
Germany
CERN, Geneva, Switzerland

Monika KÖFARAGÒ
Hungary
CERN, Geneva, Switzerland

Sergei KOLEVATOV Russia	St. Petersburg State University Saint Petersburg, Russian Federation
Kateryna KORNIEIEVA Ukraine	Joint Institute for Nuclear Research Dubna, Russian Federation
Christian P. KORTHALS-ALTES The Netherlands	CNRS-Luminy, Marseille, France
Vladimir KOVALENKO Russia	St. Petersburg State University Saint Petersburg, Russian Federation
Jens Frederik Colding KROG Denmark	University of Southern Denmark, Odense, Denmark
Andràs LÀSZLO Hungary	MTA Wigner Research Centre for Physics Budapest, Hungary
Alexander LEITHES UK	Queen Mary University of London, UK
Teresa LENZ Germany	Hamburg University, Germany
Roberto LEONARDI Italy	INFN-Università degli studi di Perugia, Italy
Stefanos LEONTSINIS Greece	CERN, Geneva, Switzerland
Heinrich LEUTWYLER Switzerland	University of Bern, Switzerland
Andrii LIASHYK Ukraine	Kyiv State University, Bogolyubov ITP, Kyiv, Ukraine
Árpád László LUKÁCS Hungary	HAS Wigner Research Centre for Physics Budapest, Hungary
Sergii MAMEDOV Ukraine	G.V. Kurdyumov Inst. for Metal Physics N.A.S. of Ukraine, Kyiv, Ukraine
Alessio MARRANI Italy	Instituut Voor Theoretische Fysica, Leuven, Belgium
Paolo MASSAROTTI Italy	University of Naples Federico II and INFN-Naples, Italy
Jack MEDLEY UK	The University of Edinburgh, UK

Thomas METCALF
UK
The University of Edinburgh, Scotland, UK

Declan Andrew MILLAR
UK
University of Southampton, UK

Peter William MILLINGTON
UK
School of Physics and Astronomy
University of Manchester, UK

Peter MINKOWSKI
Switzerland
University of Bern, Switzerland

Hashim Rasool MIR
India
Research Scholar, Dept. of Physics
Aligarh Muslin University, Aligarh, India

Akira MIYAZAKI
Japan
CERN, Geneva, Switzerland

Dimitri V. NANOPOULOS
Greece
Texas A&M University, College Station,
TX, USA

Max NEUNER
Germany
University of Heidelberg, Germany

Hermann NICOLAI
Germany
Max Planck Institute, Potsdam, Germany

Francesco NOFERINI
Italy
University of Bologna, Italy

Kontantinos A. NTEKAS
Greece
National Technical University of Athens
Zografou Campus, Greece

Alexandra OSTROVSKAYA
Russia
Faculty of Mathematics, HSE
Moscow, Russian Federation

Jochen OTT
Germany
Hamburg University, Germany

Luca PARISI
Italy
University of Salerno, Italy

Dennis V. PEREPELITSA
USA
Physics Dept., Brookhaven National Laboratory
Upton, NY, USA

Maxim PERFILOV
Russia
Moscow State University, Russian Federation

Andrey PIKELNER
Russia
Joint Institute for Nuclear Research
Dubna, Russian Federation

Dmitriy PONOMAREV
Russia
Ludwig-Maximilian University, Munich,
Germany

Fedor POPOV — MIPT and ITEP, Moscow, Russian Federation
Russia

Semyon POZDNYAKOV — St. Petersburg State University, Saint Petersburg, Russian Federation
Russia

Vladimir PROCHAZKA — The University of Edinburgh, UK
Slovak

Jérémie QUEVILLON — Laboratoire de Physique Théorique, Université Paris XI Sud, Orsay, France
France

Stefano RAGAZZI — Gran Sasso National Laboratory, L'Aquila, Italy
Italy

Sergii RASPOPOV — NSC KIPT, Kharkov, Ukraine
Ukraine

Anastasiia REBESH — Bogolyubov Institute for Theoretical Physics of the NAS of Ukraine, Kyiv, Ukraine
Ukraine

Alain RIAZUELO — Institut d'Astrophysique de Paris, France
France

Igor RUDENOK — Taras Shevchenko National University of Kyiv, Ukraine
Ukraine

Riccardo RUSSO — INFN and University of Turin, Italy
Italy

Adel SALAKH — Kiev State University, Bogolyubov ITP, Kyiv, Ukraine
Ukraine

Mykola SAVITSKYI — National University of Kyiv, Ukraine
Ukraine

Dieter SCHILDKNECHT — Werner-Heisenberg-Institut, München; Universität Bielefeld, Germany
Germany

Mario SCHRÖCK — INFN-Sezione Roma Tre, Roma, Italy
Germany

John H. SCHWARZ — CALTECH, Pasadena, CA, USA
USA

Markus SEIDEL
Germany
Universität Hamburg, Germany

Elizaveta SEMENOVA
Russia
St. Petersburg State University
Saint Petersburg, Russian Federation

Alla SEMENOVA
Russia
Petersburg Nuclear Physics Institute
Saint Petersburg, Russian Federation

Mykola SEMENYAKIN
Ukraine
Kyiv State University, Bogolyubov ITP,
Kyiv, Ukraine

Anton SHCHECHKIN
Ukraine
Physics Faculty, Kiev University,
Kyiv, Ukraine

Andrey SHESHUKOV
Russia
JINR, Dubna, Moscow Region,
Russian Federation

Anton SHEYKIN
Russia
St. Petersburg State University
Saint Petersburg, Russian Federation

Valentin SLEPUKHIN
Russia
ITEP, Moscow, Russian Federation

Aleksandra SOKOLNIKOVA
Russia
JINR, Dubna, Moscow Region,
Russian Federation

Nikita SOPENKO
Russia
MIPT and ITEP, Moscow, Russian Federation

Vladyslav SYROTENKO
Ukraine
V.N. Karazin Kharkov National University
Kharkov, Ukraine

Michael J. TANNENBAUM
USA
Brookhaven National Laboratory, Upton, NY, USA

Abdel Naser TAWFIK
Egypt
ECPT, MTI University, Cairo, Egypt

Astkhik TER-MINASYAN
Russia
Kurchatovskiy University, Moscow,
Russian Federation

Gerardus 't HOOFT The Netherlands	Utrecht University, The Netherlands
Sergii V. TROFYMENKO Ukraine	Kharkov Institute of Physics and Technology Kharkov, Ukraine
Constantino TSALLIS Brazil	National Institute of Science and Technology for Complex Systems (INCT-SC), Rio de Janeiro, Brazil
Valiyaneerilakkal UVAIS India	National Institute of Technology-Calicut Calicut, Kerala, India
Nicholas WARDLE UK	CERN, Geneva, Switzerland
Horst WENNINGER Germany	CERN, Geneva, Switzerland
Shahriyar YAFARZADE Azerbaijan	Physics Faculty, Baku State University Baku, Azerbaijan
Vahagn YEGHIKYAN Armenia	Yerevan State University, Armenia
Ivan YELETSKIKH Ukraine	Joint Institute for Nuclear Research Dubna, Russian Federation
Taras ZAGOSKIN Ukraine	V.N. Karazin Kharkov National University Kharkov, Ukraine
Jakub ZAREMBA Poland	Institute of Nuclear Physics, PAN, Krakow, Poland
Joseph ZENNAMO USA	University of Chicago, Chicago, IL, USA
Antonino ZICHICHI Italy	CERN, Geneva, Switzerland; and INFN & University of Bologna, Italy
Michele ZOCCALI Italy	Cambridge University, DAMTP, Cambridge, UK
George ZWEIG USA	MIT, Cambridge, MA, USA